DRAWDOWN

THE MOST COMPREHENSIVE PLAN EVER PROPOSED
TO REVERSE GLOBAL WARMING

STAFF

PROJECT DIRECTOR, EDITOR, WRITER: Paul Hawken
DESIGN: Janet Mumford
SENIOR WRITER: Katharine Wilkinson
WEBSITE: Chad Upham
COPY EDITOR: Christian Leahy
WRITING ASSISTANT: Olivia Ashmoore

RESEARCH DIRECTOR: Chad Frischmann
SENIOR RESEARCHER: Ryan Allard
SENIOR RESEARCHER: Kevin Bayuk
SENIOR RESEARCHER: João Pedro Gouveia
SENIOR RESEARCHER: Mamta Mehra
SENIOR RESEARCHER: Eric Toensmeier
RESEARCH COORDINATOR: Crystal Chissell

RESEARCH FELLOWS

Zak Accuardi
Raihan Uddin Ahmed
Carolyn Alkire
Ryan Allard
Kevin Bayuk
Renilde Becqué
Erika Boeing
Jvani Cabiness
Johnnie Chamberlin
Delton Chen
Leonardo Covis
Priyanka deSouza
Anna Goldstein
João Pedro Gouveia
Alisha Graves
Karan Gupta

Zhen Han
Zeke Hansfather
Yuill Herbert
Amanda Hong
Ariel Horowitz
Ryan Hottle
Troy Hottle
David Jaber
Dattakiran Jagu
Daniel Kane
Becky Xilu Li
Sumedha Malaviya
Urmila Malvadkar
Alison Mason
Mihir Mathur
Victor Maxwell

David Mead
Mamta Mehra
Ruth Metzel
Alex Michalko
Ida Midzic
S. Karthik Mukkavilli
Kapil Narula
Demetrios Papaioannou
Michelle Pedraza
Chelsea Petrenko
Noorie Rajvanshi
George Randolph
Abby Rubinson
Adrien Salazar
Aven Satre-Meloy
Christine Shearer

David Siap
Kelly Siman
Leena Tähkämö
Eric Toensmeier
Melanie Valencia
Ernesto Valero Thomas
Andrew Wade
Marilyn Waite
Charlotte Wheeler
Christopher Wally Wright
Liang Emlyn Yang
Daphne Yin
Kenneth Zame

ESSAYISTS

Janine Benyus
Anne Biklé
Pope Francis

Mark Hertsgaard
David Montgomery
Michael Pollan

Bren Smith
Peter Wohlleben
Andrea Wulf

THE BOARD OF DIRECTORS

Janine Benyus
Peter Byck
Pedro Diniz
Lisa Gautier

Paul Hawken
John Lanier
Lyn Davis Lear
Peggy Liu

Martin O'Malley
Gunhild A. Stordalen
Greg Watson
John Wick

WITH GRATITUDE TO OUR FUNDERS, DONORS, AND SUPPORTERS

Ray C. Anderson Foundation
TomKat Charitable Trust
Pedro Paulo Diniz
Lear Family Foundation
Leonardo DiCaprio Foundation
Dr. Bronner's
Overbrook Foundation
The Caldera Foundation
Interface Environmental Foundation
Natural Co+op Grocers
Jamie Wolf
Newman's Own Foundation
Leonard C. and Mildred F. Ferguson Foundation
Heinz Endowments
Paul Hawken

Justin Rosenstein
Russ and Suki Munsell
Better Tomorrow Fund
Jessica and Decker Rolph
Gautier Family
Stephen and Byron Katie Mitchell
Michael and Jena King Family Fund
Colin le Duc
Autodesk
TSE Foundation
Janine Benyus
Organic Valley
Nutiva Foundation
Leslie and Geoffrey Oelsner
Guayaki

Drawdown is a message grounded in science; it also is a testament to the growing stream of humanity who understands the enormity of the challenge we face, and is willing to devote their lives to a future of kindness, security, and regeneration. The young girl here is from the Borana Oromo people, who reside in the Nakuprat-Gotu Community Conservancy in northern Kenya. Her picture has been our talisman, calling us daily to the work that we do.

PENGUIN BOOKS

An imprint of Penguin Random House LLC
375 Hudson Street
New York, New York 10014
penguin.com

"Reciprocity" by Janine Benyus. © 2017 Janine Benyus. Published by arrangement with the author.

Excerpt adapted from *Hot: Living Through the Next Fifty Years on Earth* by Mark Hertsgaard. Copyright © 2011 by Mark Hertsgaard. Used by permission of Houghton Mifflin Harcourt Publishing Company. All rights reserved.

Excerpts from *The Hidden Half of Nature: The Microbial Roots of Life and Health* by David R. Montgomery and Anne Biklé. Copyright © 2016 by David R. Montgomery and Anne Biklé. Used by permission of W. W. Norton & Company, Inc.

Excerpt from "Why Bother?" by Michael Pollan. Published in *The New York Times Magazine*, April 20, 2008. Copyright © 2008 by Michael Pollan. Used by permission of International Creative Management. All rights reserved.

Foreword by Dr. Jonathan Foley. Published by arrangement with the author.

Excerpts from *The Hidden Life of Trees: What They Feel, How They Communicate, Discoveries from a Secret World* by Peter Wohlleben, 2016 (Greystone Books). Used by permission of the publisher.

Excerpt from *The Invention of Nature: Alexander von Humboldt's New World* by Andrea Wulf. Copyright © 2015 by Andrea Wulf. Used by permission of Alfred A. Knopf, an imprint of the Knopf Doubleday Publishing Group, a division of Penguin Random House LLC. All rights reserved.

Credits for images appear on page 240.

Library of Congress Cataloging-in-Publication Data

Names: Hawken, Paul, editor.
Title: Drawdown : the most comprehensive plan ever proposed to reverse global warming / edited by Paul Hawken.
Other titles: Draw down
Description: New York, New York : Penguin Books, [2017] | Includes index.
Identifiers: LCCN 2017007034 (print) | LCCN 2017007803 (ebook) | ISBN 9780143130444 (pbk.) | ISBN 9781524704650 (ebook)
Subjects: LCSH: Climate change mitigation. | Global warming—Prevention.
Classification: LCC TD171.75 .D73 2017 (print) | LCC TD171.75 (ebook) | DDC 363.738/746—dc23
LC record available at https://lccn.loc.gov/2017007034

Printed in the United States of America

10 9 8

Designed by Janet Mumford and Paul Hawken

DRAWDOWN

THE MOST COMPREHENSIVE PLAN EVER PROPOSED

TO REVERSE GLOBAL WARMING

EDITED BY PAUL HAWKEN

PENGUIN BOOKS

CONTENTS

We accumulated more than 5,000 references, citations, and sources in the process of researching and writing Drawdown. *Although they are too numerous to be published in the book, they may be found at www.drawdown.org/references.*

The Big Sur coastline in Northern California at the 'Golden Hour' near sunset. Here, we see the climate system in action, with the constant interplay of atmosphere, ocean, land, and biota.

FOREWORD

As a climate scientist, it's disheartening to witness world events unfold as they have over the past few decades. The clear and precise warnings we scientists have made about our planet's changing climate are materializing as predicted. Greenhouse gases are trapping heat in the earth's atmosphere, producing warmer seasons and an amped-up water cycle.

Warmer air holds more moisture, allowing for higher rates of evaporation and precipitation. Record heat waves, coupled with intense droughts, spark the perfect conditions for massive wildfires. Warming oceans trigger supercharged storms, with greater rainfall and higher storm surges. We can expect a steady rise in extreme weather events in the coming decades, potentially causing countless lost lives and significant financial losses.

Whether we like it or not—whether we choose to "believe" the science or not—the reality of climate change is upon us. It's affecting everything: not just weather patterns, ecosystems, ice sheets, islands, coastlines, and cities across the planet, but the health, safety, and security of every person alive and the generations to come. Worldwide, we're seeing related symptoms like the acidification of our oceans, which could devastate coral reefs and marine life, and the changing biochemistry of plants, including staple crops.

We know exactly why this is happening. We've known for more than a hundred years.

When we burn fossil fuels (coal, oil, and natural gas), manufacture cement, plow rich soils, and destroy forests, we release heat-trapping carbon dioxide into the air. Our cattle, rice fields, landfills, and natural gas operations release methane, warming the planet even more. Other greenhouse gases, including nitrous oxide and fluorinated gases, are seeping out of our agricultural lands, industrial sites, refrigeration systems, and urban areas, compounding the greenhouse effect. It's important to remember that climate change stems from many sources such as energy production, agriculture, forestry, cement, and chemical manufacturing; thus, the solutions must arise from those same many sources.

Beyond the damage to our planet, climate change threatens to undermine our social fabric and the foundations of democracy. We see the impacts of this in the United States in particular, where key parts of the federal government are denying the science, and are closely aligned with fossil fuel industries. While most people continue to move through their day as if nothing is wrong, others who are aware of the science are fearful, if not in despair. The climate change narrative has become a doom and gloom story, causing people to experience denial, anger, or resignation.

At times, I have been one of those people.

Thanks to *Drawdown*, I have a different perspective. Paul Hawken and his colleagues have researched and modeled the one hundred most substantive ways we can reverse global warming. These solutions reside in energy, agriculture, forests, industries, buildings, transportation, and more. They also highlight critical social and cultural solutions, such as empowering girls, reducing population growth, and changing our diets and consumption patterns. Together, these solutions not only slow climate change, they can reverse it.

Drawdown goes beyond solar panels and energy-efficient light bulbs to show that the needed solutions are far more diverse than just those associated with clean energy, and that there are many effective means to address global warming. *Drawdown* illustrates how we can make dramatic strides by reducing the emissions of more exotic greenhouse gases, like refrigerants and black carbon, lowering nitrous oxide emissions from agriculture, cutting methane emissions from cattle production, and reducing carbon dioxide emissions from deforestation. Moreover, *Drawdown* demonstrates the potential for removing carbon dioxide from the atmosphere through innovative land use practices, regenerative agriculture, and agroforestry.

But, more importantly to me, *Drawdown* illuminates ways we can overcome the fear, confusion, and apathy surrounding climate change, and take action as individuals, neighborhoods, towns and cities, states, provinces, businesses, investment firms, and non-profits. This book should become the blueprint for building a climate-safe world. By modeling solutions that are hands-on, well understood, and already scaling, *Drawdown* points to a future where we can reverse global warming and leave a better world for new generations.

We think that our climate future is harsh because news and reports have focused on what will happen if we do not act. *Drawdown* shows us what we can do. Because of that, I think this is the single most important book ever written about climate change.

Drawdown has helped restore my faith in the future, and in the capacity of human beings to solve incredible challenges. We have all the tools we need to combat climate change, and thanks to Paul and his colleagues, we now have a plan showing us how to use them.

Now let's get to work and do it.

Dr. Jonathan Foley
Executive Director, California Academy of Sciences
San Francisco

ORIGINS

The genesis of Project Drawdown was curiosity, not fear. In 2001 I began asking experts in climate and environmental fields a question: Do we know what we need to do in order to arrest and reverse global warming? I thought they could provide a shopping list. I wanted to know the most effective solutions that were already in place, and the impact they could have if scaled. I also wanted to know the price tag. My contacts replied that such an inventory did not exist, but all agreed it would be a great checklist to have, though creating one was not within their individual expertise. After several years, I stopped asking because it was not within my expertise either.

Then came 2013. Several articles were published that were so alarming that one began to hear whispers of the unthinkable: It was game over. But was that true, or might it possibly be game on? Where did we actually stand? It was then that I decided to create Project Drawdown. In atmospheric terms drawdown is that point in time at which greenhouse gases peak and begin to decline on a year-to-year basis. I decided that the goal of the project would be to identify, measure, and model one hundred substantive solutions to determine how much we could accomplish within three decades towards that end.

The subtitle of this book—the most comprehensive plan ever proposed to reverse global warming—may sound a bit brash. We have chosen that description because no detailed plan to reverse warming has been proposed. There have been agreements and proposals on how to slow, cap, and arrest emissions, and there are international commitments to prevent global temperature increases from exceeding two degrees centigrade over preindustrial levels. One hundred and ninety-five nations have made extraordinary progress in coming together to acknowledge that we have a momentous civilizational crisis on our earthly doorstep and have created national plans of action. The UN's Intergovernmental Panel on Climate Change (IPCC) has completed the most significant scientific study in the history of humankind, and continues to refine the science, expand the research, and extend our grasp of one of the most complex systems imaginable. However, there is, as yet, no road map that goes beyond slowing or stopping emissions.

To be clear, our organization did not create or devise a plan. We do not have that capability or self-appointed mandate. In conducting our research, we found a plan, a blueprint that already exists in the world in the form of humanity's collective wisdom, made manifest in applied, hands-on practices and technologies that are commonly available, economically viable, and scientifically valid. Individual farmers, communities, cities, companies, and governments have shown that they care about this planet, its people, and its places. Engaged citizens the world over are doing something extraordinary. This is their story.

In order for Project Drawdown to be credible, a coalition of researchers and scientists needed to be at its foundation. We had a tiny budget and oversized ambitions, so we sent out appeals inviting students and scholars from around the world to become research fellows. We were inundated with responses from some of the finest women and men in science and public policy. Today, the Drawdown Fellows comprise seventy individuals from twenty-two countries. Forty percent are women, nearly half have PhDs, and others have at least one advanced degree. They have extensive academic and professional experience at some of the world's most respected institutions.

Together we gathered comprehensive lists of climate solutions and winnowed them down to those that had the greatest potential to reduce emissions or sequester carbon from the atmosphere. We then compiled literature reviews and devised detailed climate and financial models for each of the solutions. The analyses informing this book were then put through a three-stage process including review by outside experts who evaluated the inputs, sources, and calculations. We brought together a 120-person Advisory Board, a prominent and diverse community of geologists, engineers, agronomists, politicians, writers, climatologists, biologists, botanists, economists, financial analysts, architects, and activists who reviewed and validated the text.

Almost all of the solutions compiled and analyzed here lead to regenerative economic outcomes that create security, produce jobs, improve health, save money, facilitate mobility, eliminate hunger, prevent pollution, restore soil, clean rivers, and more. That these are substantive solutions does not mean that they are all the best ones. There are a small handful of entries in this book whose spillover effects are clearly detrimental to human and planetary health, and we try to make that clear in our descriptions. The overwhelming majority, however, are no-regrets solutions, initiatives we would want to achieve regardless of their ultimate impact on emissions and climate, as they are practices that benefit society and the environment in multiple ways.

The final section of the main part of *Drawdown* is called "Coming Attractions" and features twenty solutions that are nascent or on the horizon. Some may succeed, while others may fail. Notwithstanding, they provide a demonstration of the ingenuity and gumption that committed individuals have brought to

Three-week-old spotted owl hatchlings on a mossy hemlock branch in northern Oregon.

address climate change. Additionally, you will find essays from prominent journalists, writers, and scientists—narratives, histories, and vignettes—that offer a rich and varied context to the specifics of the book.

We remain a learning organization. Our role is to collect information, organize it in ways that are helpful, distribute it to any and all, and provide the means for anyone to add, amend, correct, and extend the information you find here and on the drawdown.org website. Technical reports and expanded model results are available there. Any model that projects out thirty years is going to be highly speculative. However, we believe the numbers are approximately right and welcome your comments and input.

Unquestionably, distress signals are flashing throughout nature and society, from drought, sea level rise, and unrelenting increases in temperatures to expanded refugee crises, conflict, and dislocation. This is not the whole story. We have endeavored in *Drawdown* to show that many people are staunchly and un-

waveringly on the case. Although carbon emissions from fossil fuel combustion and land use have a two-century head start on these solutions, we will take those odds. The buildup of greenhouse gases we experience today occurred in the absence of human understanding; our ancestors were innocent of the damage they were doing. That can tempt us to believe that global warming is something that is happening *to* us—that we are victims of a fate that was determined by actions that precede us. If we change the preposition, and consider that global warming is happening *for* us—an atmospheric transformation that inspires us to change and reimagine everything we make and do—we begin to live in a different world. We take 100 percent responsibility and stop blaming others. We see global warming not as an inevitability but as an invitation to build, innovate, and effect change, a pathway that awakens creativity, compassion, and genius. This is not a liberal agenda, nor is it a conservative one. This is the human agenda. —Paul Hawken

The decree carved into the Rosetta Stone in Egypt in 196 B.C. is known less for its content—an affirmation of the rule of King Ptolemy V—than for its unique combination of scripts. The same text repeats in Greek, Egyptian hieroglyphics, and Egyptian demotic, respectively royal, sacred, and common languages of the time. In the 19th century, European scholars used the Rosetta Stone to crack the code of hieroglyphics, opening up understanding of Ancient Egypt. Today, the Rosetta Stone is what Richard Parkinson, professor of Egyptology at Oxford, calls "an icon of...decipherment" and "a symbol of our desire to understand each other." To convey and comprehend through language is at the heart of the human endeavor.

LANGUAGE

Confucius wrote that calling things by their proper name is the beginning of wisdom. In the world of climate change, names can sometimes be the beginning of confusion. Climate science contains its own specialized vocabulary, acronyms, lingo, and jargon. It is a language derived by scientists and policy makers that is succinct, specific, and useful. However, as a means of communication to the broader public, it can create separation and distance.

I remember my economics professor asking for a definition of Gresham's law and how I rattled off the answer mechanically. He looked at me—none too pleased, though the answer was correct—and said, now explain it to your grandmother. That was much more difficult. The answer I gave the professor would have made no sense to her. It was lingo. So it is with climate and global warming. Very few people actually understand climate science, yet the basic mechanism of global warming is pretty straightforward.

We have sought to make *Drawdown* understandable to people from all backgrounds and points of view. We endeavored to bridge the climate communication gap by the words we choose, the analogies we avoid, the jargon we stay away from, and the metaphors we employ. As much as possible, we refrain from acronyms and lesser-known climate terminology. We generally spell out *carbon dioxide* instead of abbreviating it. We write *methane*, not CH4.

Let's consider an example. In November 2016, the White House released its strategy for achieving deep decarbonization by mid-century. From our perspective, *decarbonization* is a word that describes the problem, not the goal: we decarbonized the earth by removing carbon in the form of combusted coal, gas, and oil, as well as through deforestation and poor farming practices, and releasing it into the atmosphere. When the word *decarbonization* is used, as it was by the White House, it refers to replacing fossil fuel energy with clean, renewable sources. However, the term is often employed as the overarching goal of climate action—one that is unlikely to inspire and more likely to confuse.

Another term used by scientists is "negative emissions." This term has no meaning in any language. Imagine a negative house, or a negative tree. The absence of something is nothing. The phrase refers to sequestering or drawing down carbon from the atmosphere. We call that sequestration. It is carbon positive, not negative. This is another example where climate-speak removes itself from common parlance and common sense. Our goal is to present climate science and solutions in language that is accessible and compelling to the broadest audience, from ninth graders to pipe fitters, from graduate students to farmers.

We also avoid using military language. Much of the rhetoric and writing about climate change is violent: the war on carbon, the fight against global warming, and frontline battles against fossil fuels. Articles refer to slashing emissions as if we had machetes. We understand the use of these terms because they convey the gravity of what we face and the tightening window of time to address global warming. Yet, terms such as "combat," "battle," and "crusade" imply that climate change is the enemy and it needs to be slain. Climate is a function of biological activity on earth, and physics and chemistry in the sky. It is the prevalent weather conditions over time. Climate changes because it always has and will, and variations of climate produce everything from seasons to evolution. The goal is to come into alignment with the impact we are having on climate by addressing the human causes of global warming and bringing carbon back home.

The term "drawdown" needs explanation as well. The word is conventionally used to describe the reduction of military forces, capital accounts, or water from wells. We use it to refer to reducing the amount of carbon in the atmosphere. However, there is an even more important reason for the use of the word: drawdown names a goal that has been hitherto absent in most conversations about climate. Addressing, slowing, or arresting emissions is necessary, but insufficient. If you are traveling down the wrong road, you are still on the wrong road if you slow down. The only goal that makes sense for humanity is to reverse global warming, and if parents, scientists, young people, leaders, and we citizens do not name the goal, there is little chance it will be achieved.

Last, there is the term "global warming." The history of the concept goes back to the nineteenth century when Eunice Foote (1856) and John Tyndall (1859) independently described how gases trap heat in the atmosphere and how changes in the concentration of gases would alter the climate. The term *global warming* was first used by geochemist Wallace Broecker in a 1975 *Science* article entitled "Climatic Change: Are We on the Brink of a Pronounced Global Warming?" Before that article, the term used was *inadvertent climate modification*. Global warming refers to the surface temperature of the earth. Climate change refers to the many changes that will occur with increases in temperature and greenhouse gases. That is why the U.N. climate agency is called the Intergovernmental Panel on Climate Change—the IPCC, and not the IPGW. It studies the comprehensive impacts of climate change on all living systems. What we measure and model in *Drawdown* is how to begin the reduction of greenhouse gases in order to reverse global warming. —Paul Hawken

NUMBERS

WHAT YOU WILL SEE ON THE PAGE

Behind every one of the solutions in *Drawdown* are hundreds of pages of research and rigorous mathematical models developed by some very bright minds. Each solution includes an introduction that draws on history, science, key examples, and the most current information available. Every description is supported by a detailed technical assessment available on our website for further exploration. Each entry also features a summary of output from the models, including a ranking of the solution by its emissions-reduction potential. We enumerate how many gigatons of greenhouse gases are avoided or removed from the atmosphere, as well as the total incremental cost to implement the solution, and the net cost or—in most cases—savings. In the models, we rely on peer-reviewed science for inputs. In some areas, such as land use and farming, there is a plethora of anecdotal facts and figures, some of which we refer to but we do not use in our calculations.

At the end of the book, you will find a summary table presenting the combined impact of solutions, grouped by sector.

RANKING OF SOLUTIONS

There are several ways one can rank solutions: how cost-effective they are; how quickly they can be implemented; or how beneficial they are to society. All are interesting and useful methods with which to interpret the results. For our purposes, we rank solutions based on the total amount of greenhouse gases they can potentially avoid or remove from the atmosphere. The rankings are global. The relative importance of one solution may differ depending on geography, economic conditions, or sector.

GIGATONS OF CARBON DIOXIDE REDUCED

Carbon dioxide may get the most press, but it is not the only greenhouse gas. Other heat-trapping gases include methane, nitrous oxide, fluorinated gases, and water vapor. Each has long-term impacts on global temperatures, depending on how much of it is in the atmosphere, how long it remains there, and how much heat it absorbs or radiates back out during its lifetime. Based on these factors, scientists can calculate their global warming potential, which makes it possible to have a "common currency" for greenhouse gases, translating any given gas into its equivalent in carbon dioxide.

Each solution in *Drawdown* reduces greenhouse gases by avoiding emissions and/or by sequestering carbon dioxide already in the atmosphere. The degree to which a given solution has a bearing on greenhouse gases is translated into gigatons of carbon dioxide removed between 2020 and 2050. Taken together, they represent the total reduction of greenhouse gases that could be achieved by 2050, compared to a fixed reference case, a world where very little changes.

But what is a gigaton? To appreciate its magnitude, imagine 400,000 Olympic-sized pools. That is about a billion metric tons of water, or 1 gigaton. Now multiply that by 36, yielding 14,400,000 pools. Thirty-six gigatons is the amount of carbon dioxide emitted in 2016.

TOTAL NET COST AND OPERATIONAL SAVINGS

The total cost of each solution in this book is the amount needed to purchase, install, and operate it over thirty years. By comparing this to what we typically would spend on food, fuel for cars, heating and cooling for our homes, etc., we determined the net costs or savings from investing in a given solution.

We err on the side of being conservative. That means assuming costs associated with the solution that are on the high end, and then keeping them relatively constant from 2020 to 2050. Because technologies are changing rapidly and will vary in different parts of the world, we expect the actual cost to be less and the amount of savings higher. Even taking a conservative approach, however, the solutions tend to offer an overwhelming net savings. For some solutions though, the costs and savings are incalculable, as in the cost to save a specific rainforest or support girls' education.

How much are we willing to spend to achieve results that benefit all of humanity? In the back of the book, we summarize the net cost and savings solution-by-solution for comparison. Net savings are based on the operating costs of solutions after implementation from 2020 to 2050. This calculation reveals the cost-effectiveness of the solutions presented. When considering the scale of benefits, the potential profits and savings, and the investments needed if conditions remain the same, the costs become negligible. The payback period for most solutions is relatively short in time.

TO LEARN MORE

The solutions presented in *Drawdown* are only a summary of the full research conducted to support our findings. A more detailed outline of our approach and assumptions can be found in the section "Methodology." We also provide a full description of our research at drawdown.org—how all the data were generated, sources used, and assumptions made.

As you read the book, what will become apparent is how sensible and empowering these solutions are. Rather than a lengthy technical manual, impenetrable to all save experts who have spent their lives immersed in the science behind these technologies, *Drawdown* aims to be accessible to anyone who wants to know what we, collectively, can do and the role each one of us might play. —Chad Frischmann

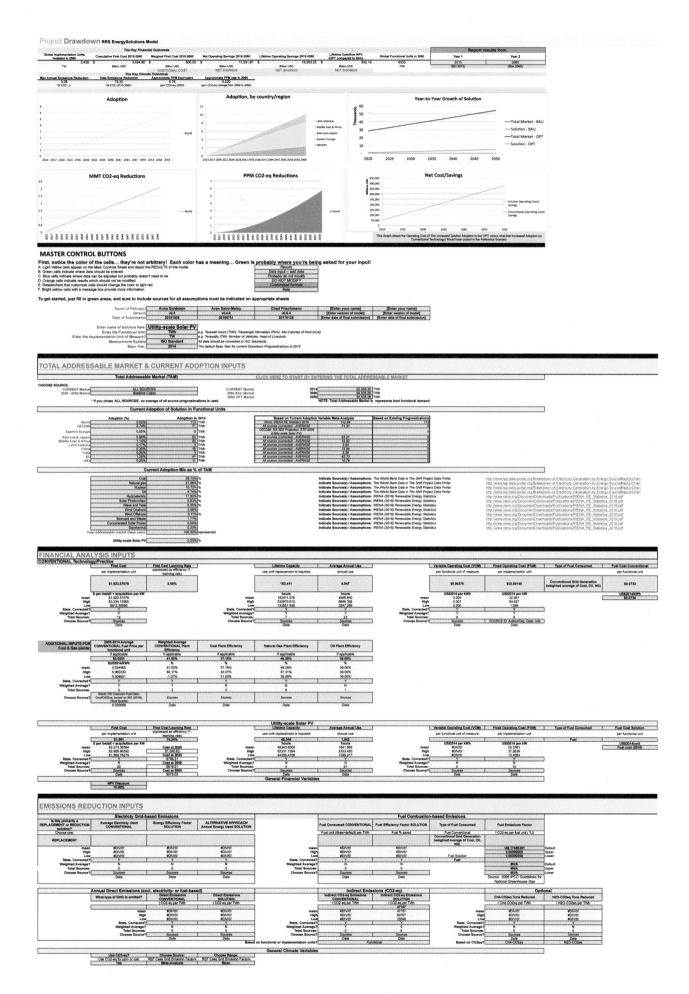

ENERGY

This section highlights the technologies and strategies supplanting energy production from fossil fuels. What were once fools' errands in the energy business, particularly wind and solar, have relentlessly defied predictions and are now competitive with coal, gas, and oil. Renewable costs are continuing to fall on a year-to-year basis, while oil, gas, and coal from new sources are significantly more difficult to extract, which will cause carbon-based fuels to rise in cost. Canada, Finland, and four other countries have banned coal, and more are preparing to. Political leadership is a wonderful thing, but its absence does not slow the renewable transition. The United States pulled out of the Kyoto Protocol in 2001, and that act had virtually no impact on the growth of the renewable energy industry. If you spend a year immersed in the economic data about energy, as we did, there is only one plausible conclusion: We are, in writer Jeremy Leggett's words, squarely in the middle of the greatest energy transition in history. The era of fossil fuels is over, and the only question now is when the new era will be fully upon us. Economics make its arrival inevitable: Clean energy is less expensive.

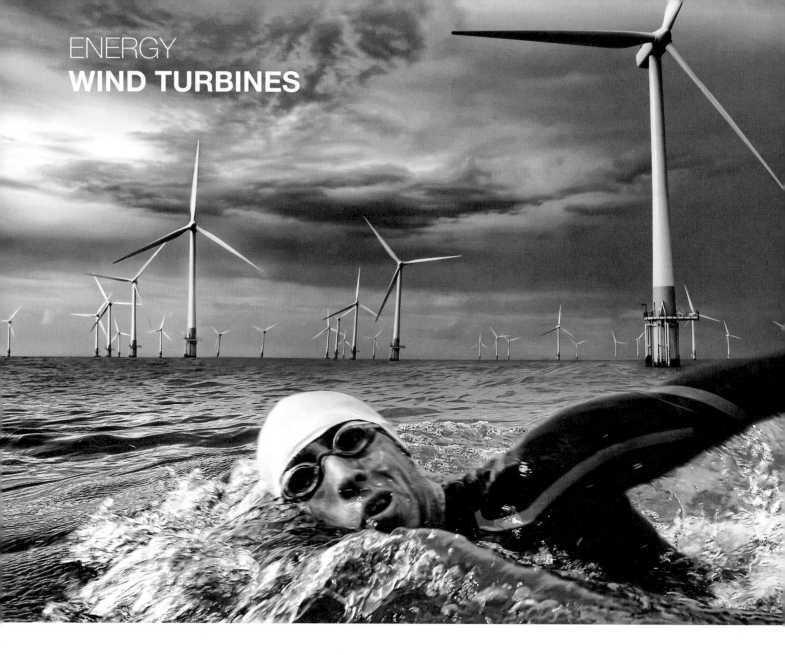

Wind never blows. Because of uneven heating of the earth's surface and the planet's rotation, it is drawn from areas of higher pressure to lower, undulating over and above the landscape like an incoming tide of air. Change is riding on that tide: Wind energy is at the crest of initiatives to address global warming in the coming three decades, second only to refrigeration in total impact.

Take the thirty-two offshore wind turbines—each double the height of the Statue of Liberty—that have been installed off the coast of Liverpool, England, at the Burbo Bank Extension. Owned by a surprising entry into the energy business—Lego, the toy maker—Burbo is an international effort: The blades are made on the Isle of Wight in the United Kingdom by a Japanese company for its Danish client, Vestas. Each turbine generates 8 megawatts of electricity. Their 269-foot blades have a sweep diameter nearly twice the length of a football field, and weigh 33 tons. A single rotation of the blades generates the electricity for one household's daily use. Altogether, the project will supply power for all 466,000 inhabitants of Liverpool.

Today, 314,000 wind turbines supply 3.7 percent of global electricity. And it will soon be much more. Ten million homes in Spain alone are powered by wind. Investment in offshore wind was $29.9 billion in 2016, 40 percent greater than the prior year.

Human beings have harnessed the power of wind for millennia, capturing breezes, gusts, and gales to send mariners and their cargo down rivers and across seas or to pump water and grind grain. The earliest recorded windmills were created around 500 to 900 AD in Persia. The technology spread to Europe during the Middle Ages, and for centuries the Dutch fostered most windmill innovation. By the late 1800s, inventors around the world were successfully converting the kinetic energy of wind into electricity. Prototypical turbines popped up in Glasgow, Scotland, Ohio, and Denmark, and the 1893 World's Columbian Exposition in Chicago featured a variety of manufacturers and their designs. In the 1920s and '30s, farms across the midwestern United States were dotted with wind turbines as a dominant energy source. In 1931, Russia launched utility-scale wind production, and the world's first megawatt turbine went online in Vermont in 1941.

84.6 GIGATONS REDUCED CO2	$1.23 TRILLION NET COST	$7.4 TRILLION NET SAVINGS

RANKING AND RESULTS BY 2050 (OFFSHORE) #22

15.2 GIGATONS REDUCED CO2	$545.3 BILLION NET COST	$762.5 BILLION NET SAVINGS

Fossil fuels sidelined wind energy during the mid-twentieth century. The oil crisis of the 1970s reignited interest, investment, and invention. This modern resurgence paved the way for where the wind industry is today with its proliferation of turbines, dropping costs, and heightened performance. In 2015, a record 63 gigawatts of wind power were installed around the world, despite a dramatic drop in fossil fuel prices. China alone brought nearly 31 gigawatts of new capacity online. Denmark now supplies more than 40 percent of its electricity needs with wind power, and in Uruguay, wind satisfies more than 15 percent of demand. In many locales, wind is either competitive with or less expensive than coal-generated electricity.

In the United States, the wind energy potential of just three states—Kansas, North Dakota, and Texas—would be sufficient to meet electricity demand from coast to coast. Wind farms have small footprints, typically using no more than 1 percent of the land they sit on, so grazing, farming, recreation, or conservation can happen simultaneously with power generation. Turbines can harvest electricity while farmers harvest alfalfa and corn. What's

more, it takes one year or less to build a wind farm, quickly producing energy and a return on investment.

Wind energy has its challenges. The weather is not the same everywhere. The variable nature of wind means there are times when turbines are not turning. Where the intermittent production of wind (and solar) power can span a broader geography, however, it is easier to overcome fluctuations in supply and demand. Interconnected grids can shuttle power to where it is needed. Critics argue that turbines are noisy, aesthetically unpleasant, and at times deadly to bats and migrating birds. Newer designs address these concerns with slower turning blades and siting practices that avoid migration paths. Yet, not-in-my-backyard sentiment—from the British countryside to the shores of Massachusetts—remains an obstacle.

Another impediment to wind power is inequitable government subsidies. The International Monetary Fund estimates that the fossil fuel industry received more than $5.3 trillion in direct and indirect subsidies in 2015; that is $10 million a minute, or about 6.5 percent of global GDP. Indirect fossil fuel subsidies include health costs due to air pollution, environmental damage, congestion, and global warming—none of which are factors with wind turbines. In comparison, the U.S. wind-energy industry has received $12.3 billion in direct subsidies since 2000. Outsize subsidies make fossil fuels look less expensive, obscuring wind power's cost competitiveness, and they give fossil fuels an incumbent advantage, making investment more attractive.

Ongoing cost reduction will soon make wind energy the least expensive source of installed electricity capacity, perhaps within a decade. Current costs are 2.9 cents per kilowatt-hour for wind, 3.8 cents per kilowatt-hour for natural gas combined-cycle plants, and 5.7 cents per kllowatt-hour for utility-scale solar. A Goldman Sachs research paper published in June 2016 stated simply, "wind provides the lowest cost source of new capacity." The cost of both wind and solar includes production tax credits; however, Goldman Sachs believes that the continuing decline in wind turbine costs will make up for the phasing out of tax credits in 2023. Wind projects built in 2016 are coming in at 2.3 cents

An athlete swims past the Sheringham Shoal Offshore Wind Farm off the coast of Norfolk, England. The wind farm consists of 88 Siemens 3.6-megawatt turbines placed over a 14-square-mile area, 11 miles from shore.

per kilowatt-hour. A Morgan Stanley analysis shows that new wind energy production in the Midwest is one-third of the cost of natural gas combined-cycle plants. And finally, Bloomberg New Energy Finance has calculated that "the lifetime cost of wind and solar is less than the cost of building new fossil fuel plants." Bloomberg predicts that wind energy will be the lowest-cost energy globally by 2030. (This accounting does not include the cost of fossil fuels with respect to air quality, health, pollution, damage to the environment, and global warming.)

Costs are going down because turbines are being built at higher elevations—meaning longer blades in locations that have more wind, a combination that has more than doubled the capacity of a given turbine to generate electricity. Onshore turbines can be made larger because assembly is far easier than on water. Turbines that generate 20 megawatts of power with tip heights taller than the Empire State Building are on the drawing boards.

Could the United States power itself with wind? The National Renewable Energy Laboratory calculates that nearly 775,000 square miles of land area is suited to 40 to 50 percent capacity factors, more than twice the average capacity factor a decade ago. (A wind turbine is rated to be able to produce a stated amount of power at a constant given wind speed, however the capacity factor takes into consideration the variability of wind speed in the actual location.) The ways and means for the United States to be fossil fuel and energy independent are here. What is often missing is political will and leadership.

Critics in Congress disparage wind power because it is subsidized, implying that the federal government is pouring money down a hole. Coal is a freeloader when it comes to the costs borne by society for environmental impacts. Putting aside the difference in emissions costs—none for wind, high for fossil fuels—the subsidy arguments do not include the difference in water usage between wind and fossil fuels. Wind power uses 98 to 99 percent less water than fossil fuel–generated electricity. Coal, gas, and nuclear power require massive amounts of water for cooling, withdrawing more water than agriculture—22 trillion to 62 trillion gallons per year. Water for many fossil fuel and nuclear plants is "free," bestowed by the federal government or the states, but it is hardly free and instead represents another unacknowledged subsidy. Who else besides the fossil fuel and nuclear power industries can take trillions of gallons of water in the United States and not pay for it?

China's rise as the world's wind leader demonstrates that consistent government commitment to scaling wind energy can accelerate a declining cost curve, especially if government support remains constant regardless of shifting political winds. A predictable environment is key to the industry's development. On the policy side, portfolio standards can mandate a share of renewable generation. Grants, loans, and tax incentives can encourage construction of more wind capacity and ongoing innovation, on such technologies as vertical-axis turbines and offshore systems. Where governments do support wind energy, such as in the European Union, political action is failing to keep up with the growth of renewable wind energy. In Germany in 2015, bottlenecks in the grid caused 4,100 gigawatt-hours of wind electricity to be wasted—enough energy to power 1.2 million homes for a year. Concerns that wind would be unable to supply enough energy for Europe are being replaced by worries that grid integration and utility and distributed energy storage systems will not keep up with demand.

Wind energy, like other sources of energy, is part of a system. Investment in energy storage, transmission infrastructure, and distributed generation is essential to its growth. Technologies and infrastructure to store excess power are developing quickly now. Power lines to connect remote wind farms to areas of high demand are being built. For the world, the decision is simple: Invest in the future or in the past. ◉

IMPACT: *An increase in onshore wind from 3 to 4 percent of world electricity use to 21.6 percent by 2050 could reduce emissions by 84.6 gigatons of carbon dioxide. For offshore wind, growing from .1 percent to 4 percent could avoid 15.2 gigatons of emissions. At a combined cost of $1.8 trillion, wind turbines can deliver net savings of $8.2 trillion over three decades of operation. These are conservative estimates, however. Costs are falling annually and new technological improvements are already being installed, increasing capacity to generate more electricity at the same or lower cost.*

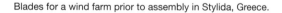

Blades for a wind farm prior to assembly in Stylida, Greece.

ENERGY
MICROGRIDS

AN ENABLING TECHNOLOGY—COST AND SAVINGS
ARE EMBEDDED IN RENEWABLE ENERGY

The "macro" grid is a massive electrical network of energy sources that connects utilities, energy generators, storage, and 24-7 control centers monitoring supply and demand. Anything that is plugged in taps into the grid's centralized power—electricity that is available from large fossil-fuel plants, day or night and rain or shine. This setup made sense when power generation was concentrated. Today, it hinders society's transition from dirty energy produced in a few places to clean energy produced everywhere.

Enter microgrids. A microgrid is a localized grouping of distributed energy sources, such as solar, wind, in-stream hydro, and biomass, together with energy storage or backup generation and load management tools. This system can operate as a stand-alone entity or its users can plug into the larger grid as needed. Microgrids are nimble, efficient microcosms of the big grid, designed for smaller, diverse energy sources. By bringing together renewables and storage, microgrids provide reliable power that can augment the centralized model or operate independently in an emergency situation.

Microgrids will play a critical role in the advancement of a flexible and efficient electric grid. The use of local supply to serve local demand reduces energy lost in transmission and distribution, increasing efficiency of delivery compared to a centralized grid. When coal is burned to boil water to turn a turbine to generate electricity, two-thirds of the energy is dispersed as waste heat and in-line losses.

Microgrid installations in grid-connected regions offer several key advantages. Civilization is dependent on electricity; losing access due to outages or blackouts is a critical risk. In developed countries, economic losses from such events can be many billions of dollars per year. Associated social costs include increased crime, transportation failures, and food wastage, in addition to the environmental cost of diesel-fueled backup power. Studies indicate that as overall demand for electricity increases, owing in part to use of air conditioning and electric vehicles, existing power systems become more frail and blackouts more frequent. By virtue of being localized systems, microgrids are more resilient and can be more responsive to local demand. In the event of disruption, a microgrid can focus on critical loads that require uninterrupted service, such as hospitals, and shed non-critical loads until adequate supply is restored.

In low-income countries, the advantages are greater. Globally, 1.1 billion people do not have access to a grid or electricity. More than 95 percent of them live in sub-Saharan Africa and Asia, a majority in rural areas where highly polluting kerosene lamps are still the main source of lighting and meals are cooked on rudimentary stoves. While the connection between electrification and human development has been clear, progress has remained slow due to the high cost of extending the grid to

This is the Solar Settlement in Freiburg, Germany. A 59-home community, it is the first in the world to have a positive energy balance, with each home producing $5,600 per year in solar energy profits. The way to positive energy is designing homes that are extraordinarily energy efficient, what designer Rolf Disch calls PlusEnergy.

remote regions. In rural parts of Asia and Africa, populations are best supplied with electricity from microgrids (and in remote locations by stand-alone solar).

Establishing microgrids in low-income rural areas is easier than operating them in energy-rich high-income locales. In many places, the business models of large utilities are not compatible with distributed energy and storage. They have sunk costs in a system of generation and delivery that is becoming outmoded. Where utilities are resistant, monopoly, not technology, is the biggest challenge for microgrids. Lessons could cross-pollinate: large grids need to be less rigid and adapt to a changing world; microgrids need to adopt robust technical standards for long-term success. In the age of technological disruption, working out a partnership of technologies makes good sense. •

IMPACT: *We model the growth of microgrids in areas that currently do not have access to electricity, using renewable energy alternatives such as in-stream hydro, micro wind, rooftop solar, and biomass energy, paired with distributed energy storage. It is assumed that these systems replace what would otherwise be the extension of a dirty grid or the continued use of off-grid oil or diesel generators. Emissions impacts are accounted for in the individual solutions themselves, preventing double counting. For higher-income countries the benefits of microgrid systems fall under "Grid Flexibility."*

ENERGY
GEOTHERMAL

Ours is an active planet. A constant flow of heat moves up toward the earth's crust, generating tectonic plate movement, earthquakes, volcanoes, and mountain making. About a fifth of the earth's internal heat is primordial, lingering from the planet's formation 4.6 billion years ago. The balance is generated by ongoing radioactive decay of potassium, thorium, and uranium isotopes in the crust and mantle. The heat energy generated is about 100 billion times more than current world energy consumption. Geothermal energy—literally "earth heat"—creates underground reservoirs of steamy hot water. The geysers of Yellowstone National Park are iconic evidence of a geothermal hot pot simmering beneath us, which occasionally gushes *en plein air*. The hot springs scattered across Iceland's fire-and-ice landscape are another.

Hot water and steam within hydrothermal reservoirs can be piped to the surface and drive turbines to produce electricity—a feat first accomplished in Larderello, Italy, on July 15, 1904, when five lightbulbs were lit by a mechanical device powered by geothermal steam, the invention of Prince Piero Ginori Conti. More than a century later, the Larderello plant still runs, and most of the world's 13 gigawatts of geothermal electricity generation are located along boundaries between tectonic plates, where liquid bodies made themselves apparent on the surface in some way. Another 22 gigawatts of direct geothermal supplies heat for district heating, spas, greenhouses, industrial processes, and other uses.

Geothermal energy is earth energy and depends on heat, an underground reservoir, and water or steam to carry that heat up to the earth's surface. Although prime geothermal conditions are found on less than 10 percent of the planet, new technologies dramatically expand production

Iceland's Svartsengi ("Black Meadow") geothermal power plant, located on the Reykjanes Peninsula in Iceland, was the first geothermal plant designed to both create electricity and provide hot water for district heating. With six different plants, it generates 75 megawatts of electricity, enough to supply 25,000 homes. Its "waste" hot water is piped to the Blue Lagoon Geothermal Spa, visited by 400,000 guests annually.

Maintenance engineer with protective clothing repairs a
pipe connection that is leaking 221° Fahrenheit steam.

potential in areas where useful resources were previously un-known. Conventionally, locating hydrothermal pools is the first step; however, pinpointing subsurface resources has been a challenge and limitation for geothermal power. It is difficult to know where reservoirs are and expensive to drill wells to find out. But new exploration techniques are opening up larger territories.

One of these new approaches is enhanced geothermal systems (EGS), which typically targets deep underground cavities and creates hydrothermal pools where they do not currently exist. EGS uses engineering to make use of areas that contain ample heat but little or no water, adding it in rather than relying on nature's supply. By injecting high-pressure water into the earth, EGS techniques fracture and break up hot rock, making it more permeable and accessible. Once the rock is porous, water can be pumped in via one borehole, heated underground, then returned to the surface via another. After using it for electricity production, injection wells pump spent water back down into the reservoir. Or, in the case of Iceland's Blue Lagoon geothermal spa, the Svartsengi power plant's wastewater becomes bathwater for residents and tourists alike. With recirculation, the cycle repeats.

These innovations could dramatically increase the geographic reach of geothermal energy and, in certain locales, help address a critical challenge for renewables: providing baseload or readily dispatchable power. Wind power dwindles when winds are not blowing. Solar power takes the night off. With subterranean resources flowing 24-7, without interlude, geothermal production can take place at all hours and under almost any weather conditions. Geothermal is reliable, efficient, and the heat source itself is free.

In the process of pursuing its potential, geothermal's negatives need to be managed. Whether naturally occurring or pumped in, water and steam can be laced with dissolved gases, including carbon dioxide, and toxic substances such as mercury, arsenic, and boric acid. Though its emissions per megawatt hour are just 5 to 10 percent of a coal plant's, geothermal is not without greenhouse impact. In addition, depleting hydrothermal pools can cause soil subsidence, while hydrofracturing can produce microearthquakes. Additional concerns include land-use change that can cause noise pollution, foul smells, and impacts on viewsheds.

In twenty-four countries around the world, tackling these drawbacks is proving worthwhile because geothermal power can provide reliable, abundant, and affordable electricity, with low operational costs over its lifetime. In El Salvador and the Philippines, geothermal accounts for a quarter of national electric capacity. In volcanic Iceland, it is one-third. In Kenya, thanks to the activity beneath Africa's Great Rift Valley, fully half of the country's electricity generation is geothermal—and growing. Though less than .5 percent of national electricity production, U.S. geothermal plants lead the world with 3.7 gigawatts of installed capacity.

There is opportunity to pursue geothermal with greater steam and in more places. According to the Geothermal Energy Association, 39 countries could supply 100 percent of their electricity needs from geothermal energy, yet only 6 to 7 percent of the world's potential geothermal power has been tapped. Theoretical projections based on geologic surveys of Iceland and the United States indicate that undiscovered geothermal resources could supply 1 to 2 terawatts of power or 7 to 13 percent of current human consumption. However, that number is significantly lower when capital requirements and other costs and constraints are factored in.

The world's geothermal vanguards point the way forward. They also underscore the importance of government involvement in growing generation. Even with a viable location in hand, geothermal plants can be expensive to bring online. The up-front costs of drilling are especially steep, particularly in less certain, more complex environments. That is why public investment, national targets for its production, and agreements that guarantee power will be purchased from companies that develop it have a crucial role to play in expansion. These measures all help to rein in the level of risk for investing. While hot new technologies such as enhanced geothermal systems advance, continued development of traditional geothermal generation remains indispensable, especially in Indonesia, Central America, and East Africa—places where the planet is most active and "earth heat" is abundant. ●

IMPACT: *Our calculations assume geothermal grows from .66 percent of global electricity generation to 4.9 percent by 2050. That growth could reduce emissions by 16.6 gigatons of carbon dioxide and save $1 trillion in energy costs over thirty years and $2.1 trillion over the lifetime of the infrastructure. By providing baseload electricity, geothermal also supports expansion of variable renewables.*

ENERGY
SOLAR FARMS

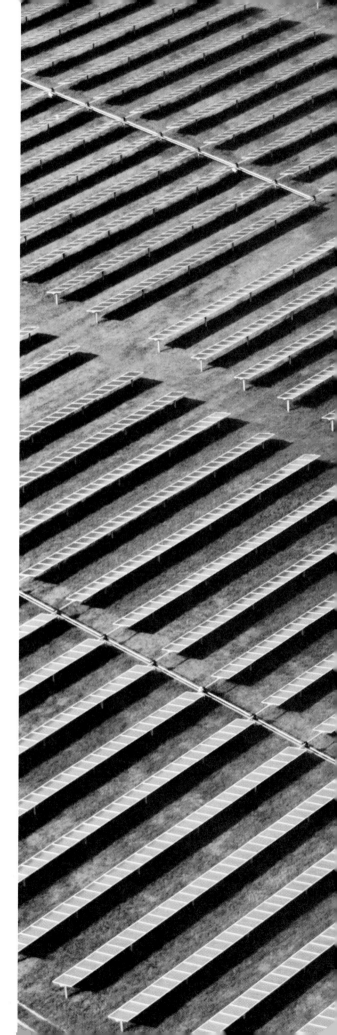

Any scenario for reversing global warming includes a massive ramp-up of solar power by mid-century. It simply makes sense; the sun shines every day, providing a virtually unlimited, clean, and free fuel at a price that never changes. Small, distributed clusters of rooftop panels are the most conspicuous evidence of the renewables revolution powered by solar photovoltaics (PV). The other, less obvious iteration of the PV phenomenon is large-scale arrays of hundreds, thousands, or in some cases millions of panels that achieve generating capacity in the tens or hundreds of megawatts. These solar farms operate at a utility scale, more like conventional power plants in the amount of electricity they produce, but dramatically different in their emissions. When their entire life cycle is taken into account, solar farms curtail 94 percent of the carbon emissions that coal plants emit and completely eliminate emissions of sulfur and nitrous oxides, mercury, and particulates. Beyond the ecosystem damage those pollutants do, they are major contributors to outdoor air pollution, responsible for 3.7 million premature deaths in 2012.

The first solar PV farms went up in the early 1980s. Now, these utility-scale installations account for 65 percent of additions to solar PV capacity around the world. They can be found in deserts, on military bases, atop closed landfills, and even floating on reservoirs, where they perform the additional benefit of reducing evaporation. If Ukrainian officials have their way, Chernobyl, the site of a mass nuclear meltdown in 1986, will house a 1-gigawatt solar farm, which would be one of the world's largest. Whatever the site, *farm* is an appropriate term for these expansive solar arrays because photovoltaics are literally a means of energy harvesting. The silicon panels that make up a solar farm harvest the photons streaming to earth from the sun. Inside a panel's hermetically sealed environment, photons energize electrons and create electrical current—from light to voltage, precisely as the name suggests. Beyond particles, no moving parts are required.

Silicon PV technology was discovered by accident in the 1950s, alongside the invention of the silicon transistor that is present in almost every electronic device used today. That work happened under the auspices of the United States' Bell Labs, accelerated by a search for sources of distributed power that could work in hot, humid, remote locations, where batteries might fail and the grid would not reach. Silicon, the Bell scientists found, was a major improvement over the selenium that had been standard for experimental solar panels since the late 1800s. It achieved more than a tenfold rise in efficiency of converting light to electricity. In the 1954 debut of the Bell "solar battery," a tiny panel of silicon cells powered a twenty-one-inch Ferris wheel and then a radio transmitter. Small as they were, the demos duly impressed the press. The *New York Times* proclaimed it might mark "the beginning of a new era, leading eventually to the realization of one of mankind's most cherished dreams—the harnessing of the almost limitless energy of the sun for the uses of civilization."

A solar farm owned by the Sacramento Municipal Utility District in California, the first municipal district to meet the state's mandated renewable energy standards. The utility sells SolarShares in the solar farms to its ratepayers so that they may harvest a monetary return from the renewable energy revolution in California.

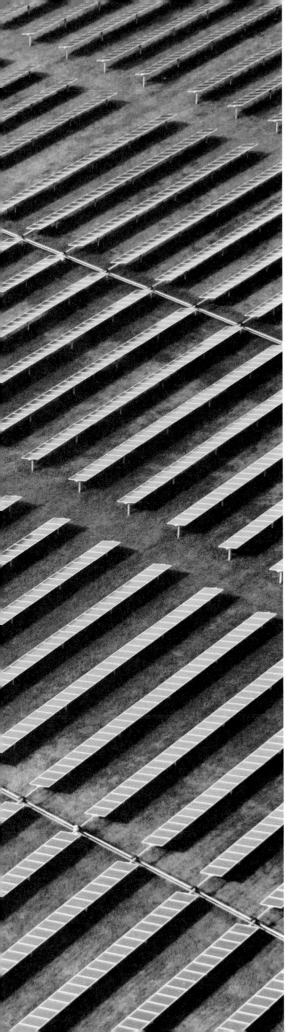

36.9 GIGATONS REDUCED CO2	-$80.6 BILLION NET COST	$5.02 TRILLION NET SAVINGS

At that time, photovoltaics were so expensive (more than $1,900 per watt in today's currency), their only sensible use was in satellites. Up to space they went, but almost nowhere else. Ironically, the first major purchaser of solar cells for use on earth was the oil industry, which needed a distributed energy source for its rigs and extraction operations. Since then, public investment, tax incentives, technology evolution, and brute manufacturing force have chipped away at the cost of creating PV, bringing it down to sixty-five cents per watt today. The decline in price has always outpaced predictions, and drops will continue. Informed predictions about the cost and growth of solar PV indicate that it will soon become the least expensive energy in the world. It is already the fastest growing. Solar power is a solution, but it might be fair to say it is a revolution as well. Constructing a solar farm is also getting cheaper, and it is faster than creating a new coal, natural gas, or nuclear plant. In many parts of the world, solar PV is now cost competitive with or less costly than conventional power generation. Developers are bidding select projects at pennies per kilowatt-hour, which would have been unthinkable a few years ago. Thanks to plunging hard and soft costs, alongside zero fuel use and modest maintenance requirements over time, the growth of large-scale solar has outpaced the most bullish expectations.

Compared to rooftop solar, solar farms enjoy lower installation costs per watt, and their efficacy in translating sunlight into electricity (known as efficiency rating) is higher. When their panels rotate to make the most of the sun's rays, generation can improve by 40 percent or more. At the same time, no matter where solar panels are placed, they are subject to the diurnal and variable nature of solar radiation and its misalignment with electricity use, peaking midday while demand peaks a few hours later. That is why as solar generation continues to grow, so should complementary renewables that are constant, such as geothermal, and that have rhythms different from the sun, such as wind, which tends to pick up at night. Energy storage and more flexible, intelligent grids that can manage the variability of production from PV farms will also be integral to solar's success.

The International Renewable Energy Agency already credits 220 million to 330 million tons of annual carbon dioxide savings to solar photovoltaics, and they are less than 2 percent of the global electricity mix at present. Could solar meet 20 percent of global energy needs by 2027, as some University of Oxford researchers calculate? Thanks to complementary government interventions and market progress, there are many promising signs: costs reaching "grid parity" with fossil fuel generation and dropping, the typical solar panel factory churning out hundreds of megawatts of solar capacity each year, and panels lasting easily for twenty-five years, if not decades more. In 2015, solar PV met almost 8 percent of electricity demand in Italy and more than 6 percent in Germany and Greece, leaders in the solar revolution. PV has had a long history of surpassing expectations and taking unexpected leaps forward. Hand in hand with distributed solar and supported by the right enabling technologies, the "new era" cited by the *New York Times* in 1954 is becoming reality. ●

IMPACT: *Currently .4 percent of global electricity generation, utility-scale solar PV grows to 10 percent in our analysis. We assume an implementation cost of $1,445 per kilowatt and a learning rate of 19.2 percent, resulting in implementation savings of $81 billion when compared to fossil fuel plants. That increase could avoid 36.9 gigatons of carbon dioxide emissions, while saving $5 trillion in operational costs by 2050 — the financial impact of producing energy without fuel.*

ENERGY
ROOFTOP SOLAR

The year was 1884, when the first solar array appeared on a rooftop in New York City. Experimentalist Charles Fritts installed it after discovering that a thin layer of selenium on a metal plate could produce a current of electricity when exposed to light. How light could turn on lights, he and his solar-pioneering contemporaries did not know, for the mechanics were not understood until the early twentieth century when, among other breakthroughs, Albert Einstein published his revolutionary work on what are now called photons. Though the scientific establishment of Fritts's day believed power generation depended on heat, Fritts was convinced that "photoelectric" modules would wind up competing with coal-fired power plants. The first such plant had been brought online by Thomas Edison just two years earlier, also in New York City.

Today, solar is replacing electricity generated from coal as well as from natural gas. It is replacing kerosene lamps and diesel generators in places where people lack access to the power grid, true for more than a billion people around the world. While society grapples with electricity's pollution in some places and its absence in others, the mysterious waves and particles of the sun's light continuously strike the surface of the planet with an energy more than ten thousand times the world's total use. Small-scale photovoltaic systems, typically sited on rooftops, are playing a significant role in harnessing that light, the most abundant resource on earth. When photons strike the thin wafers of silicon crystal within a vacuum-sealed solar panel, they knock electrons loose and produce an electrical circuit. These subatomic particles are the only moving parts in a solar panel, which requires no fuel.

While solar photovoltaics (PV) provide less than 2 percent of the world's electricity at present, PV has seen exponential growth over the past decade. In 2015 distributed systems of less than 100 kilowatts accounted for roughly 30 percent of solar PV capacity installed worldwide. In Germany, one of the world's solar leaders, the majority of photovoltaic capacity is on rooftops, which don 1.5 million systems. In Bangladesh, population 157 million, more than 3.6 million home solar systems

An Uros mother and her two daughters live on one of the 42 floating islands made of totora reeds on Lake Titicaca. Their delight upon receiving their first solar panel is infectious. Installed at an elevation of 12,507 feet, the panel will replace kerosene and provide electricity to her family for the first time. As high tech as solar may be, it is a perfect cultural match: The Uru People know themselves as Lupihaques, Sons of the Sun.

24.6 GIGATONS
REDUCED CO2

$453.1 BILLION
NET COST

$3.46 TRILLION
NET SAVINGS

have been installed. Fully 16 percent of Australian homes have them. Transforming a small section of rooftop into a miniature power station is proving irresistible.

Roof modules are spreading around the world because of their affordability. Solar PV has benefited from a virtuous cycle of falling costs, driven by incentives to accelerate its development and implementation, economies of scale in manufacturing, advances in panel technology, and innovative approaches for end-user financing—such as the third-party ownership arrangements that have helped mainstream solar in the United States. As demand has grown and production has risen to meet it, prices have dropped; as prices have dropped, demand has grown further. A PV manufacturing boom in China has helped unleash a torrent of inexpensive panels around the world. But hard costs are only one side of the expense equation. The soft costs of financing, acquisition, permitting, and installation can be half the cost of a rooftop system and have not seen the same dip as panels themselves. That is part of the reason rooftop solar is more expensive than its utility-scale kin. Nonetheless, small-scale PV already generates electricity more cheaply than it can be brought from the grid in some parts of the United States, in many small island states, and in countries including Australia, Denmark, Germany, Italy, and Spain.

The advantages of rooftop solar extend far beyond price. While the production of PV panels, like any manufacturing process, involves emissions, they generate electricity without emitting greenhouse gases or air pollution—with the infinite resource of sunlight as their sole input. When placed on a grid-connected roof, they produce energy at the site of consumption, avoiding the inevitable losses of grid transmission. They can help utilities meet broader demand by feeding unused electricity into the grid, especially in summer, when solar is humming and electricity needs run high. This "net metering" arrangement, selling excess electricity back to the grid, can make solar panels financially feasible for homeowners, offsetting the electricity they buy at night or when the sun is not shining.

Numerous studies show that the financial benefit of rooftop PV runs both ways. By having it as part of an energy-generation portfolio, utilities can avoid the capital costs of additional coal or gas plants, for which their customers would otherwise have to pay, and broader society is spared the environmental and public health impacts. Added PV supply at times of highest electricity demand can also curb the use of expensive and polluting peak generators. Some utilities reject this proposition and posit contradictory claims of rooftop PV being a "free rider," as they aim to block the rise of distributed solar and its impact on their revenue and profitability. Others accept its inevitability and are trying to shift their business models accordingly. For all involved, the need for a grid "commons" continues, so utilities, regulators, and stakeholders of all stripes are evolving approaches to cover that cost.

The first solar array installed by Charles Fritts in 1884 in New York City. Fritts built the first solar panels in 1881, reporting that the current was "continuous, constant and of considerable force not only by exposure to sunlight but also to dim, diffused daylight, and even to lamplight."

Off the grid, rooftop panels can bring electricity to rural parts of low-income countries. Just as mobile phones leapfrogged installation of landlines and made communication more democratic, solar systems eliminate the need for large-scale, centralized power grids. High-income countries dominated investment in distributed solar until 2014, but now countries such as Chile, China, India, and South Africa have joined in. It means rooftop PV is accelerating access to affordable, clean electricity and thereby becoming a powerful tool for eliminating poverty. It is also creating jobs and energizing local economies. In Bangladesh alone, those 3.6 million home solar systems have generated 115,000 direct jobs and 50,000 more downstream.

Since the late nineteenth century, human beings in many places have relied on centralized plants that burn fossil fuels and send electricity out to a system of cables, towers, and poles. As households adopt rooftop solar (increasingly accompanied and enabled by distributed energy storage), they transform generation and its ownership, shifting away from utility monopolies and making power production their own. As electric vehicles also spread, "gassing up" can be done at home, supplanting oil companies. With producer and user as one, energy gets democratized. Charles Fritts had this vision in the 1880s, as he looked out over the roofscape of New York City. Today, that vision is increasingly coming to fruition. ●

IMPACT: *Our analysis assumes rooftop solar PV can grow from .4 percent of electricity generation globally to 7 percent by 2050. That growth can avoid 24.6 gigatons of emissions. We assume an implementation cost of $1,883 per kilowatt, dropping to $627 per kilowatt by 2050. Over three decades, the technology could save $3.4 trillion in home energy costs.*

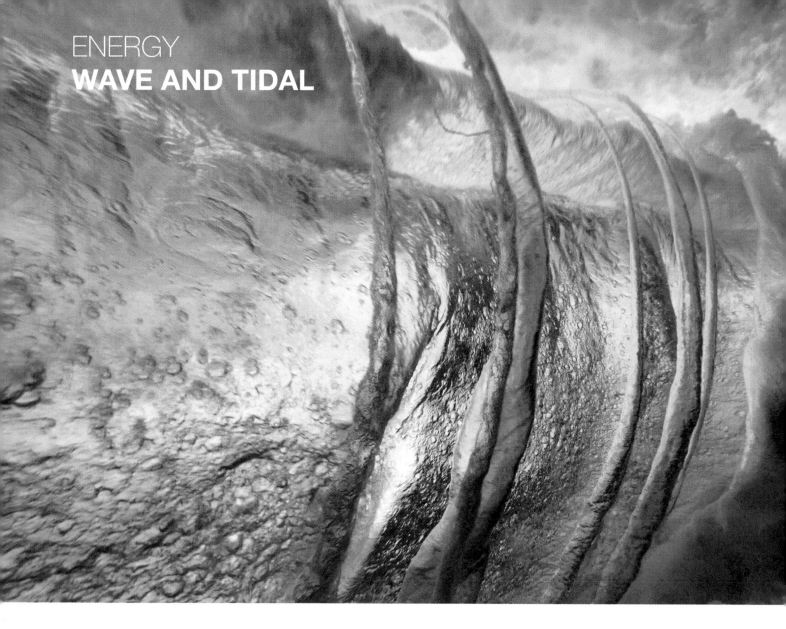

ENERGY
WAVE AND TIDAL

The oceans are in constant motion, rippling, swirling, swelling, retreating. As wind blows across the surface, waves are formed. As the gravitational forces of earth, moon, and sun interact, tides are created. These are among the most powerful and constant dynamics on earth.

Wave- and tidal-energy systems harness natural oceanic flows to generate electricity. A variety of companies, utilities, universities, and governments are working to realize the promise of consistent and predictable ocean energy, which currently accounts for a fraction of global electricity generation. Early technologies date back more than two centuries, with modern designs emerging in the 1960s, thanks especially to the work of Japanese naval commander Yoshio Masuda and his 1947 invention of the oscillating water column (OWC). As a wave or tide rises within an OWC, air is displaced and pushed through a turbine, creating electricity. With the ongoing movement of ocean waters, air is compressed and decompressed continuously. It is the same principle used in whistling buoys, which draw on compressed air to create noise near treacherous shoals or outcroppings. Today, there are several OWC power plants in the world.

The appeal of wave and tidal energy is its constancy: No energy storage is required. And while communities often resist the presence of wind turbines along ridges or shorelines for violating viewsheds, the idea of underwater, out-of-sight wave and tidal systems has proven to be more acceptable to coastal citizens. (Though they can pose concerns for local fishermen, whose livelihoods depend on the same waters.)

When it comes to energy generation, not all waves and tides are created equal. East-west trade winds blow at 30 to 60 degrees latitude, giving the west coasts of all continents the greatest wave activity. Surfing destinations are often wave-energy hot spots. Key locations for vigorous tidal energy are the northeastern coast of the United States, the western coast of the United Kingdom, and the shoreline of South Korea. Many experts also point to smaller islands as candidates for wave and tidal energy, given isolated geographies and limited energy resources.

While the ocean's perpetual power makes wave and tidal energy possible, it also creates obstacles. Operating in harsh and complex marine environments is a challenge—from designing the most effective systems to building installations for their

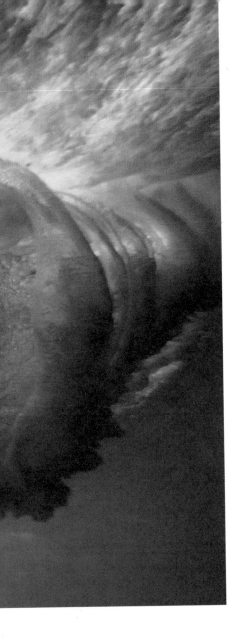

9.2 GIGATONS REDUCED CO2	$411.8 BILLION NET COST	-$1 TRILLION NET SAVINGS

The Annapolis Royal Generating Station is a 20-megawatt power station located on the Annapolis River in Nova Scotia. Built in 1984, it remains the only tidal generating station in North America and takes advantage of the highest tidal range in the world. The difference in height between high and low tides can be over 50 feet. Currently in-stream turbines are being tested nearby, a simpler design with far less environmental impact.

implementation to maintaining them over time. Salt water corrodes equipment, and waves are more multidimensional than a gust of wind—moving up, down, and in all other directions when there are turbulent conditions. It is also critical to ensure marine ecosystems are not harmed by discharges of sound or substance, or by trapping or killing sea life. All told, these dynamics make operating in salt water more exacting and expensive than operating on solid ground.

Marine technologies are still in early development, lagging decades behind solar and wind. Tidal energy is more established than wave, with more projects in operation today. They are ideally suited for natural bays, inlets, or lagoons—places where ocean water enters and exits in circadian cycles—harnessing the incoming and outgoing tides to generate electricity. Some resemble dams, inside which rising or retreating tides drive turbines. More experimental in-stream systems function like underwater wind turbines, with tides moving blades to produce electricity.

Across the world, a variety of wave-energy technologies are being tested and honed, in pursuit of the ideal design for converting waves' kinetic energy into electricity. Some look like yellow buoys bobbing up and down on the ocean's surface. Others resemble large red snakes riding the waves, or long arms waving back and forth. Still others are fully submerged floating discs that incorporate electricity generation right there in the sea. It is not yet clear which technology is most effective. But whatever their shape and form, these systems tap into the upward and downward, the incoming and outgoing movement of waves to power generation. Oscillation is the key, so the higher the wave, the greater its power potential tends to be.

The opportunity of marine-based energy is massive, but realizing it will require substantial investment and expanded research. Proponents believe wave power could provide up to 25 percent of U.S. electricity and 30 percent or more in Australia. In Scotland, that number may be upwards of 70 percent. Wave and tidal energy is currently the most expensive of all renewables, and with the price of wind and solar dropping rapidly, that gap will likely widen. However, as this technology evolves and policy comes into place to support implementation, marine renewables may follow a similar path, attracting private capital investment and the interest of large companies such as General Electric and Siemens. On a trajectory like that, wave and tidal energy could also become cost competitive with fossil fuels. ●

IMPACT: *There are not many projections of wave and tidal energy to 2050. Building on those few, we estimate that wave and tidal energy can grow from .0004 percent of global electricity production to .28 percent by 2050. The result: reducing carbon dioxide emissions by 9.2 gigatons over thirty years. Cost to implement would be $412 billion, with net losses of $1 trillion over three decades, but the investment would pave the way for longer-term expansion and emissions reductions.*

ENERGY
CONCENTRATED SOLAR

So far, concentrated solar power (CSP) "has been a tale of two countries, Spain vs. the U.S." That is how the International Energy Agency sums up the beginning of the story of CSP, also known as solar thermal electricity. The first plants came online in California in the 1980s, and still run today. Instead of capturing energy from the sun's light and converting it directly into electricity like photovoltaics do, they rely on the core technology of conventional fossil fuel generation: steam turbines. The difference is that rather than using coal or natural gas, CSP uses solar radiation as its primary fuel—free and clear of carbon. Mirrors, the essential component of any CSP plant, are curved or angled in specific ways to concentrate incoming solar rays to heat a fluid, produce steam, and turn turbines. As of 2014, this technology was limited to just 4 gigawatts worldwide. Roughly half was in Spain, the one country where CSP is significant enough to show up in national generation statistics, at about 2 percent. Because of CSP's unique advantages, it will grow and those stats will shift. Morocco's giant Noor Ouarzazate Solar Complex, on the edge of the Sahara, is already changing the solar thermal landscape and will be the world's largest when complete.

CSP plants rely on immense amounts of direct sunshine—direct normal irradiance (DNI). DNI is highest in hot, dry regions where skies are clear, typically between latitudes of 15 and 40 degrees. Optimal locales range from the Middle East to Mexico, Chile to Western China, India to Australia. According to a 2014 study in the journal *Nature Climate Change*, the Mediterranean basin and the Kalahari Desert of Southern Africa have the greatest potential for large, interconnected networks of CSP, with the potential to supply power at a cost comparable to that of fossil fuels. In many regions best suited to making solar thermal power, technical generation capacity (the electricity they could be capable of producing) far surpasses demand. With advances in transmission lines, they could supply local populations *and* export power to places where CSP is more constrained.

Rather ironically, the recent success of solar photovoltaic (PV) has limited the growth of solar thermal electricity. PV panels have become so cheap with such speed that CSP has been sidelined; steel and mirrors have not seen the same price plunge. But as PV comes to comprise a greater fraction of the generation mix, it may shift from a damper to a boost. That is because CSP has

the very advantage photovoltaics struggle with and need: energy storage. Unlike PV panels and wind turbines, CSP makes heat before it makes electricity, and the former is much easier and more efficient to store. Indeed, heat can be stored twenty to one hundred times more cheaply than electricity. In the past decade, it has become relatively standard to build CSP plants with storage in the form of molten salt tanks. Warmed with excess heat during the day, molten salt can be kept hot for five to ten hours, depending on the DNI of a particular site, then used to generate electricity when the sun's rays soften. That capacity is crucial for the hours when people remain awake, consuming electricity, but the sun has gone down. Even without molten salt, CSP plants can store heat for shorter periods of time, giving them the ability to buffer variations in irradiance, as can happen on cloudy days—something PV panels cannot do. More flexible and less intermittent than other renewables, CSP is easier to integrate into the conventional grid and can be a powerful complement to solar PV. Some plants pair the two technologies, strengthening the value of both.

Compared to wind and PV generation, the major downside of CSP, to date, is that it is less efficient, in terms of both energy and economics. Solar thermal plants convert a smaller percentage of the sun's energy to electricity than PV panels do, and they are highly capital intensive, particularly because of the mirrors used. Experts anticipate that the reliability of CSP will hasten its growth, however, and as the technology scales, costs could fall quickly. Efficiency of energy conversion is also projected to improve. (Technologies currently under development are already proving it.)

Other downsides require attention as well. Solar thermal typically relies on natural gas as a production backup or, in some cases, a consistent production boost, with accompanying carbon dioxide emissions. The use of heat often implies the use of water for cooling, which can be a scarce resource in the hot, dry places ideal for CSP. Dry cooling is possible, but it is less efficient and more expensive. Lastly, by concentrating channels of intense heat, CSP plants have killed bats and birds, which literally combust in midair. One company, Solar Reserve, has developed an effective strategy to stop bird deaths; spreading that practice for mirror operation will be critical as more plants come online.

Human beings have long used mirrors to start fires. The Chinese, Greeks, and Romans all developed "burning mirrors"— curved mirrors that could concentrate the sun's rays onto an object, causing it to combust. Three thousand years ago, solar igniters were mass-produced in Bronze Age China. They're how the ancient Greeks lit the Olympic flame. In the sixteenth century, Leonardo da Vinci designed a giant parabolic mirror to boil water for industry and to warm swimming pools. Like so many technologies, using mirrors to harness the sun's energy has been lost and found repeatedly, enchanting experimentalists and tinkerers through the ages—and once again today.

IMPACT: *CSP comprised .04 percent of world electricity generation in 2014. Despite slow adoption in recent years, this analysis assumes CSP could rise to 4.3 percent of world electricity generation by 2050, avoiding 10.9 gigatons of carbon dioxide emissions. Implementation costs are high at $1.3 trillion, but net savings could be $414 billion by 2050 and $1.2 trillion over the lifetime of the technology. An additional benefit of CSP is that it can easily integrate energy storage, allowing for extended use after dark.*

The Crescent Dunes Solar Energy Project is a 110-megawatt solar thermal plant located near Tonopah, Nevada. It also is a molten salt storage plant, capable of holding 1.1 billion kilowatt-hours of energy. 10,347 heliostats circle a 640-foot tower at the center and have a combined surface area of 1.28 million square feet. The $1 billion plant produces electricity at 13.5 cents per kilowatt-hour, higher than wind and solar farms to be sure. However, Tonopah provides steady baseload power, which in turn enables intermittent energy from renewable wind and solar to be seamlessly integrated into the grid.

ENERGY
BIOMASS

How does the world get from one powered by fossil fuels to one that runs entirely on energy from the wind, sun, earth's heat, and water's movement? Part of the answer is biomass energy generation. It is a "bridge" solution from status quo to desired state—imperfect, riddled with caveats, and probably necessary. Necessary because biomass energy can produce electricity on demand, helping the grid meet predictable changes in load and complementing variable sources of power, like wind and solar. Biomass can aid the shift away from fossil fuels and buy time for flexible grid solutions to come online, while utilizing wastes that might otherwise become environmental problems. In the near-term, substituting biomass for fossil fuels can prevent carbon stocks in the atmosphere from rising.

Photosynthesis is an energy conversion and storage process; solar energy is captured and stored as carbohydrates in biomass. Under the right conditions and over millions of years, biomass left intact would become coal, oil, or natural gas—the carbon-dense fossil fuels that, at present, dominate electricity production and transportation. Or, it can be harvested to produce heat, create steam for electricity production, or be processed into oil or gas. Rather than releasing fossil-fuel carbon that has been stored for eons far belowground, biomass energy generation trades in carbon that is already in circulation, cycling from atmosphere to plants and back again. Grow plants and sequester carbon. Process and burn biomass. Emit carbon. Repeat. It is a continuous, neutral exchange, so long as use and replenishment remain in balance. Energy efficiency and cogeneration are integral to ensure that, in any given year, carbon from biomass combustion is equal to or less than the carbon uptake of replanted vegetation. When this balance is achieved, the atmosphere sees net zero new emissions.

There is an if: Biomass energy is a viable solution *if* it uses appropriate feedstock, such as waste products or sustainably grown, appropriate energy crops. Optimally, it also uses a low-emission conversion technology such as gasification or digestion. Using annual grain crops such as corn and sorghum for energy production depletes groundwater, causes erosion, and requires high inputs of energy in the form of fertilizer and equipment operation. The sustainable alternative is perennial crops or so-called short-rotation woody crops. Perennial herbaceous grasses such as switchgrass and *Miscanthus* can be harvested for fifteen years before replanting becomes necessary, and they require fewer inputs of water, and labor. Woody crops such as shrub willow, eucalyptus, and poplar are able to grow on "marginal" land not suited to food production. Because they grow back after being cut close to the ground, they can be harvested repeatedly for ten to twenty years. These woody crops circumvent the deforestation that comes with using forests as fuel and sequester carbon more rapidly than most other trees can, but not if they replace already

forested lands. Care needs to be taken with both *Miscanthus* and eucalyptus, however, as they are invasive.

Another important feedstock is waste from wood and agricultural processing. Scraps from saw mills and paper mills are valuable biomass. So are discarded stalks, husks, leaves, and cobs from crops grown for food or animal feed. While it is important to leave crop residues on fields to promote soil health, a portion of those agricultural wastes can be diverted for biomass energy production. Many such organic residues would either decompose on-site or get burned in slash piles, thus releasing their stored carbon regardless (albeit perhaps over longer periods of time). When organic matter decomposes, it often releases methane and when it is burned in piles, it releases black carbon (soot). Both methane and soot increase global warming faster than

7.5 GIGATONS REDUCED CO2	$402.3 BILLION NET COST	$519.4 BILLION NET SAVINGS

This is a single-pass, cut-and-chip harvester reaping fast-growing willow for a carbon-neutral biomass plant, part of Germany's *Energiewende* or "energy turnaround." Germany currently produces 7 percent of its energy from biomass. When the total cost of harvesting and processing wood is calculated, it is not carbon neutral. The industry exists because of significant government subsidies.

trees. The trees will grow back, but over decades—a lengthy and uncertain lag time to achieve carbon neutrality. When biomass energy relies on trees, it is not a true solution.

Biomass is controversial. To some, biomass is a friend; to others, a foe. A considerable academic effort is under way to more accurately assess its environmental and social impacts. Debates center around three main issues: life-cycle carbon emissions (as previously described), indirect land-use change and deforestation, and impacts on food security. Often, the latter two debates are constructed as forests versus fuel and food versus fuel. In reality, managing land, cultivating food, and producing biomass feedstock interact dynamically—and not always in line with conventional wisdom. The three can be mutually reinforcing or play out to one another's detriment, so *how* biomass feedstocks are approached within a given local context matters enormously. At present, biomass fuels 2 percent of global electricity production, more than any other renewable. In some countries—Sweden, Finland, and Latvia among them—bioenergy is 20 to 30 percent of the national generation mix, almost entirely provided for by trees. Biomass energy is on the rise in China, India, Japan, South Korea, and Brazil. Reaching greater scale in more places requires investment in biomass production facilities and infrastructure for collection, transport, and storage. It is crucial to manage, through regulation, the drawbacks of biomass energy. Pelletizing native forests for biomass continues to be a giant step backward. However, extracting invasive species from forests accompanied with appropriate ecological safeguards can be a good source of biomass energy. That approach is being tested in India by the government of the state of Sikkim, which is making "bio-briquettes" for clean cookstoves. Additionally, smallholder farmers need to be protected from displacement by industrial-scale approaches to biomass generation. Most important to bear in mind is that biomass—carefully regulated and managed—is a bridge to reach a clean energy future, not the destination itself. •

IMPACT: *Biomass is a "bridge" solution, phased out over time in favor of cleaner energy sources. This analysis assumes all biomass is derived from perennial bioenergy feedstock—not forests, annuals, or waste—and replaces coal and natural gas in electricity production. By 2050, biomass energy could reduce 7.5 gigatons of carbon dioxide emissions. As clean wind and solar power become more available in a flexible grid, the need for biomass energy will decline.*

carbon dioxide; simply preventing them from being emitted can yield a significant benefit, beyond putting the embodied energy of biomass to productive use.

In the United States, a majority of the more than 115 biomass electricity generation plants under construction or in the permitting process plan on burning wood as fuel. Proponents state that these plants will be powered by branches and treetops left over from commercial logging operations, but these claims do not stand up to scrutiny. In the states of Washington, Vermont, Massachusetts, Wisconsin, and New York, the amount of slash generated by logging operations falls far short of the amount needed to feed the proposed biomass burners. In Ohio and North Carolina, utilities have been more forthright and admit that biomass electricity generation means cutting and burning

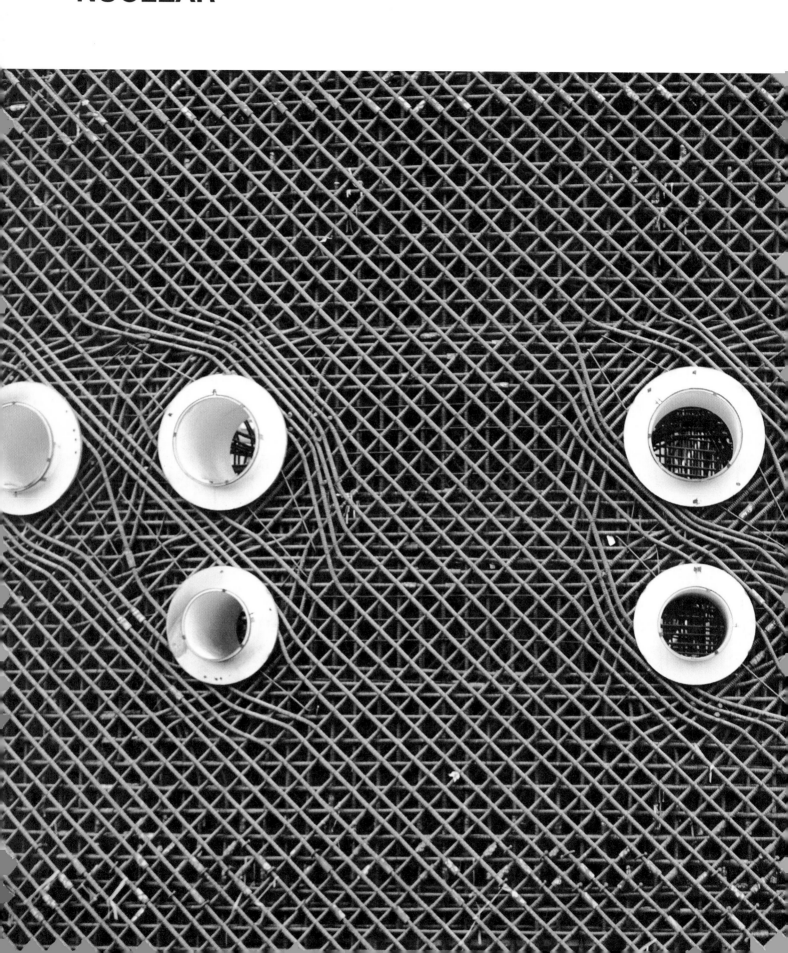

16.09 GIGATONS REDUCED CO2	$.88 BILLION NET COST	$1.7 TRILLION NET SAVINGS

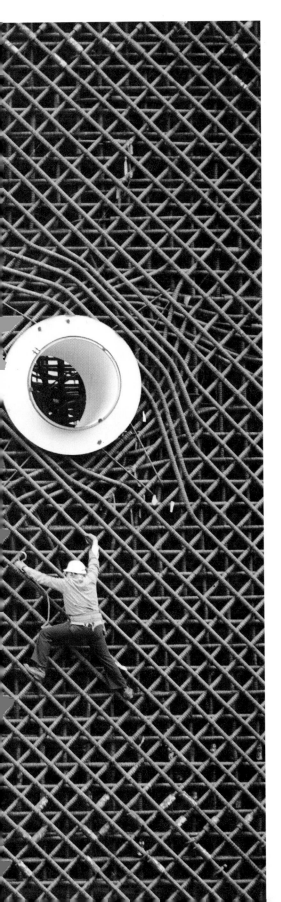

I n effect, nuclear power plants boil water. Nuclear fission splits atomic nuclei and releases the energy that binds the protons and neutrons together. The energy released by radioactivity is used to heat water, which in turn is used to power turbines. It is the most complex process ever invented to create steam. However, nuclear power has a low carbon footprint, which is why it is seen by some people to be a critical global warming solution; many others believe that it is not now, nor will it ever be, cost-effective compared with other low-carbon options. The almost-universal method used to power steam turbines is gas- or coal-fired power. Greenhouse gases emitted to generate electricity are calculated to be ten to a hundred times higher for coal than for nuclear.

Currently, nuclear power generates about 11 percent of the world's electricity and contributes about 4.8 percent to the world's total energy supply. There are more than 440 operating nuclear reactors in 30 countries, and 60 more are under construction. Of the 30 countries countries with operative nuclear power plants, France has the highest nuclear contribution to its electrical energy supply, at over 70 percent.

Nuclear reactors are broadly classified by generation. The oldest, Generation 1, first came online in the 1950s and are now almost entirely decommissioned. The majority of current nuclear capacity falls into the Generation 2 category. (Chernobyl consisted of both Gen 1 and Gen 2. The four Fukushima Daiichi reactors are Gen 2, as are all of the reactors in the United States and France.) Generation 2 distinguishes itself from its predecessor by the use of water (as opposed to graphite) to slow down nuclear chain reactions and the use of enriched, as opposed to natural, uranium for fuel. The Generation 3 reactors, five of which are in operation worldwide and several more under construction, along with Generation 4 reactors, which are currently being researched, constitute what is called "advanced nuclear." In theory, advanced nuclear has standardized designs that reduce construction time and achieve longer operating lifetimes, improved safety features, greater fuel efficiency, and less waste.

What makes the future of nuclear energy difficult to predict is its cost. While the cost of virtually every other form of energy has gone down over time, a nuclear power plant's is four to eight times higher than it was four decades ago. According to the U.S. Department of Energy, advanced nuclear is the most expensive form of energy besides conventional gas turbines, which are comparatively inefficient. Onshore wind is a quarter of the cost of nuclear power.

For those who argue against nuclear because of cost, timing, and safety reasons, the counterargument at one time was the unremitting pace of new coal-fired plant construction. Hundreds of coal-fired plants were being built or planned, primarily in south and east Asia, with three-fourths of them slated to be built by China, India, Vietnam, and Indonesia. If the coal boom is not stopped, global warming will increase far beyond any reasonable limit. This is why climate reporting focuses primarily on energy, and it is why proponents of nuclear are frustrated at the sluggish pace of new plant construction. Licensing, permitting, and financing have brought nuclear plants to a near standstill in the United States, while Germany is shutting its plants down and decommissioning. On the other hand, China has thirty-seven nuclear plants operative and twenty under construction. It is committed to peak carbon dioxide in 2030 with a reduction of its carbon footprint from that date forward.

To give a sense of the scale of a nuclear power plant, this image shows a worker climbing a lattice of steel rods at one of the original Hanford Site nuclear reactors.

Discussion of nuclear power goes right to the heart of the climate dilemma with respect to carbon emissions: Is an increase in the number of nuclear power plants, with all their flaws and inherent risks, worth the risk? Or, as some proponents insist, will there be a total meltdown of climate by limiting their use? Nuclear power has been the subject of contentious disagreements by proponents and critics. The arguments for and against are fascinating, complex, and polarized. Take the following three scientists, widely respected in the environmental community, who do not agree:

According to physicist Amory Lovins, "Nuclear power is the only energy source where mishap or malice can destroy so much value or kill many faraway people; the only one whose materials, technologies, and skills can help make and hide nuclear weapons; the only proposed climate solution that [creates] proliferation, major accidents, and radioactive-waste dangers. . . . [N]uclear power is continuing its decades-long collapse in the global marketplace because it's grossly uncompetitive, unneeded, and obsolete—so hopelessly uneconomic that one need not debate whether it is clean and safe; it weakens electric reliability and national security; and it worsens climate change compared with devoting the same money and time to more effective options."

James Hansen, the NASA scientist who put the United States on notice in his 1988 congressional testimony on climate change, takes another perspective. He authored an open letter with three other climate leaders stating, "Renewables like wind and solar and biomass will certainly play roles in a future energy economy, but those energy sources cannot expand fast enough to deliver cheap and reliable power at the scale the global economy requires. While it may be theoretically possible to stabilize the climate without nuclear power, in the real world there is no credible path to climate stabilization that does not include a substantial role for nuclear power." Their proposal would require building 115 reactors per year for thirty-five years.

Joseph Romm, one of the most respected climate writers and bloggers, does not buy it. Nuclear reactors are too expensive and unwieldy and, given the still-plummeting cost of wind and solar, have priced themselves out of the market. Romm summarizes the perspective of the International Energy Agency (IEA): nuclear can play "an important but limited role. In the IEA's estimation, nuclear can grow from its current 11 percent of generated electricity to 17 percent by 2050.

There seem to be two different worlds here, not one. Nuclear is expensive, and the highly regulated industry in the European Union and the United States may continue to be over-budget and slow. The French company Areva is ten years behind schedule and $5.4 billion over budget on the Olkiluoto reactor in Finland. In Normandy, a $3.4 billion pressurized-water reactor slated for start-up in 2012 will not commence construction until 2018, at a revised cost of $11.3 billion. On the other side of the globe, the largest emitter of carbon in the world is building nuclear reactors more rapidly, motivated in no small part because its cities are extraordinarily polluted from cars and coal-fired power plants. The Chinese nuclear power industry is self-sufficient, in a position to export, and able to complete new plants within two to three years. Yet even where nuclear seems to be "working," there

is a dramatic shift to renewables. China currently leads the world in installed renewable energy capacity, has canceled plans for dozens of coal-fired plants, and is committing to a combined wind and solar capacity of 320 gigawatts by 2020.

Or maybe there is another possibility. Can nuclear power plants be redesigned to be smaller, lighter, safer, and cheaper? That is a question dozens of start-ups are working on. Generation 3 reactors notwithstanding, the nuclear reactor world is stuck on large, expensive, hugely complex systems that are better than those in the past, but that repeat the past. Do large, centralized power plants of any sort make sense in a world of inexpensive renewables, distributed storage, and advanced batteries? Nearly fifty companies are competing to solve the nuclear problem, creating what could be called Generation 4 reactors. These technologies include molten-salt reactors, high-temperature gas reactors, pebble-bed modular reactors, and fusion reactors (hydrogen-boron reactors). There are new reactor designs that address some of the main criticisms and concerns about nuclear energy. These

reactors are being designed to shut down quickly and safely with no one in attendance ("walk-away safety"). They employ better coolants and can scale down to plants one five-hundredth the size of conventional nuclear. They reduce construction time to one or two years. The world may soon have better choices when it comes to nuclear energy than it has had in the past, but it may be too late given the accelerating cost and construction advantages of renewable energy technologies. •

IMPACT: *Nuclear's complicated dynamics around safety and public acceptance will influence its future direction—of expansion or contraction. We assume its share of global electricity generation will grow to 13.6 percent by 2030, but slowly decline to 12 percent by 2050. With a longer lifetime than fossil fuel plants resulting in fewer facilities overall, installation of nuclear power plants could cost an additional $900 million, despite the high implementation cost of $4,457 per kilowatt. Net operating savings over thirty years could reach $1.7 trillion. This scenario could result in 16.1 gigatons of carbon dioxide emissions avoided.*

EDITOR'S NOTE: *One hundred solutions are featured in Drawdown. Of those, almost all are no-regrets solutions society would want to pursue regardless of their carbon impact because they have many beneficial social, environmental, and economic effects. Nuclear is a regrets solution, and regrets have already occurred at Chernobyl, Three Mile Island, Rocky Flats, Kyshtym, Browns Ferry, Idaho Falls, Mihama, Lucens, Fukushima Daiichi, Tokaimura, Marcoule, Windscale, Bohunice, and Church Rock. Regrets include tritium releases, abandoned uranium mines, mine-tailings pollution, spent nuclear waste disposal, illicit plutonium trafficking, thefts of fissile material, destruction of aquatic organisms sucked into cooling systems, and the need to heavily guard nuclear waste for hundreds of thousands of years.*

Steam rises from the Grafenrheinfeld nuclear power plant in Germany. The plant had been in operation since 1981 and ceased operation in June 2015. Germany is withdrawing from nuclear energy and hopes to cease all nuclear power generation by 2022.

ENERGY
COGENERATION

RANKING AND RESULTS BY 2050

#50

3.97 GIGATONS
REDUCED CO2

$279.3 BILLION
NET COST

$567 BILLION
NET SAVINGS

U.S. coal-fired or nuclear power plants are about 34 percent efficient in terms of producing electricity, which means two-thirds of the energy goes up the flue and heats the sky. All told, the U.S. power-generation sector throws away an amount of heat equivalent to the entire energy budget of Japan. Put your hand behind the tailpipe of your car when the engine is running. It is the same principle, only worse—75 to 80 percent of the energy generated by an internal combustion engine is wasted heat. Coal and single-cycle gas generating plants are the best candidates for capturing wasted energy through cogeneration.

Cogeneration puts otherwise-forfeited energy to work, heating and cooling homes and offices or creating additional electricity. Cogeneration systems, also known as combined heat and power (CHP), capture excess heat generated during electricity production and use that thermal energy at or near the site for district heating and other purposes. The opportunity to reduce emissions and save money through cogeneration is significant because of the inherent low efficiency of electrical generation.

Many of the cogeneration systems currently online are found in the industrial sector. In the United States, 87 percent of them are used in energy-intensive industries such as chemical, paper, and metal manufacturing and food processing. In countries such as Denmark and Finland, cogeneration makes up a significant part of electricity production largely because of its use in district heating systems.

In countries with a high-CHP share in total generation, such as Denmark and Finland, the need to address energy security played a decisive role. Denmark's progress came in large part from specific government policies, while Finland's was more market driven. Finland's large paper and forestry industries are naturally motivated to utilize biomass-based cogeneration given the on-site availability of this wood energy resource. Moreover, the cold climate in the country has provided a basis for a healthy return on investment in heat supply infrastructure. As of 2013, 69 percent of Finland's district heating is provided by cogeneration systems.

Denmark's approach to energy supply is policy driven. Although the use of CHP in the country dates back to 1903, it was the 1970s oil crisis that spurred the use of this technology. Since that time, policies have compelled local authorities to identify opportunities for energy-efficient heat production, helped to move power generation from centralized plants to a decentralized network, and incentivized the use of cogeneration generally, and renewable-based systems particularly, through tax policy. Additionally, Denmark has actively participated in United Nations climate change negotiations and made advances to reduce greenhouse gas emissions. Currently around 80 percent of district heating and more than 60 percent of electricity demand is met by CHP, and there are now microcogeneration units available to households. Usually fueled by natural gas, they can be a fuel cell or heat generator that provides electricity, heating, ventilation, and air-conditioning. They are very efficient, but their price and other factors inhibit adoption.

The United States has long lagged behind Europe on cogeneration, in part because of pushback from utilities— notoriously so twenty years ago, when CHP plans at the Massachusetts Institute of Technology were challenged by the local utility. Litigation followed, with the university finally winning in the courts. Such obstruction is rare in today's energy-conscious environment, and MIT's state-of-the-art cogeneration system is nearing completion.

From a financial viewpoint, the adoption of cogeneration systems makes sense for many industrial and commercial uses, as well as for some residential uses. Cogeneration makes it possible for users that do not have access to renewable energy to produce more energy with the same amount, and cost, of fuel. In addition to clear financial benefits, adoption will reduce greenhouse gas emissions to the extent cogeneration reduces reliance on fossil fuels for heating and electricity. Moreover, it will play a substantial role in the ushering in of smart, distributed, and renewable-based energy networks. Because distributed systems are necessarily placed close to the site of generation, they reduce the need for transmission lines. Cogeneration systems are easily adaptable to user preference and thus allow for a variety of energy sources. Additionally, cogeneration systems can help to reduce water usage and thermal water pollution when compared to separate combustion-based heat and power systems, decreasing demand pressure on another vital natural resource.

IMPACT: *In our analysis cogeneration refers to on-site CHP from natural gas in commercial, industrial, and transportation sectors. In 2014, industrial cogeneration using natural gas comprised approximately 3.2 percent of global power generation and 1.7 percent of heat generation. If adoption grows to 5.4 percent of power and 3.3 percent of heat by 2050, 4 gigatons of carbon dioxide emissions can be avoided. At an average installation cost of $1,851 per kilowatt, total installation would cost $279 billion. By replacing grid-based electricity and on-site heat generation with more efficient and less costly technology, the growth in cogeneration could produce operational savings of $567 billion over thirty years and lifetime savings of $1.7 trillion.*

ENERGY
MICRO WIND

With capacity of 100 kilowatts or less, micro wind turbines are akin to the windmills of yore—standing solo in a Kansas cornfield, meeting the electricity needs of a family or small farm or business. They are often used to pump water, charge batteries, and provide electrification in rural locations. Typically, only one is installed at a particular location, on as little as an acre of land, in contrast to the large, sweeping groupings found at commercial wind farms.

When the electric grid was still sparse in many rural U.S. states, on-site wind energy was often used to fill the gap. It is playing a similar role in developing countries today, where these small-scale systems can bring power to the 1.1 billion people around the world without access to electricity, predominantly in rural parts of sub-Saharan Africa and developing Asia. Micro wind turbines are a notable technology for expanding electrification, giving people a way to light their homes or cook their evening meals, which has wide-ranging benefits for well-being and economic development. At the same time, micro wind in high-income countries can be paired with utility-scale renewables, augmenting production. Though the locations may vary widely, micro wind turbines achieve the same climate benefit: energy production without creating greenhouse gases.

Depending on its speed, wind contains a certain amount of kinetic energy. The efficiency with which a turbine extracts power from the wind is called its capacity factor. For small-scale wind turbines, real-world capacity is typically 25 percent or lower. Siting is critical to maximizing their output, but the technology for doing so is in its infancy compared to that for the commercial wind industry. At the same time, micro wind turbines are able to avoid challenges that plague their utility-scale brethren. Being smaller in scale means they avoid aesthetic issues—claims of ruining bucolic views along ridgelines or off coasts—and noise grievances, as many are nearly inaudible.

At present, the major demand for micro wind turbines is for off-grid use. That means they are often installed with a diesel generator to supply electricity when the breeze does not blow. From a carbon perspective, relying on a fossil fuel complement is not ideal. There are already some combined solar photovoltaic and micro wind systems on the market, which is one fruitful alternative. Improved battery storage technology could also boost the viability of small-scale wind. Where these turbines *are* linked up to the grid, owners may be able to send their unneeded electrons out to the larger network for financial return through net metering.

Experts estimate that a million or more micro wind turbines are currently in use around the world, with the majority whirling in China, the United States, and the United Kingdom. The key factor for growing that number is cost in both low- and

This is a VisionAIR5 vertical axis wind turbine that is quieter than a human whisper at low speeds. The turbine is 10.5 feet high and is rated at 3.2 kilowatts of power. The minimum wind speed required is 9 miles per hour and it can withstand speeds up to 110 miles per hour.

high-income countries alike. Currently, the price per kilowatt of small-scale wind is much higher than that of utility-scale turbines, and payback periods can be long, in part because they are installed individually. Acquiring micro-wind technology is beyond the reach of many. Public-support schemes, such as feed-in-tariffs, tax credits, capital subsidies, and net metering, can shift that equation—and have in places where it is thriving. Until small-scale turbine manufacturers can reach economies of scale, end-user cost is likely to remain a challenge. Continued evolution of turbine technology itself also will play an important role in reducing price.

Integrating micro turbines into large structures within the built environment is showing unique promise. Structures that enable turbine placement at high elevation, such as skyscrapers, can take advantage of stronger, steadier breezes. That is one reason visitors to the Eiffel Tower can now find vertical axis turbines on its second level, four hundred feet above the ground, overlooking the Champ de Mars. Their design enables them to utilize wind coming from any direction, producing electricity to power the tower's restaurants, shop, and exhibits. A symbol of engineering innovation, the Eiffel Tower is an appropriate perch for technologies that can help propel a clean energy future.

IMPACT: *Increase micro wind fivefold to 1 percent of global electricity generation by 2050, and it can deliver .2 gigatons of emissions reductions. Like in-stream hydro, micro wind turbines allow for the extension of clean, renewable electricity in areas without grid access.*

Alexander von Humboldt
ANDREA WULF

Though little known or studied today, Alexander von Humboldt (b. September 14, 1769) was a legend in his lifetime, and remains one of the most important scientists in history. More places and species are named after Humboldt than after any other human being. His one hundredth birthday was celebrated all over the world with festivities and parades. More than 25,000 people gathered in Central Park to pay homage, 10,000 in Pittsburgh, 15,000 in Syracuse, 80,000 in Berlin, with thousands more in Buenos Aires, Mexico City, London, and Sydney. As people around the world become more aware of how vulnerable living systems are to global warming, Humboldt's insights and writings seem more than prescient. He was the first person to describe the phenomenon and cause of human-induced climate change, in 1800 and again in 1831, based on observations generated during his travels.

Humboldt's first journey, in 1799, took him on a five-year odyssey through Latin America—an expedition that transformed his thinking and that of the rest of the world. It was here that Humboldt created the idea of isotherms, the lines delineating changes in barometric pressure and temperature on weather maps. His concept of climatic zones came about from his near ascent of Chimborazo, a 20,564-foot inactive volcano in Ecuador. He had taken a trunk full of instruments and measured, described, scrutinized, and drawn the plants, animals, forests, people, and lands encountered with an almost perfect recall, giving him an encyclopedic ability to compare any species with another he had previously seen. During his five-year immersion in largely unspoiled wilderness, Humboldt realized that nature is intricately interconnected in ways that surpass human knowledge. And he saw that

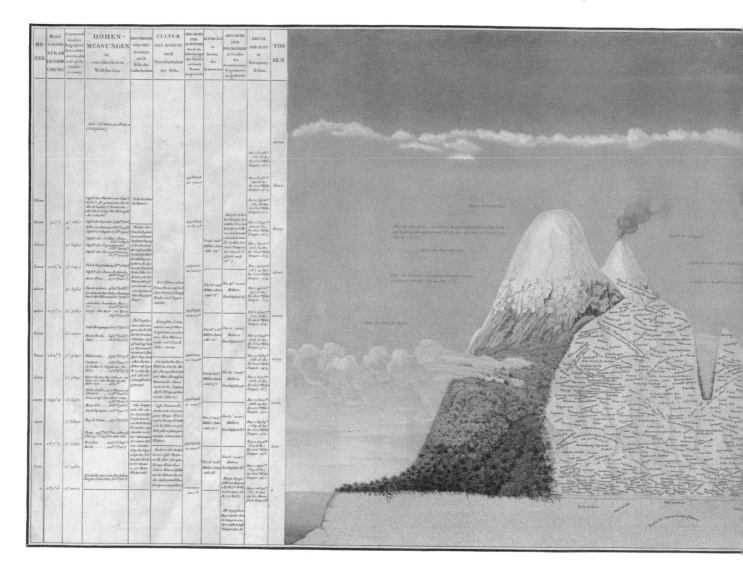

living systems, and indeed the whole of the planet, are highly vulnerable to disturbances by human beings. The principles of the web of life variously described by Darwin, Muir, Emerson, and Thoreau arose directly from Humboldt's Latin America expedition and his subsequent writings.

In 1829, the sixty-year-old Humboldt set off on his last journey, a wide-ranging expedition to Russia arranged after receiving welcoming invitations from Czar Nicholas I and foreign minister Count Georg von Cancrin. In twenty-five weeks, his party traveled 9,614 miles. When he returned, he described precisely and prophetically what could happen to a civilization if it did not recognize how sensitive our atmosphere is to changes on the ground. In this wonderful excerpt from Andrea Wulf's brilliant biography, she describes his return to Moscow and St. Petersburg at the end of his journey. —PH

It was now the end of October and the Russian winter was almost upon them. Humboldt was expected first in Moscow and then in St. Petersburg to report on his expedition. He was happy. He had seen deep mines and snow-capped mountains as well as the largest dry steppe in the world and the Caspian Sea. He had drunk tea with the Chinese commanders at the Mongolian border as well as fermented mare's milk with the Kyrgyz. Between Astrakhan and Volgograd, the learned khan of the Kalmyk choir sang Mozart overtures. Humboldt had watched Saiga antelopes chasing across the Kazakh Steppe, snakes sunbathing on a Volga island and a naked Indian fakir in Astrakhan. He had correctly predicted the presence of diamonds in Siberia, had against his instructions talked to political exiles, and had even met a Polish man who had been deported to Orenburg and who proudly showed Humboldt his copy of *Political Essay of New Spain*. During the previous months Humboldt had survived an anthrax epidemic and had lost weight because he found the Siberian food indigestible. He had plunged his thermometer into deep wells, carried his instruments across the Russian Empire, and taken thousands of measurements. He and his team returned with rocks, pressed plants, fish in vials, and stuffed animals, as well as ancient manuscripts and books for Wilhelm.

As before, Humboldt was not just interested in botany, zoology, or geology but also in agriculture and forestry. Noting the rapid disappearance of the forests around the mining centers, he had written to Cancrin about the "lack of timber" and advised him against using steam engines to drain flooded mines because doing so would consume too many trees. In the Baraba Steppe, where the anthrax epidemic had raged, Humboldt had noted the environmental impact of intense husbandry. The region was (and is) an important agriculture center of Siberia, and the farmers there had drained swamps and lakes to turn the land into fields and pastures. This had caused a considerable desiccation of the marshy plains which would continue to increase, Humboldt concluded.

Humboldt was searching for the "connections which linked all phenomena and all forces of nature." Russia was the final chapter in his understanding of nature—he consolidated, confirmed, and set into relation all the data he had collected over the past decades. Comparison not discovery was his guiding theme. Later, when he published the results of the Russian expedition in two books, Humboldt wrote about the destruction of forests and of humankind's long-term changes to the environment. When he listed the three ways in which the human species was affecting the climate, he named deforestation, ruthless irrigation, and, perhaps most prophetically, the "great masses of steam and gas" produced in the industrial centers. No one but Humboldt had looked at the relationship between humankind and nature like this before. ●

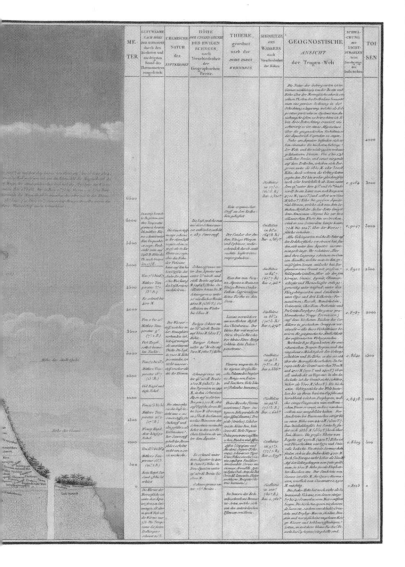

Humboldt's first and most stunning depiction of nature as an interconnected whole was his so-called *Naturgemälde*, a German term that can mean "painting of nature" but that also implies a sense of unity or wholeness. It was, as Humboldt later explained, a "microcosm on one page." In today's parlance, this is probably the first infographic ever created, another first by Humboldt.

ENERGY
METHANE DIGESTERS

The same year Thomas Jefferson penned the U.S. Declaration of Independence, Italian physicist Alessandro Volta discovered methane gas. Intrigued by the flammable air rising up from muddy waters along Lake Maggiore, he captured some and recorded his findings from ensuing experiments in a series of letters to friend and fellow curious mind Carlo Campi. "No, sir, no air is more combustible than the air from marshy soil," Volta wrote on November 21, 1776, beginning to fathom the connection between the gas and decaying vegetation. He went on to engage the fiery power of methane in a pistol of his own design. But it was not until a century later that scientists came to understand that microbes were responsible for the creation of Volta's combustible air. That microbial wisdom is now being used to manage the planet-warming methane emissions that arise from organic waste—creating clean energy in the process.

Agricultural, industrial, and human digestion processes create an ongoing (and growing) stream of organic waste. Around the world, people grow crops, raise animals, make foodstuffs, and nourish themselves. Every one of those activities creates by-products, from residues to excrement. Even with best efforts to reduce, there is no way around waste. Some spoilage, for instance, is inevitable. And, as the saying goes, shit happens. Without thoughtful management, organic wastes can emit fugitive methane gases as they decompose. Molecules of methane that make their way into the atmosphere create a warming effect up to thirty-four times stronger than carbon dioxide over a one-hundred-year time horizon. But that need not be the case. One option is to control their decomposition in sealed tanks called anaerobic digesters, which facilitate the natural processes Volta found along Maggiore's marshy shores. They harness the power of microbes to transform scraps and sludge and produce two main products: biogas, an energy source, and solids called digestate, a nutrient-rich fertilizer.

Harnessing organic waste as an energy resource has a long history. Just before the turn of the twentieth century, sewage-gas lamps illuminated the streets of Exeter, England. A full millennium before, biogas warmed Assyrian bathwater. During his years in ancient China, Venetian explorer Marco Polo encountered covered sewage tanks that produced cooking fuel. An asylum for lepers near Mumbai installed a biogas system in 1859, also for lighting. Today, anaerobic digestion is used around the world at backyard, farmyard, and industrial scales, and is on the rise. Thanks to a supportive regulatory environment, Germany leads the way among established economies with nearly eight thousand methane digesters as of 2014—almost 4,000 megawatts of installed capacity in total. Their adoption is increasing in the United States as well, particularly as attention to methane emissions grows. Small-scale digesters dominate in Asia. More than 100 million people in rural China have access to digester gas.

Whatever size or shape digesters take, the dynamics within are the same. As organic wastes are mixed within an airtight, oxygen-less tank, bacteria and other microbes break them down into their component parts, step by step. Over the course of days or weeks, biogas wafts off the top, while solid digestate falls to the bottom, concentrating nutrients such as nitrogen. Biogas is a blend of methane and carbon dioxide that can be used raw or further purified into biomethane, akin to natural gas. The digestion process unfolds continuously, so long as feedstock supplies are sustained and the microorganisms remain happy.

Additional emissions savings result from how a digester's versatile outputs are put to use. Those end uses tend to depend on the scale of production. At the household level, largely in rural and unelectrified areas in Asia and Africa, biogas is utilized for cooking, lighting, and heating, while digestate enriches home gardens and small agricultural plots. Importantly, biogas can reduce demand for wood, charcoal, and dung as fuel sources and therefore their noxious fumes, which impact both planetary and human health. When produced at industrial scales, biogas can displace dirty fossil fuels for heating and electricity generation. When cleaned of contaminants, it also can be used in vehicles that would otherwise rely on natural gas. On the solids side, digestate supplants fossil fuel–based fertilizers while improving soil health. In addition to reducing greenhouse gases, methane digesters reduce landfill volumes and water-polluting effluent, and eradicate odors and pathogens.

Around the same time Volta was combusting gas, the phrase "Waste not, want not" came into fashion. The Latin root of the word waste, *vastus*, means "uncultivated." The opportunity for digesting organic wastes is, indeed, largely uncultivated. In the face of an ongoing stream of animal and human excrement and organic waste from food production and consumption—and a tandem surge of energy demand—we would do well to take the opportunity to waste not, want not to heart. ●

IMPACT: *Our analysis includes both small and large methane digesters. We project that by 2050, small digesters can replace 57.5 million inefficient cookstoves in low-income economies, while large digesters can grow to 69.8 gigawatts of installed capacity. The cumulative result: 10.3 gigatons of carbon dioxide emissions avoided at a cost of $217 billion.*

ENERGY
IN-STREAM HYDRO

4 GIGATONS REDUCED CO2	$202.5 BILLION NET COST	$568.4 BILLION NET SAVINGS

Kinetic energy is energy in motion. The world's waterways brim with it, as gravity draws water across watersheds, through rivulets and creeks, down larger tributaries, and into rivers flowing seaward. For millennia we have harnessed that energy, first to turn waterwheels and power machinery, then, in the nineteenth century, to generate electricity. Today, hydropower conjures images of massive, landscape-shattering dams: the Three Gorges on upper tributaries of the Yangtze River in China, the Hoover on the Colorado River in the United States, and the Itaipu on the Paraná River, between Paraguay and Brazil. To maximize the kinetic energy available for electricity generation, dams use the vertical distance or "head"—water falls from the top of their structures to their base, rushing over turbine blades with high volume and velocity. Hydroelectric dams produce enormous amounts of electricity. But they also swallow up vast swaths of natural and human habitat—the Three Gorges alone displaced 1.2 million people—while impacting water movement and quality, sediment patterns, and fish migration.

These drawbacks have shifted attention from grand dams to smaller, in-stream turbines that are akin to an updated waterwheel. Placed within a free-flowing river or stream, in-stream turbines can capture hydrokinetic energy without creating a reservoir and its repercussions. The underwater analogue to wind turbines activated by the breeze, their blades rotate as water moves past. No barriers, diversions, or storage are required, only limited structural support, and no emissions ensue. In-stream hydro can produce renewable energy that is ecologically sound. The presence of a submerged apparatus with moving parts will always have some impact on the life of a river or stream, and concerns persist about harming fish populations and impeding their migration. Careful design and installation are of utmost importance.

Though water flows can shift season-to-season and year-to-year, hydrokinetic turbines offer a relatively continuous supply of energy. They must be kept free of debris, but upkeep is minimal and initial costs are low. Because in-stream hydro can function in smaller waterways, where currents' powerful, concentrated energy is often untapped, it is a strong candidate for providing electrification in remote areas. From native communities in rural Alaska to rice fields needing irrigation, this technology is being tested and adopted where expensive and dirty diesel generators have been the conventional source of power. Waterways fed by Himalayan snowmelt are hotbeds of in-stream activity, with the potential to propel rural economic development. In urban environments, in-stream turbines target another hydrokinetic resource: city water mains. In Portland, Oregon, 3.5-foot-wide turbines fit perfectly inside underground pipes. As water rushes down from the Cascade Range to the city, it also generates power for the local utility—without harming flow. This subcategory of in-stream technologies is called conduit hydropower.

Mini hydroelectric power station with 12 kilowatts of installed power produces around 33,000 kilowatt-hours of electricity per year in Bruton, Somerset, England.

According to a national assessment of U.S. hydrokinetic resources, the in-stream energy that is technically recoverable is more than 100 terawatt-hours per year. Roughly 95 percent of it is located in the Mississippi, Alaska, Pacific Northwest, Ohio, and Missouri hydrologic regions. The technology needed to seize that opportunity is fairly new and rare, likened by some to the status of wind power fifteen years ago. Small players populate the industry, but their efforts benefit from the similarities between in-stream and tidal energy and the surge of research and investment in the latter. As entrepreneurs and engineers develop in-stream technologies and governments support those efforts, it is important to bear in mind that not all "run-of-river" projects actually let the river run. Some have diverted waterways' currents, impairing their vitality; others have been stacked up so closely that flooding results when waters run high. If potential missteps are managed and in-stream hydro harnesses river power properly, an ancient form of energy could well be important for our future.

IMPACT: *If in-stream hydro grows to supply 3.7 percent of the world's electricity by 2050, it can reduce 4 gigatons of carbon dioxide emissions and save $568.4 billion in energy costs. Communities in remote mountainous areas are among the last regions in need of electrification; in-stream hydro offers them a reliable and economical method of generating electricity.*

WASTE-TO-ENERGY

Some call this a solution, while others call it pollution. It is certainly the latter. Waste-to-energy is detailed here as a transitional strategy for a world that wastes too much. In *Drawdown*, there are several solutions that we call *regrets solutions*, and this is one of them. A regrets solution has a positive impact on overall carbon emissions; however, the social and environmental costs are harmful and high.

The waste incineration industry in the United States arose from the collapse of the nuclear industry in the 1970s and 1980s. Companies that benefited from building nuclear plants got into a business called "resource recovery," also nicknamed "trash to cash." This solution does not eliminate waste: It releases the energy contained in plastic, paper, foodstuffs, and junk, and leaves a residual ash. In other words, it changes the form of the waste. Some of the heavy metals and toxic compounds latent within the trash are emitted into the air, some are scrubbed out, and some remain in the resulting ash. At that time, a hundred tons of municipal waste created thirty tons of fly ash, a granular substance laden with toxins. The ash goes to landfills lined with plastic to ensure that leachates from the ash do not seep into groundwater. How long the plastic liners last is not known. The amount of ash generated today is much lower due to newer techniques.

There are four methods used by industry to convert waste to energy: incineration, gasification, pyrolysis, and plasma. Waste-to-energy also refers to smaller conversion facilities sited at government agencies, companies, or hospitals that use one of these techniques to dispose of medical, manufacturing, or radioactive waste, as well as tires, sewage sludge, laboratory chemicals, or neighborhood garbage.

So why feature waste-to-energy *in Drawdown* at all? In a sustainable world, waste would be composted, recycled, or reused; it would never be thrown away because it would be designed at the outset to have residual value, and systems would be in place to capture it. Yet cities and land-scarce countries such as Japan face a dilemma: What is to be done with their trash—a veritable Tower of Babel comprising tens of thousands

of different materials and chemicals? Landfilling requires extensive tracts of land, which countries like Japan do not have or cannot afford. If landfill sites are available, burying waste creates methane gas from the decomposition of organic matter, a greenhouse gas that is up to thirty-four times more powerful than carbon dioxide over a one-hundred-year period. Waste-to-energy plants create energy that might otherwise be sourced from coal- or gas-fired power plants. Their impact on greenhouse gases is positive when compared to methane-creating landfills.

Today, the United States burns more than 30 million tons of garbage annually—roughly 13 percent of its total generated waste. The nation's initial foray into incineration was a toxicological disaster. One study conducted in the 1980s of a New Jersey incinerator showed the following results: If 2,250 tons of trash were incinerated daily, the annual emissions would be 5 tons of lead, 17 tons of mercury, 580 pounds of cadmium, 2,248 tons of nitrous oxide, 853 tons of sulfur dioxide, 777 tons of hydrogen chloride, 87 tons of sulfuric acid, 18 tons of fluorides, and 98 tons of particulate matter small enough to lodge permanently in the lungs. The study also showed varying amounts of the persistent toxic pollutant dioxin, depending on the amount of paper and wood involved in incineration. Essentially, inert hazardous waste goes into an incinerator and bioavailable hazardous and toxic emissions come out.

Modern incinerators address these concerns in part. Employing considerably higher temperatures and equipped with scrubbers and filters, almost all traces of pollutants can be captured—but not all. For cities and urban communities, the allure of waste-to-energy plants is compelling. In Europe, more than 480 waste-to-energy plants exist, burning roughly a quarter of all waste. Sweden is among the leaders, importing 800,000 tons of garbage from other countries, at considerable cost in carbon emissions, to fuel its district heating plants—the most extensive network in the world. The Swedes assert that they are very careful about the trash they import: It has to be well sorted with all of the recyclables, including food, removed. Landfills are banned, so if it is not recycled, it is burned.

In a modern Swedish waste-to-energy plant the remaining ash is filtered, removing any metal bits, which are also recycled. Tile or ceramic pieces are gathered to use for gravel in road construction. The use of electric filters negatively charges and removes any particulate matter, and the remaining smoke is considered toxin-free and almost entirely consisting of water and carbon dioxide. Because of higher temperatures, there is a significant reduction in total fly ash. The small remainder goes to landfills. The Swedish municipal association believes that for every ton of garbage, imported or domestic, there is an equivalent savings of 1,100 pounds of carbon dioxide if compared to the garbage being landfilled.

As a strategy for managing our trash, waste-to-energy is better than the landfill alternative when state-of-the-art facilities are employed. In Europe, despite the market for trash (the Germans, Danes, Dutch, and Belgians also are in the business of importing garbage), the rate of recycling, including green waste, is going up, and a 50 percent recycling directive is in place for the year 2020. There is also a strategy for addressing the whole waste stream as effectively as possible: Where more rubbish could be reduced, reused, recycled, or composted, it should be.

Waste-to-energy continues to evoke strong feelings. Its champions point to the land spared from dumps and to a cleaner-burning source of power. One ton of waste can generate as much electricity as one-third of a ton of coal. But opponents continue to decry pollution, however trace, as well as high capital costs and potential for perverse effects on recycling or composting. Because incineration is often cheaper than those alternatives, it can win out with municipalities when it comes to cost. Data shows high recycling rates tend to go hand in hand with high rates of waste-to-energy use, but some argue recycling could be higher in the absence of burning trash. These are among the reasons that construction of new plants in the United States has been at a near standstill for many years, despite evolution in incineration technology.

There is even greater cause for concern in low-income countries, where waste-to-energy can resemble the early toxic incinerators. Public health is particularly an issue in China and East Asia. That is where waste-to-energy is seeing its most rapid market growth, but also where pollution regulation and enforcement are weak. The Green Climate Fund, established by the United Nations, invests in waste-to-energy plants in low-income countries but requires waste sorting, recycling, and removal of toxics.

While some agencies and investors believe waste-to-energy is a renewable source of energy, it is not. Truly renewable resources, like solar and wind, cannot be depleted. There is nothing renewable about burning plastic athletic shoes, CDs, Styrofoam peanuts, and auto upholstery. Waste is certainly a repeatable resource at this point, but that is only because we generate so very much.

Drawdown includes waste-to-energy as a bridge solution: It can help move us away from fossil fuels in the near-term, but is not part of a clean energy future. Even when incineration facilities are state-of-the-art (and many are not), they are not truly clean and toxin-free. The Scotgen gasification incinerator in Dumfries, Scotland, was supposed to be advanced but proved to be one of the country's worst polluters and dioxin emitters. The government shut it down in 2013. Although it may be technically possible to eliminate all dioxin releases, the reality is that measurable breaches of dioxin limits occur at waste-to-energy sites throughout the world. Thus, there are many reasons to oppose plants, especially existing facilities that do not meet the highest standard. But there is another reason we list this as a regrets solution. Waste-to-energy can impede emergence of something better: zero-waste practices that eliminate the need for landfills and incinerators altogether. If this sounds starry-eyed or impractical, know that ten large corporations have committed to zero waste to landfill, including Interface, Subaru, Toyota, and Google.

Zero waste is a growing movement that wants to go upstream, not down, in order to change the nature of waste and the ways in which society recaptures its value. It is saying, in essence, that material flows in society can imitate what we see in forests and grasslands where there truly is no waste that is not feedstock for some other form of life. It relies on green chemistry and material innovation that has the end in mind, not just the beginning. Like solar and wind energy, technologies that were once impractical and unaffordable, zero waste is an engineering and design revolution, which will make waste so valuable that the last thing you would want to do is burn or bury it. Rossano Ercolini of Lucca, Italy, is one of the leaders of the Zero Waste International Alliance. The teacher was galvanized to action when a proposed incinerator was to be built near his school. He successfully stopped that one, and did not pause there. Through his efforts to promote recycling and waste reduction, 117 other Italian municipalities have shut down their waste-to-energy plants and committed to zero waste. That is a true solution, with nothing to regret. ●

IMPACT: *The risks of waste-to-energy are significant, but it has some benefits: 1.1 gigatons of carbon dioxide emissions can be avoided by 2050, primarily due to reduced methane emissions from keeping waste out of landfills. Considering the disadvantages, this is a "bridge" solution—one that will decline as preferable waste-management solutions, including zero waste, composting, and recycling, become more widely adopted globally. Island nations, with limited available space, may continue to use waste-to-energy as an alternative to landfilling—employing more advanced technologies, such as plasma gasification, to limit the negative impacts. At a $36 billion cost to implement, savings over thirty years could be $20 billion.*

ENERGY
GRID FLEXIBILITY

During John Muir's first summer exploring the Sierra Nevada, he wrote in his journal, "When we try to pick out anything by itself, we find it hitched to everything else in the Universe." For more than a century, people have used this quote to describe the interconnectedness of ecosystems and the planetary ripple effects of everything from food to transport. It is also useful for describing the phenomenon of the grid: the dynamic web of electricity production, transmission, storage, and consumption that 85 percent of the world relies on. Increasingly, the phrase "global energy transition" gets bandied about, usually to describe a wholesale shift from fossil fuels to clean, renewable sources of energy. While this shift in sources is the crux of the matter when it comes to greenhouse gas emissions, broader change is afoot: a transformation of the entire grid system.

Some sources of renewable power have constancy akin to that of fossil fuel–generated electricity: geothermal steam, rushing water, or combusted biomass, to name three. Producing electricity from the wind and sun, however, is an intermittent endeavor. With everyday rhythms and variations in wind, they vary from minute to minute, day to day, and season to season. The month of November in Germany, for example, has notoriously low wind and sun, so extra production must come from elsewhere. In addition to variability, solar and wind generation is diverse, ranging from centralized and utility-scale to small and distributed, such as solar on rooftops. Integrating geothermal into the grid is a standard procedure, but the current grid was not designed for wind. Utilities and regulators around the world are grappling with the question: In a rapidly shifting landscape, how can the grid best align electricity supply and end-user demand, keeping lights on and costs in check?

AN ENABLING TECHNOLOGY—COST AND SAVINGS
ARE EMBEDDED IN RENEWABLE ENERGY

a small scale, batteries are the key, including those within electric vehicles. Demand-response technologies, such as web-connected smart thermostats and appliances, can adjust consumers' energy draw on the grid in real time to avoid times of peak demand.

Transmission and distribution networks—the connective tissue between generation and consumption—need to be strong to be flexible. Where grid connections span larger geographies, they encompass broader patterns of wind and sunshine: If the air is still in one place, it is likely moving in another. At any given moment, then, the total output of renewables is less variable. In Spain, the grid operator Red Eléctrica de España controls almost all of the country's wind power production. Working in the aggregate, it can control wind generation to specific levels within fifteen minutes. Interconnection with neighboring power systems, as exists in northwestern Europe, creates additional opportunities for production spillover and backup supply.

There are various operational practices that aid flexibility. When weather and electricity generation go hand in hand, as they do with wind and solar power, forecasting and prediction may be a utility's most important tool. In Denmark, predictions are still made a day in advance, but they also are updated in real time. Comparing forecasts to actual wind output throughout the day and night results in predictability being continually refined. Grid operators can adjust how far in advance generation is scheduled and the length of time for each segment of production. When necessary, suppliers can be required to curtail electricity generation, and negative prices may be used to discourage overproduction, however economically undesirable those measures may be.

By 2050, 80 percent renewable generation could be a global reality. In many grids around the world, renewable energy is already reaching 20 to 40 percent share, including variable renewables as well as constant. So far, the balancing act is working well—better, in fact, than many predicted. More and more jurisdictions soon will be pursuing advanced grid flexibility, integrating the mix of measures that works best for a particular context. Renewable sources in tandem with more flexible grids will make the global energy transition possible. While photovoltaic panels and towering turbines may garner most of the attention, flexibility is the means for renewables to become the dominant form of energy on the planet.

IMPACT: *We do not model grid flexibility because it is a complicated, dynamic system, and it is nearly impossible to account for all local factors at a global scale. However, to grow beyond a 25 percent share of generation, variable renewable energy sources require grid flexibility. The emissions reductions from this solution are counted in the variable renewable solutions that could not reach their full potential without it.*

The answer is *flexibility*. For electricity supply to become predominantly or entirely renewable, the grid needs to become more adaptable than it is today. The front-runners of renewable energy integration, such as California, Denmark, Germany, and South Australia, are showing that grid flexibility stems from a variety of measures—on both the supply and demand sides, as well as utility operations—and looks different in different places. A number of the solutions profiled in this book support a more pliable grid. Constant renewables, such as methane captured from landfills, are valuable complements to wind and solar photovoltaics. Combined-heat-and-power or cogeneration plants can be accessed quickly, especially if they store excess heat in large water tanks. A variety of utility-scale storage measures will be increasingly important, from the long-standing technology of pumped hydro to newer arrivals such as molten salt and compressed air. At

ENERGY STORAGE (UTILITIES)

About eleven thousand years ago, when we humans shifted from hunter-gatherer mode to permanent settlements and agriculture, we started learning about storage. We had no choice, really, because those first crops yielded temporary surpluses that had to be protected from mice and humidity. Earthen, wooden, then ceramic granaries were the early answers. Nowadays we excel at storage. If we make it, we contain it . . . with one notable exception. The most fundamental commodity in the industrialized world—electricity—is one for which storage in volume has not been considered. What is the hedge against brownouts, blackouts, and inefficiency? In the absence of large-scale energy storage, utilities rely on highly polluting "peaker" plants that they rev up to meet high demand. As we seek to reduce emissions from electricity production and enable the shift to variable renewable sources of power, storage is doubly vital.

Since utilities first delivered electricity to paying customers in San Francisco in 1879, the business plan has been to generate sufficient power to meet demand in real time. When the grid could not produce, the lights and motors went out. In some countries, that still happens regularly. As economies shift to variable renewables, management of the power grid with energy storage systems is critical. This includes daily, multiday, and longer-term or seasonal storage. When solar and wind power supplied a small fraction of the total electricity in the grid, their variability was not a major problem; traditional fossil fuel–powered plants could adjust for any shortfalls without undue stress. As renewables begin to account for 30 to 40 percent of total power, the variability becomes more complicated for the grid to cope with reliably and economically. In May 2016, Germany set a global record as the country ran on 88 percent renewable power for several hours, much of it from solar PV. The U.S. renewables record may have been set one February evening in Texas in 2015, when forty-odd wind farms accounted for 45 percent of the grid's total power generation. Unless renewable energy can be used or exported, peaks in production create surpluses that have to be thrown away

AN ENABLING TECHNOLOGY—COST AND SAVINGS ARE EMBEDDED IN RENEWABLE ENERGY

because conventional power plants cannot be turned off. One way to overcome surplus is through high voltage direct current (HVDC) power lines that can extend energy for thousands of miles with small line losses. Additionally, there are a suite of energy-storage technologies that address precisely these issues.

How does a utility store large amounts of electricity? One option is pumping water from lower reservoirs into higher ones, ideally fifteen hundred feet higher. The water is released back down into the lower reservoir as needed and runs through power-generating turbines. Utilities pump the water at night, when electrical power is in surplus, and bring it down again when demand and prices peak. In an example, General Electric has teamed up with a German company to create energy when there is no wind. The project requires a sloping topography with four wind turbines working in concert to generate energy to pump water from a reservoir at a lower elevation to a reservoir at a higher elevation. When wind is lacking or demand is high, the water flowing downhill powers a conventional hydroelectric

plant. All told, there are more than two hundred pumped storage systems in the world at present, accounting for 99 percent of global storage capacity. It is an opportunity that works when the topography obliges.

Nevada is experimenting with energy storage by rail. Here, where there is no water, gravity can still be enlisted. The system takes its cues from the myth of Sisyphus, forever pushing his boulder up a hill. When power is abundant, mining railcars freighted with 230 tons of rock and cement are sent up to a rail yard three thousand feet higher. The railcars are equipped with 2-megawatt generators that act as an engine on the way up. On the way down, a regenerative braking system converts rolling resistance to electrical power.

The technology at the core of both solutions is more than a century old. When the railcars are parked at elevation, they can sit there for a year and not lose any power, while reservoirs evaporate. Both systems share a key advantage: how quickly they can respond to demand. The ramp-up time to full power is seconds, whereas fossil fuel plants take minutes or hours. The grid needs storage at speed.

Concentrated solar power plants are also at the forefront of energy storage, where molten salt is used to hold heat until it is needed to generate electricity. A mix of sodium and potassium nitrate, these salts melt at temperatures above 435 degrees Fahrenheit and can absorb heat reflected by concentrated solar mirrors. Molten salt remains hot for five to ten hours, and returns as much as 93 percent of the energy absorbed. Now a common element of concentrated solar plants, molten salt storage allows generators to keep going hours after sunset.

Then, there are batteries at scale. Some utilities are installing banks of lithium-ion batteries to help meet peak demand. By 2021, Los Angeles plans to take its natural gas peaker plant offline, replacing it with eighteen thousand batteries that will be charged by wind power at night and solar in the morning, while energy needs are low. And dozens of start-ups and established companies are racing to create low-cost, low-toxicity, and safe (no spontaneous ignition) batteries that will revolutionize energy storage from flashlights to utilities—batteries of the future.

IMPACT: *Taken on its own, the production of energy storage does not reduce emissions; instead, energy storage enables adoption of wind and solar energy. No carbon impact numbers are included above in order to prevent double counting with the variable renewable energy solutions themselves. As with other forms of grid flexibility, the costs and total growth are not modeled directly.*

Plus and minus signs indicate the poles on the new energy storage facility at the Fraunhofer Institute in Magdeburg, Germany. During a full-scale test, the entire Fraunhofer research center was supplied with energy from the battery. The lithium-based storage system has an available capacity of 0.5 megawatts per hour and an output of one megawatt. The storage battery is housed in a 26-ton transportable container. This type of equipment is designed to stabilize intermittent and variable energy.

ENERGY STORAGE (DISTRIBUTED)

There is an energy transition under way, one as radical as the adoption of coal, oil, and gas at the beginning of the Industrial Revolution. Most would describe the transformation as the shift away from carbon-based fuels to renewable energy, and they would be right—in part. Another part of the breakthrough will be distributed energy storage—the ability to retain small or large amounts of energy produced where you live or work. If global warming is "the transformation that transforms everything," as sociology and human geography professor Karen O'Brien has observed, distributed energy storage may be the transformation that transforms the energy industry.

Where does your electricity come from? When energy is centrally generated and distributed from large power-generating plants—gas, coal, nuclear, hydro—it feeds into high-voltage transmission lines that crisscross the country into step-down transformers that flow into regional power grids and, finally, your home or place of work. Distributed energy systems turn this sequence on its head. No longer passive consumers, customers can become producers and buy or sell power to the grid when they choose. They can avoid peak demand charges and enable a more resilient grid, preventing demand spikes that can cause brownouts or grid failure.

The wind and sun have their own timetables, making renewable energy variable. That poses a critical challenge for utilities that need to closely monitor supply and demand. The capacity to turn on backup power-generating plants at a moment's notice—lest the grid go down—is critical. Creating a distributed energy storage system, or grid independence, requires affordable storage, and until now, prices for batteries have been prohibitively expensive. That is changing. There are two basic sources of storage: stand-alone batteries and electric vehicles. Storage costs are measured in kilowatt-hours. From $1,200 per kilowatt-hour in 2009, the cost has dropped to roughly $200 in 2016. Companies are now predicting $50 per kilowatt-hour in a few years. For $1,200 per kilowatt-hour, you can purchase a 24-kilowatt-hour energy storage system and get a car thrown in for free—the all-electric Nissan LEAF.

Whether in a car, garage, or the basement of an office building, distributed energy storage is coming faster than expected. Just as every prediction of cost and growth in solar was underestimated for the past two decades, the predictions around battery prices keep missing the mark. In 2012, the global consultancy McKinsey & Company predicted $200-per-kilowatt-hour batteries by 2020, but both General Motors and Tesla achieved that in 2016.

At current cost, a $500 billion investment in distributed energy systems would save U.S. businesses and households $4 trillion in peak-demand utility billing over the next thirty years. Battery cost could halve in the next four years, further amplifying those gains. If storage is used to enable more reliance on renewables there will be substantial climate benefits. If storage is just used to shift peak demand to nights in systems that rely heavily on coal, there will be little benefit.

Not so long ago, solar photovoltaics had high carbon costs. So much coal-fired energy was required for the glass, aluminum, gases, installation, and 3,600-degree Fahrenheit sintering ovens, it would have been fair to call solar panels coal extenders. Today, the energy costs of making solar have dropped significantly. Batteries seem to be following suit; plummeting costs will likely be accompanied by less energy-intensive manufacturing methods. As that occurs, an entirely new energy grid will come online—one that promises to be more resilient and democratic—powered by sensors, apps, and software yet to be invented.

IMPACT: *Distributed energy storage is an essential supporting technology for many solutions. Microgrids, net zero buildings, grid flexibility, and rooftop solar all depend on or are amplified by the use of dispersed storage systems, which facilitate uptake of renewable energy and avert the expansion of coal, oil, and gas electricity generation. Adoption of distributed storage varies depending on whether it is used in an urban or rural setting; those dynamics are not explicitly modeled.*

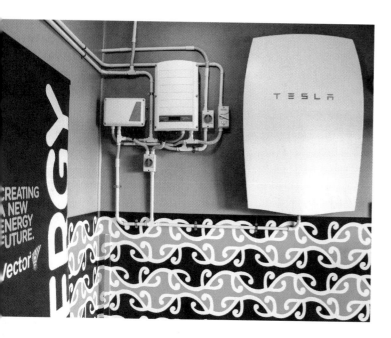

A Tesla Powerwall being installed and celebrated at the Rongomai School in Auckland, New Zealand. The primary school specializes in curriculum designed around Maori cultural values. The battery allows the school to be powered after hours and into the evening from its solar array.

AN ENABLING TECHNOLOGY—COST AND SAVINGS
ARE EMBEDDED IN RENEWABLE ENERGY

For as long as people have bathed, they have sought ways to heat bathwater. During the nineteenth century, the most rudimentary solar-heating technology exposed dark-colored metal tanks to the sun. It worked but was not robust. In 1891, American inventor and manufacturer Clarence Kemp patented a design that improved performance dramatically by using the greenhouse effect. The Climax—the world's first commercial solar water heater—placed iron water tanks inside an insulated, glass-covered box, thereby increasing the tanks' ability to collect and retain solar heat. "Using one of nature's generous forces," Kemp's advertisements proclaimed, the Climax could provide "hot water at all hours of the day and night. No delay. Always charged. Always ready." A residential model cost $25.

At the turn of the twentieth century, solar water heating (SWH) spread across Southern California, as other entrepreneurs worked to improve on Kemp's invention. William Bailey's Day and Night model added a separate storage tank to the rooftop solar heat collector, and revolutionized the industry. As Miami boomed in the 1920s, so did solar collectors—some still operating atop art deco buildings today. During the 1930s, they became standard on public housing in the American South. Cheap energy in the post–World War II years stymied the industry in the United States, but the concept took off in Israel, Japan, and parts of South Africa and Australia. Throughout its history, SWH has

risen and fallen based on the price of energy, as well as government intervention to support it.

Today, China is home to more than 70 percent of the world's SWH capacity, but the technology is in use in many countries and almost every climate, without freezing in winter or overheating in summer. In Cyprus and Israel, where the use of SWH has been mandated since the 1980s, 90 percent of homes have systems. Residential continues to be the primary application for sun-warmed water, though large-scale installations are on the rise. Some systems use tubes, while others employ flat plates; some rely on pumps, while others are passive. As Bailey found, good storage tanks are fundamental. All told, SWH is considered to be "one of the most effective technologies to convert solar energy into thermal energy," with payback periods as short as two to four years, depending on specifics of system, location, and alternatives.

What is also true today is that water heating is a major energy use. Hot water for showers, laundry, and washing dishes consumes a quarter of residential energy use worldwide; in commercial buildings, that number is roughly 12 percent. SWH can reduce that fuel consumption by 50 to 70 percent. But it has yet to be widely tapped as a resource because of up-front costs and complexity of installation, which are higher than gas and electric boilers. Increasingly, SWH gets considered alongside solar photovoltaics, when it comes to roof space, investment, and potential synergies or trade-offs between the two. To achieve uptake at the level Cyprus and Israel have accomplished, governments can require or incent use in new construction—and more and more they are. If the United States maximized its potential for SWH, the country could reduce natural gas consumption by 2.5 percent and electricity use by 1 percent, and avoid producing 57 million tons of carbon each year—as much as 13 coal-fired power plants or 9.9 million cars. With national ambitions for growth in Malawi, Morocco, Mozambique, Jordan, Italy, Thailand, and beyond, clearly SWH has not come close to reaching its zenith, even 125 years after the original Climax was first devised.

IMPACT: *If solar water heating grows from 5.5 percent of the addressable market to 25 percent, the technology can deliver emissions reductions of 6.1 gigatons of carbon dioxide and save households $774 billion in energy costs by 2050. In our calculations of up-front costs, we assume solar water heaters supplement and do not replace electric and gas boilers.*

A solar water array in Esbjerg, Denmark, used for house and district heating, employs buffer tanks for thermal storage. Esbjerg, a port city on the Jutland Peninsula, runs almost entirely on renewable energy and is at the center of Denmark's offshore wind and wave energy industries.

FOOD

Think of the causes of global warming, and fossil fuel energy probably comes to mind. Less conspicuous are the consequences of breakfast, lunch, and dinner. The food system is elaborate and complex; its requirements and impacts are extraordinary. Fossil fuels power tractors, fishing vessels, transport, processing, chemicals, packaging materials, refrigeration, supermarkets, and kitchens. Chemical fertilizers atomize into the air, forming the powerful greenhouse gas nitrous oxide. Our passion for meat involves over 60 billion land animals that require nearly half of all agricultural land for food and pasture. Livestock emissions, including carbon dioxide, nitrous oxide, and methane, are responsible for an estimated 18 to 20 percent of greenhouse gases annually, a source second only to fossil fuels. If you add to livestock all other food-related emissions—from farming to deforestation to food waste—what we eat turns out to be the number one of the greatest causes of global warming along with the energy supply sector. This section profiles techniques, behaviors, and practices that can transform a source into a sink: Instead of releasing carbon dioxide and other greenhouse gases into the atmosphere, food production can capture carbon as a means to increase fertility, soil health, water availability, yields, and ultimately nutrition and food security.

66.11 GIGATONS
REDUCED CO2

GLOBAL COST AND SAVINGS DATA
TOO VARIABLE TO BE DETERMINED

The Buddha, Confucius, and Pythagoras. Leonardo da Vinci and Leo Tolstoy. Gandhi and Gaudí. Percy Bysshe Shelley and George Bernard Shaw. Plant-based diets have had no shortage of notable champions, long before omnivore Michael Pollan famously simplified the conundrum of eating: "Eat food. Not too much. Mostly plants." "Mostly plants" is the key, although some argue all. Shifting to a diet rich in plants is a demand-side solution to global warming that runs counter to the meat-centric, highly processed, often-excessive Western diet broadly on the rise today.

That Western diet comes with a steep climate price tag. The most conservative estimates suggest that raising livestock accounts for nearly 15 percent of global greenhouse gases emitted each year; the most comprehensive assessments of direct and indirect emissions say more than 50 percent. Outside of the innovative, carbon-sequestering managed grazing practices described in this book, the production of meat and dairy contributes many more emissions than growing their sprouted counterparts—vegetables, fruits, grains, and legumes. Ruminants such as cows are the most prolific offenders, generating the potent greenhouse gas methane as they digest their food. In addition, agricultural land use and associated energy consumption to grow livestock feed produce carbon dioxide emissions, while manure and fertilizer emit nitrous oxide. If cattle were their own nation, they would be the world's third-largest emitter of greenhouse gases.

Overconsumption of animal protein also comes at a steep cost to human health. In many places around the world, the protein eaten daily goes well beyond dietary requirements. On average, adults require 50 grams of protein each day, but in 2009, the average per capita consumption was 68 grams per day—36 percent higher than necessary. In the United States and Canada, the average adult consumes more than 90 grams of protein per day. Where plant-based protein is abundant, human beings do not need animal protein for its nutrients (aside from vitamin B12 in strict vegan diets), and eating too much of it can lead to certain cancers, strokes, and heart disease. Increased morbidity and health-care costs go hand in hand.

With billions of people dining multiple times a day, imagine how many opportunities exist to turn the tables. It is possible to eat well, in terms of both nutrition and pleasure, while eating lower on the food chain and thereby lowering emissions. According to the World Health Organization, only 10 to 15 percent of one's daily calories need to come from protein, and a diet primarily of plants can easily meet that threshold.

A groundbreaking 2016 study from the University of Oxford modeled the climate, health, and economic benefits of a worldwide transition to plant-based diets between now and 2050. Business-as-usual emissions could be reduced by as much as 70 percent through adopting a vegan diet and 63 percent for a vegetarian diet (which includes cheese, milk, and eggs). The model also calculates a reduction in global mortality of 6 to 10 percent. The potential health impact on millions of lives translates into trillions of dollars in savings: $1 trillion in annual health-care costs and lost productivity, and upwards of $30 trillion when accounting for the value of lives lost. In other words, dietary shifts could be worth as much as 13 percent of worldwide gross domestic product in 2050. And that does not begin to include avoided impacts of global warming.

Similarly, a 2016 World Resources Institute report analyzes a variety of possible dietary modifications and finds that "ambitious animal protein reduction"—focused on reducing overconsumption of animal-based foods in regions where people devour more than 60 grams of protein and 2,500 calories per day—holds the greatest promise for ensuring a sustainable future for global food supply and the planet. "In a world that is on a course to demand more than 70 percent more food, nearly 80 percent more animal-based foods, and 95 percent more beef between 2006 and 2050," its authors argue, altering meat consumption patterns is critical to achieving a host of global goals related to hunger, healthy lives, water management, terrestrial ecosystems, and, of course, climate change.

The case for a plant-rich diet is robust. That said, bringing about profound dietary change is not simple because eating is profoundly personal and cultural. Meat is laden with meaning, blended into customs, and appealing to taste buds. The complex and ingrained nature of people's relationship with eating animal protein necessitates artful strategies for shifting demand. For individuals to give up meat in favor of options lower on the food chain, those options should be available, visible, and tempting. Meat substitutes made from plants are a key way to minimize disruption of established ways of cooking and eating, mimicking the flavor, texture, and aroma of animal protein and even replicating its amino acids, fats, carbohydrates, and trace minerals. With nutritious alternatives that appeal to meat-centric palates and practices, companies such as Beyond Meat and Impossible Foods are actively leading that charge, proving that it is possible to swap out proteins in painless or pleasurable ways. Select plant-based alternatives are now making their way into grocery store meat cases, a market evolution that can interrupt habitual behaviors around food. Between rapidly improving products, research at top universities, venture capital investment, and mounting consumer interest, experts expect markets for nonmeats to grow rapidly.

Vertumnus by the painter Giuseppe Arcimboldo, created 1590–91, symbolizing the Roman god of metamorphoses.

In addition to meat imitation, the celebration of vegetables, grains, and pulses in their natural form can update norms around these foods, elevating them to main acts in their own right, as opposed to sideshows. Omnivorous chefs are making the case for eating widely and with pleasure *without meat*. They include Mark Bittman, journalist and author of *How to Cook Everything Vegetarian*, and Yotam Ottolenghi, restaurateur and author of *Plenty*. Initiatives such as Meatless Monday and VB6 (vegan before six p.m.), as well as stories that highlight athletic heroes who eat plant-based diets (such as Tom Brady of the New England Patriots), are helping to shift biases around reduced meat consumption. Debunking protein myths and amplifying the health benefits of plant-rich diets can also encourage individuals to change their eating patterns. Instead of being the exception, vegetarian options should become the norm, especially at public institutions such as schools and hospitals.

Beyond promoting "reducetarianism," if not vegetarianism, it is also necessary to reframe meat as a delicacy, rather than a staple. First and foremost, that means ending price-distorting government subsidies, such as those benefiting the U.S. livestock industry, so that the wholesale and resale prices of animal protein more accurately reflect their true cost. In 2013, $53 billion went to livestock subsidies in the thirty-five countries affiliated with the Organisation for Economic Co-operation and Development alone. Some experts are proposing a more pointed intervention: levying a tax on meat—similar to taxes on cigarettes—to reflect its social and environmental externalities and dissuade purchases. Financial disincentives, government targets for reducing the amount of beef consumed, and campaigns that liken meat eating to tobacco use—in tandem with shifting social norms around meat consumption and healthy diets—may effectively conspire to make meat less desirable.

However they are achieved, plant-rich diets are a compelling win-win for society. Eating with a lighter footprint reduces emissions, of course, but also tends to be healthier, leading to lower rates of chronic disease. Simultaneously, it does less damage to freshwater resources and ecosystems—for example, the forests bulldozed to make way for cattle ranching and the immense aquatic "dead zones" created by farm runoff. With billions of animals currently raised on factory farms, reducing meat and dairy consumption reduces suffering that is well documented, often extreme, and commonly overlooked. Plant-based diets also open opportunities to preserve land that might otherwise go into livestock production and to engage current agricultural land in other, carbon-sequestering uses. As Zen master Thich Nhat Hanh has said, making the transition to a plant-based diet may well be the most effective way an individual can stop climate change. Recent research suggests he is right: Few climate solutions of this magnitude lie in the hands of individuals or are as close as the dinner plate. ●

IMPACT: *Using country-level data from the Food and Agriculture Organization of the United Nations, we estimate the growth in global food consumption by 2050, assuming that lower-income countries will consume more food overall and higher quantities of meat as economies grow. If 50 percent of the world's population restricts their diet to a healthy 2,500 calories per day and reduces meat consumption overall, we estimate at least 26.7 gigatons of emissions could be avoided from dietary change alone. If avoided deforestation from land use change is included, an additional 39.3 gigatons of emissions could be avoided, making healthy, plant-rich diets one of the most impactful solutions at a total of 66 gigatons reduced.*

Green chilies going on sale at the Sadarghat Market in Dhaka, Bangladesh.

FOOD
FARMLAND
RESTORATION

14.08 GIGATONS
REDUCED CO2

$72.2 BILLION
NET COST

$1.34 TRILLION
NET SAVINGS

Around the world, farmers are walking away from lands that were once cultivated or grazed because those lands have been "farmed out." Agricultural practices depleted fertility, eroded soil, caused compaction, drained groundwater, or created salinity by over-irrigation. Because the lands no longer generate sufficient income, they are abandoned. Other contributing causes include a changing climate, desertification as in China and the Sahel in Africa, and the results of farming on fragile, steeply sloped land. On the socio-economic side there is migration, the lure of higher income in cities, lack of market access, and high production costs for small-holders when competing with industrial agriculture. Whatever the case, for many, it is cheaper to walk away from the land than to work it.

These abandoned lands are not lying fallow; they are forgotten. Measuring how extensive they are and how quickly they are growing is complex, and different approaches yield different numbers. A comprehensive study out of Stanford University estimates that there are 950 million to 1.1 billion acres of deserted farmland around the world—acreage once used for crops or pasture that has not been restored as forest or converted to development. Ninety-nine percent of that abandonment occurred in the past century.

The quantity of forsaken lands continues to grow, even as the world strains to create more food. To feed a growing population and protect forests from deforestation for fresh farmland, restoring abandoned cropland and pastureland to health and long-term productivity is key. Bringing abandoned lands back into productive use can also turn them into carbon sinks. Like an empty bowl, degraded land can theoretically take up more carbon than fertile ground, as plants draw carbon from the atmosphere and send it back into depleted soils. Where soils are left to erode and diminish further, abandoned farmlands can be a source of greenhouse gas emissions. According to Professor Rattan Lal of the Ohio State University, the world's cultivated soils have lost 50 to 70 percent of their original carbon stock, which combines with oxygen in the air to become carbon dioxide.

Restoration can mean the return of native vegetation, the establishment of tree plantations, or the introduction of regenerative farming methods. In general, the more degraded the land, the more intensive the restoration efforts initially need to be. In less extreme cases, simply allowing natural processes to play out over time—passive restoration—will return the land to a healthy ecosystem. Passive approaches require little money but lots of time. Active restoration is often labor intensive, yet necessary for cultivation to revive. Its costs are higher, but so is its speed to productivity, carbon storage, and ecosystem services.

In the Gulu District of Uganda villagers learn permagardening, which integrates water-saving practices, soil fertility, companion-planting knowledge, and enriched raised beds.

The two strategies need not be mutually exclusive; combining them can aid cost-effectiveness.

Presently, there are few financial incentives to induce farmland restoration. Costs are not inconsequential, and because change is slow, returns on investment lag. For this solution to take root, formal schemes to finance regeneration will be a necessary stimulus to action, helping landowners make changes without (sometimes literally) having to bet the farm. The world's abandoned farmland offers an opportunity to improve food security, farmers' livelihoods, ecosystem health, and carbon draw-down simultaneously. Lal estimates that farmland soils could reabsorb 88 billion to 110 billion tons of carbon, all the while enhancing tilth, fertility, biodiversity, and the water cycle.

The default mode of all land is regeneration. That can be a slow process, but in the hands of skilled practitioners, the economic, social, and ecological benefits of farmland restoration can be greatly accelerated. At the moment, too much former farmland is something someone, for some reason, has abandoned—figuratively thrown away. The world, and many generations of farmers to come, would reap rewards from restoring and reactivating these neglected terrestrial assets.

IMPACT: *Currently, 1 billion acres of farmland have been abandoned due to land degradation. We estimate that by 2050 424 million acres could be restored and converted to regenerative agriculture, or other productive, carbon-friendly farming systems, for a combined emissions impact of 14.1 gigatons of carbon dioxide. This solution could provide a financial return of $1.3 trillion over three decades on an investment of $72 billion, while producing an additional 9.5 billion tons of food.*

REDUCED FOOD WASTE

One of the great miracles of life on this planet is the creation of food. The alchemy human beings do with seed, sun, soil, and water produces figs and fava beans, pearl onions and okra. It can include raising animals for their flesh or yield and transforming raw ingredients into chutney or cake or capellini. For more than a third of the world's labor force, the production of food is the source of their livelihoods, and all people are sustained by consuming it.

Yet a third of the food raised or prepared does not make it from farm or factory to fork. That number is startling, especially when paired with this one: Hunger is a condition of life for nearly 800 million people worldwide. And this one: The food we waste contributes 4.4 gigatons of carbon dioxide equivalent into the atmosphere each year—roughly 8 percent of total anthropogenic greenhouse gas emissions. Ranked with countries, food waste would be the third-largest emitter of greenhouse gases globally, just behind the United States and China. A fundamental equation is off-kilter: People who need food are not getting it, and food that is not getting consumed is heating up the planet.

Losing food to one waste heap or another is an issue in both high- and low-income countries, though the drivers differ. In places where income is low and infrastructure is weak, food loss is typically unintended and structural in nature—bad roads, lack of refrigeration or storage facilities, poor equipment or packaging, a challenging combination of heat and humidity. Wastage occurs earlier in the supply chain, rotting on farms or spoiling during storage or distribution.

In regions of higher income, unintentional losses tend to be minimal; willful food waste dominates farther along the supply chain. Retailers reject food based on bumps, bruises, coloring—aesthetic objections of all sorts. Other times, they simply order or serve too much, lest they risk shortages or unhappy customers. Similarly, consumers spurn imperfect spuds in the produce section, overestimate how many meals they will cook in a week, toss out milk that has not gone bad, or forget about leftover lasagna in the back of the fridge. In too many places, kitchen efficiency has become a lost art.

Basic laws of supply and demand also play a role. If a crop is unprofitable to harvest, it will be left in the field. And if a product is too expensive for consumers to purchase, it will idle in the storeroom. As ever, economics matter. Regardless of the reason, the outcome is much the same. Producing uneaten food squanders a whole host of resources—seeds, water, energy, land, fertilizer, hours of labor, financial capital—and generates greenhouse gases at every stage—including methane when organic matter lands in the global rubbish bin.

There are numerous and varied, but often invisible, dumps of food all around us. The interventions that can address key waste points in the food chain are also numerous and varied. The United Nations' Sustainable Development Goals speak to this chain of "orphaned" food, calling for halving per capita global food waste at the retail and consumer levels by 2030, as well as reducing food losses along production and supply chains, including those that occur postharvest. The root of the problem has many offshoots.

70.53 GIGATONS
REDUCED CO2

GLOBAL COST AND SAVINGS DATA
TOO VARIABLE TO BE DETERMINED

In lower-income countries, improving infrastructure for storage, processing, and transportation is essential. That can be as simple as better storage bags, silos, or crates. Strengthening communication and coordination between producers and buyers is also paramount for keeping food from falling through the cracks. Given the world's many smallholder farmers, producer organizations can help with planning, logistics, and closing capacity gaps.

In higher-income regions, major interventions are needed at the retail and consumer levels. Most important is to preempt food waste before it happens, for greatest reduction of upstream emissions, followed by reallocation of unwanted food for human consumption or another reuse. Standardizing date labeling on food packages is an essential step. Currently, "sell by," "best before," and the like are largely unregulated designations, indicating when food should taste best. Though not focused on safety, these markers confuse consumers about expiration. Consumer education is another powerful tool, including campaigns celebrating "ugly" produce and efforts such as Feeding the 5000—large public feasts made entirely from nearly wasted food.

National goals and policies can encourage widespread change. In 2015, the United States set a food-waste target, aligned with the Sustainable Development Goals. The same year, France passed a law forbidding supermarkets from trashing unsold food and requiring that they pass it on to charities or animal feed or composting companies instead. Italy followed suit. Entrepreneurs are capitalizing on wasted food—from turning homely fruits and veggies into juice to growing mushrooms from used coffee grounds to morphing brewery waste into animal feed. Of course, from an emissions perspective, the most effective efforts are those that avert waste, rather than finding better uses for it after the fact.

Given the complexity of the supply chain that food travels, waste reduction depends on the engagement of diverse actors: food businesses, environmental groups, antihunger organizations, and policy makers. Also critical are the world's 7.4 billion eaters—especially those who live where food waste is greatest: the United States and Canada, Australia and New Zealand, industrialized Asia, and Europe. Whether on the farm, near the fork, or somewhere in between, efforts to reduce food waste can address emissions and ease pressure on resources of all kinds, while enabling society more effectively to supply future food demand. ●

IMPACT: *After taking into account the adoption of plant-rich diets, if 50 percent of food waste is reduced by 2050, avoided emissions could be equal to 26.2 gigatons of carbon dioxide. Reducing waste also avoids the deforestation for additional farmland, preventing 44.4 gigatons of additional emissions. We used forecasts of regional waste estimates from farm to household. This data shows that up to 35 percent food in high-income economies is thrown out by consumers; in low-income economies, however, relatively little is wasted at the household level.*

Left: This is the back end of a processing plant for vegetables in Burscough, Lancashire, UK. If you wonder why you have never seen a crooked carrot in your local market, commercial or natural, this is why. Vegetables are ruthlessly sorted to conform to "quality standards" set by the food chain, and this is the result. Some is carted off to piggeries, some as you can see is already rotting in the water.

Right: Feeding the 5000 is a program developed by founder Tristram Stuart to illustrate the scope of food waste. It is a public event wherein five thousand people are provided a free lunch from ingredients that would otherwise have been thrown away. The event has been held in London, Paris, Dublin, Sydney, Amsterdam, Washington, D.C., and Brussels.

CLEAN COOKSTOVES

Preparing food is at the core of family, culture, and community. Experts debate how long humans have been cooking with fire, but it is likely hundreds of thousands of years. Cooking with heat has a host of benefits: The food is safer, more items become edible, and the taste is richer. Today, we revere chefs such as René Redzepi, Alice Waters, Alain Ducasse, and Madhur Jaffrey for honing the culinary arts and taking them to new heights, yet 3 billion people around the world continue to cook roti, tortillas, and stews hunched over open fires or the most rudimentary of stoves. As human population has swelled, so has the impact of these stoves, with atmospheric repercussions.

The cooking fuels used by 40 percent of humanity are wood, charcoal, animal dung, crop residues, and coal. As these solids burn, often inside homes or in areas with limited ventilation, they release plumes of smoke and soot liable for 4.3 million premature deaths each year. Those most likely to be around the fire are women and the children at their side, inhaling toxic particulate matter and suffering from resulting lung, heart, and eye conditions. Globally, household air pollution is the leading environmental cause of death and disability, ahead of unsafe water and lack of sanitation, and it is responsible for more premature deaths than HIV/AIDS, malaria, and tuberculosis combined.

The harm caused by cooking with these solid fuels extends beyond homes and families to the earth's climate. Traditional cooking practices comprise 2 to 5 percent of annual greenhouse gas emissions worldwide. They stem from two sources. First, unsustainable harvesting of fuel drives deforestation and forest degradation, releasing carbon dioxide. Second, burning fuels during the cooking process emits carbon dioxide, methane, and pollutants from incomplete combustion that include carbon monoxide and black carbon. The latter are known as short-lived climate pollutants—materials that cause warming but do not remain in the atmosphere long.

Black carbon is especially harmful to the climate, as well as to health. This particulate matter is highly light absorbent, soaking up a million times more energy than an equal amount of carbon dioxide. So while black carbon only remains in the atmosphere for eight to ten days, versus decades to centuries for carbon dioxide, it can cause considerable impact during that time. Some researchers point to black carbon as the second-largest driver of climate change, after carbon dioxide. At the same time, its potency, prevalence, and short life span mean that reducing black carbon emissions can have almost immediate impacts on warming. Because household fuel combustion produces roughly a quarter of black carbon emissions, along with other greenhouse gases, clean cookstoves are a key lever for curbing them.

A wide range of "improved" cookstove technologies exists, with an equally wide range of impacts on emissions. Basic efficient stoves offer a small improvement by reducing biomass consumption. Intermediate chimney rocket stoves offer significant fuel savings but, at best, have a limited impact on black carbon—some produce more of it. Advanced biomass stoves that use

---

A woman prepares food on an improved cookstove in her home in the Indian state of Gujarat. The cookstove is made of lightweight metal with a metal alloy combustion chamber. This technology maximizes the lifetime of the stove, quality control, safety, and heat transfer, while minimizing emissions.

gasification technology are the most promising. By forcing gases and smoke from incomplete combustion back into the stove's flame, some cut emissions by an incredible 95 percent, but they are more expensive and can require more advanced pellet or briquette fuels. Those are among the reasons why just 1.5 million households use gasifier stoves at present, mostly in China and India. Solar cookers are an exceedingly clean option, but because they require sunlight and do not work for all foods, they are limited to a supporting role. In the face of this diversity of technologies and impacts, organizations such as the Gold Standard foundation play a vital role, verifying which cookstoves significantly reduce greenhouse gas emissions and would check climate change if dispatched at scale.

At the helm of efforts to make clean cookstoves a worldwide phenomenon is the public-private Global Alliance for Clean Cookstoves (GACC), launched by the United Nations Foundation in 2010. The GACC aims to create a flourishing global market for household cooking technologies that are effective, efficient, and healthy for people and planet. It and its partner organizations have set out to see 100 million such stoves adopted by 2020, with universal adoption achieved by 2030. The GACC reports being ahead of schedule: As of 2015, some 28 million households around the world were cooking on clean cookstoves (though not necessarily those with the greatest impact on greenhouse gases). This global effort builds on decades of work, beginning in earnest in India during the 1950s and first taken to scale with national programs in the 1970s and '80s. The greatest need at this time lies in Asia and sub-Saharan Africa.

The sheer size and breadth of the opportunity is striking, as is the confluence of positive impacts that can result. In many places, women and girls bear the brunt of gathering fuel and preparing meals, so better cooking devices can help redress gender inequities, minimize safety risks during wood collection, and free up time for education or generating income. Healthier eyes, hearts, and lungs relieve the burden of disease and death; well-being rises. More efficient fuel combustion reduces pressure on forests and curtails air pollution and greenhouse gas emissions. The sum total of these impacts means clean cookstoves can help root out poverty and boost livelihood. As the GACC asserts, "The global community cannot reach its goals of eradicating poverty and addressing climate change without addressing the way millions of people cook."

A diverse host of actors is responding to this multidimensional opportunity, from international nongovernmental organizations (NGOs), donors, and carbon financiers to government agencies, researchers, and social entrepreneurs. But success has proven complex and often elusive. In the past, too many stoves were designed and tested in laboratory settings, and they did not translate well to life. Nuanced needs and wants were crudely understood—even those as basic as cooking with more than one pot at a time. Local materials were not up to par for fabrication. Stove durability was poor, and repair issues were not anticipated. Concerned with supply, manufacturers often overlooked demand. Moreover, many "improved" stoves did little to reduce emissions or exposure to smoke and soot. The need to accelerate a next generation of well-made, culturally attuned, low-pollution stoves is clear.

Though cookstoves may seem simple, taking them from concept to reality is as much an art as cooking itself. Family dynamics, from finances to education to gender roles, affect decisions about stoves, which must meet a suite of needs. These include preparing traditional dishes in traditional pots and achieving desired flavor; working with locally available fuels; saving on the cost of fuel or time spent obtaining it; making cooking easy, efficient, and safe; and, of course, affordability. As with any technology, happy early adopters are key, and unhappy early adopters can be hard to win back. That is why the most successful design efforts are not just created *for* but *with* end users, cocreating the ideal technology. When it comes to stoves, context really counts, so testing stoves in the field, for both technical and sociocultural performance, is critical. Locally attuned, human-centered designs are most likely to win hearts and minds and shift prevailing habits—and, most important, majority share of cooking time.

Clean cooking can lead to swift change for the climate. Some researchers place the emissions-reduction opportunity in the realm of 1 gigaton of carbon dioxide or its equivalent per year. Scaling the development and adoption of affordable, suitable, and durable cooking technologies is essential to the realization of what is possible. The GACC and leading experts are working to develop international standards that can ensure stoves meet baseline performance, inform government policy and philanthropic initiatives, and help consumers make more informed choices. Even the best technology cannot succeed without strong financing and distribution—areas equally in need of innovation. Funds for research and development, targeted subsidies, distribution support, educational efforts, and special loans are already helping; many millions more are needed. As funding continues to grow, interventions can target priority areas, such as countries where wood-fuel use per capita is highest, to achieve greater impact in the interim. The world's constellation of efforts to make clean stoves is where the future of cooking matters most. ●

IMPACT: *As of 2014, clean cookstoves comprised just 1.3 percent of the addressable market. If adoption grows to 16 percent by 2050, reductions in emissions will amount to 15.8 gigatons of carbon dioxide. The additional benefits to the health of millions of households is not calculated here.*

FOOD
MULTISTRATA AGROFORESTRY

Strata are horizontal layers. The word's Latin root means "something spread out or laid down," like a blanket. These layers are one of the defining features of forests, from undergrowth to understory, from canopy to emergents—the tallest trees that peek out from the top of a tropical forest's shady density and into the bright light above. Each layer rising up from the forest floor teems with life and activity. Multistrata agroforestry takes its cues from this natural structure, blending an overstory of taller trees and an understory of one or more layers of crops. Think of this as the Manhattan of food production, maximizing both horizontal and vertical space. If natural forests grow food for the species within them, multistrata agroforestry sets out to cultivate food for humans as well. The blend of plants varies by region and culture, but the spectrum includes macadamia and coconut, black pepper and cardamom, pineapple and banana, coffee and cacao, as well as useful materials such as rubber and timber.

Because multistrata agroforestry mimics the structure of forests, it can deliver similar environmental benefits. Multistrata systems can prevent erosion and flooding, recharge groundwater, restore degraded land and soils, support biodiversity by providing habitat and corridors between fragmented ecosystems, and absorb and store significant amounts of carbon. Thanks to the many layers of vegetation supporting sequestration in both soil and biomass, an acre of multistrata agroforestry can achieve rates of carbon sequestration that are comparable to those of afforestation and forest restoration—2.8 tons per acre per year, on average—with the added benefit of producing food. At times, the sequestration rates for multistrata agroforestry plots can out-sequester nearby natural forests.

At present, there are almost 250 million acres of multistrata agroforestry in the world, primarily in the tropics. That number has held steady in recent decades. It includes shade-grown varieties of two of the world's most beloved goods: coffee and cacao (for chocolate). Cacao plants grow in the shade on nearly 20 million acres. Shade-grown coffee accounts for almost 15 million acres. All coffee was once grown under canopy, the conditions the classic varietal arabica thrives in. But in an effort to increase yields, many farmers have shifted to full-sun operations, planting the less flavorful robusta variety instead. There are short-term yield gains, but they come at a cost: Full-sun coffee farms are monocultures that rapidly deplete their soil resources. Multistrata coffee plants live two to three times longer than sun grown, and shade farms can sustain for hundreds of years. They have better natural pest control, fertilization, and water absorption, all of which save farmers money. Needing fewer chemical inputs, if any, they are safer places for workers too, because of less exposure to toxic material. Shade-grown coffee is a higher-quality product, which gives it the potential to command higher prices. The same is true for chocolate made from shade-grown cacao.

Home gardens are another important approach to multistrata agroforestry. Dating back to 13,000 BC, these are small plots comprising dense, diverse layers of trees and crops, planted where people live. The two oldest Sanskrit epic poems, *The Ramayana* and *The Mahabharata*, contain illustrations of a precursor to the home garden called Ashok Vatika. They have been an important part of the "life scape" in Java, Indonesia, and Kerala, India, for millennia. Today, Indonesia alone is home to more than 12 million home garden acres. Given their proximity to the kitchen, home gardens have a central purpose of feeding families, and they can produce medicinal plants and items to take to market. Because they generate food security, nourishment, and income, on top of ecological benefits, home gardens have been dubbed "the epitome of sustainability" by agroforestry expert P. K. Nair. Though their origins lie in rural, tropical, subsistence-oriented areas, they are sprouting as an urban phenomenon, and temperate home gardens are increasingly taking root.

Whether the crop being grown is coffee, cacao, fruit, vegetables, herbs, fuel, or plant remedies, the benefits of multistrata

This image shows part of Fazenda da Toca, a 5,700-acre farm managed by Pedro Diniz in Itirapina, Brazil. Employing regenerative farming and agroforestry practices, the Diniz family has created the Institute Toca, which offers education and training in agroecology. The program is based on the teachings of Ernst Gotsch, one of the world's leading experts in agroforestry. By creating an agricultural system that mimics the forests, they have been able to regenerate sandy dirt into rich loam, create on-farm fertility without the use of compost or manure, and greatly increase water retention.

RANKING AND RESULTS BY 2050

#28

9.28 GIGATONS
REDUCED CO2

$26.8 BILLION
NET COST

$709.8 BILLION
NET SAVINGS

agroforestry are clear. It is well suited to steep slopes and degraded croplands, places where other cultivation might struggle. Where they provide firewood, multistrata systems can relieve pressures from natural forests. One study suggests every acre of agroforestry can prevent deforestation of five to twenty forest acres. In addition to providing long-term economic stability for farmers, thanks largely to multiple crops growing on unique time lines, these approaches may help farmers adapt to the impacts of climate change, including drought and extreme weather events.

Despite these clear advantages, multistrata agroforestry is too often lumped into more general agricultural categories, undercutting the attention it deserves. In addition to issues of awareness and understanding, multistrata agroforestry has other challenges. The costs to establish such a complex system are high and without immediate returns. Though they are quite profitable once established, that investment may be out of reach for resource-poor farmers. That same complexity makes mechanization difficult, if not impossible. Tending and cultivating by hand means higher labor costs. And though resilience and longevity are superior, yields can be lower than with conventional approaches, as crops compete for water, light, and nutrients.

Multistrata agroforestry requires a humid climate and cannot be implemented everywhere, but where it can, it promises a sizable impact. In addition to their high rates of carbon sequestration, these systems of cultivation are among the most energy efficient in the world. According to one study of traditional Pacific multistrata agroforestry, just 0.02 calories of energy produce 1 calorie of food. That kind of caloric efficiency, alongside maximizing production on small plots, makes multistrata agroforestry ideal for smallholders who live in densely populated areas. Market incentives and payment for ecosystem services could help farmers overcome financial barriers and help realize the multilayered benefits of multistrata systems for people and climate.

IMPACT: *Multistrata agroforestry can be integrated into some existing agricultural systems; others can be converted or restored to it. If adopted on another 46 million acres by 2050, from 247 million acres currently, 9.3 gigatons of carbon dioxide could be sequestered. Average sequestration rate of 2.8 tons of carbon per acre per year is strong, as is financial return: $710 billion in net profit by 2050, on a $27 billion investment.*

FOOD
IMPROVED RICE CULTIVATION

Vietnamese poet Phan Van Tri writes of rice grains: "They leave rice fields to travel far and wide: who doesn't count on them for sustenance? . . . Time after time, their forebears saved the realm—for centuries their breed has fed our folk." Rice has, in fact, been part of human life for thousands of years. Most likely domesticated in China first, today the grain is nearly universal—white, brown, and sticky; noodles, cakes, and vinegar; pilaf, paella, and porridge. Rice provides a full one-fifth of calories consumed worldwide, more than wheat or corn, and is the essential staple in the daily diet of 3 billion people, many of them poor and food insecure.

Presently, rice cultivation is responsible for at least 10 percent of agricultural greenhouse gas emissions and 9 to 19 percent of global methane emissions. Flooded rice paddies are perfect environments for methane-producing microbes that feed on decomposing organic matter, a process known as methanogenesis. Higher ambient temperatures where rice is cultivated increase emissions, which suggests that methane releases from rice paddies will increase as the planet gets hotter. Methane does not persist in the atmosphere as long as carbon dioxide does, but over a century, its global warming potential is up to thirty-four times greater. Thus, the world faces a multifaceted challenge: to find and adopt ways to produce rice that are efficient, dependable, and sustainable, meeting the growing demand for this staple food without causing warming.

It "was discovered almost by accident." That is how French Jesuit priest and agronomist Henri de Laulanié described the origins of the System of Rice Intensification (SRI), a key approach to improve rice production, which he and smallholder farmers developed on Madagascar in the 1980s. Under atypical time constraints, a group of agricultural students transplanted rice seedlings much earlier than usual. It was an unanticipated first step toward a holistic system that lowers the inputs required for rice production—seeds, water, and fertilizer—while dramatically increasing crop yields.

Three decades later, the *New York Times* described SRI as emphasizing "the quality of individual plants over the quantity" and applying "a less-is-more ethic to rice cultivation." Thanks in large part to the evangelizing efforts of Norman Uphoff, of Cornell University, that ethic is now practiced by 4 million to 5 million farmers around the world, especially in Asia. They include Sumant Kumar, a farmer in the village of Darveshpura in northeast India, who achieved a world-record yield of 24.7 tons of rice on his 2.5-acre (1-hectare) plot in 2012—eclipsing the 4.5 or 5.5 tons that are typical for a piece of land that size.

SRI is not the only approach to sustainable rice production, but it seems to be the most promising. Kumar and his friends engaged in a simple set of practices knit together in a compelling way:

RANKING AND RESULTS BY 2050
(IMPROVED RICE CULTIVATION)

#24

11.34 GIGATONS
REDUCED CO2

NO ADDITIONAL
COSTS REQUIRED

$519.1 BILLION
NET SAVINGS

RANKING AND RESULTS BY 2050
(SYSTEM OF RICE INTENSIFICATION)

#53

3.13 GIGATONS
REDUCED CO2

NO ADDITIONAL
COSTS REQUIRED

$677.8 BILLION
NET SAVINGS

1. *Planting*. Rather than bedding out three-week-old seedlings by the handful—bunched close together—SRI calls for transplanting single seedlings when they are eight to ten days old and using a square grid that gives each one wider berth. Doing so creates more access to sunshine and canopy space aboveground, and more room for the roots to spread below.

2. *Watering*. Most conventional rice fields are continuously flooded, enabling methanogenesis, but SRI specifies more purposeful, intermittent watering. Temporary draining midway through a growing season or alternating between wet and dry conditions is more favorable to soil microbes and root systems that like to breathe, while disrupting the waterlogged conditions that methane-producing microbes favor. Research shows mid-season drainage alone reduces methane emissions by 35 to 70 percent.

3. *Tending*. Weeds can be a challenge in the absence of flooding, which SRI addresses with a rotating hoe used by hand, also aerating the soil. In tandem, applying organic compost helps to enhance soil fertility and carbon sequestration. Reducing or avoiding synthetic fertilizers protects both soil and waterways.

It all adds up to creating the ideal environment for rice to grow, fed with more sunshine, more air, and more nutrients. The result: plants that are larger and healthier, with stronger root systems, aided by more abundant, thriving soil microorganisms. Not only are yields 50 to 100 percent higher than conventional rice production, but seed use drops by 80 to 90 percent and water inputs by 25 to 50 percent. This reduction in water use makes SRI not just a means of mitigating global warming but also a good approach for adapting to a warming world. SRI plants also prove more resistant to drought, flooding, and storms—phenomena heightened by climate change.

While these practices improve the productivity of a farmer's land, labor, and capital, the labor inputs required can be higher than in conventional rice cultivation, mostly in the early years, when a farmer is learning SRI. As Uphoff explains, "It's not intrinsically labor-intensive; it's initially labor-intensive." Farm incomes can double when SRI gets adopted. Despite its spread to some forty countries and millions of smallholder farms, some scientists dispute yield and income claims, citing insufficient peer-reviewed research. That body of literature is growing, but SRI may continue to face this challenge, at least in the near term. SRI's defenders suggest that the grassroots, democratic, and holistic nature of the movement may actually be the reason for critique. Farmers, those most intimately in dialogue with the earth, are the innovators and experts—not agribusiness, nor academia. SRI disrupts the mechanistic, chemically intensive approach to food production upon which so many companies depend for their income.

SRI is not the only means for achieving improved rice production. There are four general and increasingly common techniques, best used in combination, that focus on water, nutrients, plant varieties, and tillage. Mid-season water drainage and alternate wetting and drying improve aerobic conditions. More balanced application of both organic and inorganic nutrients reduces methane emissions while supporting yields. Rice varieties or *cultivars* that are less water-loving can be used in more aerobic environments. Techniques for seeding rice without tilling the ground also have positive effect.

The advantage and the burden of SRI and other improved rice-production techniques is that they hinge largely on behavior change, shifting the way farmers manage their plants, water, soil, and nutrients. On one hand, that means it is exceedingly doable for smallholder farmers, who need not buy anything before putting SRI into practice (a striking difference from conventional approaches to agricultural intensification). The main technical challenge they face is controlling water application. On the other hand, many rice farming methods have been in place for centuries; they are embedded in families, villages, and cultures. Shifting entrenched customs requires a comprehensive approach to cultivate necessary knowledge and skills, help farmers see what results are possible, and implement incentives that make the prospect of change compelling. In SRI's early days, de Laulanié and his collaborators founded an educational organization, Tefy Saina, meaning "to improve the mind" in Malagasy. There is a message in that name: On-the-ground knowledge sharing and peer-to-peer training continues to be indispensable. Deepening and spreading those efforts can help low-emissions rice cultivation take root worldwide. It was not de Laulanié's original purpose, but his work may prove to be indispensable in tackling global warming. ●

IMPACT: *Our analysis includes both SRI and improved rice production, which involves improved soil, nutrient management, water use, and tillage practices. SRI has been adopted largely by smallholder farmers and has much higher yield benefits compared to improved rice production. We calculate that SRI can expand from 8.4 million acres to 133 million acres by 2050, both sequestering carbon and avoiding methane emissions that together total 3.1 gigatons of carbon dioxide or its equivalent over thirty years. With increased yields, 477 million additional tons of rice could be produced, earning farmers an additional $678 billion in profit by 2050. If improved rice production grows from 70 million acres to 218 million acres over thirty years, another 11.3 gigatons of carbon dioxide emissions can be reduced. Farmers could realize $519 billion in additional profits.*

FOOD
SILVOPASTURE

Cows and trees do not belong together—so says conventional wisdom. And why should it not? In Brazil and elsewhere, headlines condemn ranching as a driver of mass deforestation and attendant climate change. But the practice of silvopasture challenges this assumption of mutual exclusivity and could help shape a new era for the acreage dedicated to livestock and their food.

From the Latin for "forest" and "grazing," silvopasture is just that: the integration of trees and pasture or forage into a single system for raising livestock, from cattle and sheep to deer and ducks. Rather than seeing trees as a weed to be removed, silvopasture integrates them into a sustainable and symbiotic system. It is one approach within the broader umbrella of agroforestry and revives an ancient practice, now common on over 350 million acres worldwide. The *dehesa* system of silvopasture, famous for the jamón ibérico it yields, has been cultivated on the Iberian Peninsula for more than forty-five hundred years. More recently, silvopasture has taken root in Central America, thanks to the work of champions such as the Center for Research in Sustainable Systems of Agriculture (CIPAV), based in Cali, Colombia. In many places in the United States and Canada, livestock and trees can be found intermingling.

That intermingling takes a variety of forms. Trees may be clustered, evenly spaced, or used as living fencing. Animals may graze in grassy alleys between rows of arboreal growth. Most silvopastoral systems are similar in spacing to a savanna ecosystem. They can be created by planting trees in open pasture, letting those that sprout mature, or by thinning a woodland or plantation canopy to allow for forage growth. But whatever the design, trees, animals, and their forage are just the most obvious aspects of a silvopastoral system. Soil is the other essential component—and key to the potential silvopasture has for addressing climate change.

Experts around the world are engaged in an ongoing and fiery debate about how best to manage pastures to counterbalance the methane emissions of livestock, especially cattle, and sequester carbon in the soil under-hoof. Cattle and other ruminants require 30 to 45 percent of the world's arable land, and livestock produce roughly one-fifth of greenhouse gas emissions, depending on specifics of analysis.

Research to date suggests silvopasture far outpaces any grassland technique. That is because silvopastoral systems sequester carbon in both the biomass aboveground and the soil below. Pastures that are strewn or crisscrossed with trees sequester five to ten times as much carbon as those of the same size that are treeless. Moreover, because the livestock yield on a silvopasture plot is higher (as explored below), it may curtail the need for additional pasture space and thus help avoid deforestation and subsequent carbon emissions. Some studies show that ruminants can better digest silvopastoral forage, emitting lower amounts of methane in the process.

Carbon aside, the benefits of silvopasture are considerable; these practices have spread precisely because of their demonstrable financial benefits for farmers and ranchers. Options for piecing together a silvopastoral system are many and can work across all scales, from smallholder farms to corporate ranch operations. From a financial and risk perspective, silvopasture is useful for its diversification. Livestock, trees, and any additional forestry products, such as nuts, fruit, mushrooms, and maple syrup, all come of age and generate income on different time horizons—some more regularly and short-term, some at much longer intervals. Because the land is diversely productive, farmers are better insulated from financial risk due to weather events.

The integrated, symbiotic system of silvopasture proves to be more resilient for both animals and trees. In a typical treeless pasture, livestock may suffer from extreme heat, cutting winds, and mediocre forage. But silvopasture provides distributed shade and wind protection, as well as rich food. With better nutrition and shelter from the elements, animal health goes up, as does the production of milk, meat, and offspring. Yield results vary depending on the exact silvopasture system employed, but they regularly surpass that of a comparable grass-only pasture by 5 to 10 percent. At the same time, livestock function as weed control, reducing trees' competition for moisture, sunlight, and nutrients. Their manure also provides natural fertilizer.

Silvopasture can cut farmers' costs by reducing the need for feed, fertilizer, and herbicides. Because the integration of trees into grazing lands enhances soil fertility and moisture, farmers find themselves with healthier, more productive land over time.

Though the advantages of silvopasture are clear, its growth has been limited by both practical and cultural factors. These systems are more expensive to establish, requiring higher upfront costs in addition to the necessary technical expertise. In Colombia, for example, farmers look at an investment of $400–800 per acre, a steep short-term expense. There is less incentive to plant trees and then protect them as they grow where pastures are plentiful, fire poses a risk, or landownership is unclear. Layered on these challenges is the stubborn belief that trees and pastures are not compatible—that trees inhibit the growth of pasture fodder rather than enrich it. In many places, cleared plots and companionless grass are the norm, and farmers may ridicule one another for shifting to an alternate approach. Silvopasture requires rethinking the ecology of land.

These social impediments make peer-to-peer engagement and direct experience of silvopasture's benefits key accelerants. Fellow farmers are often more trusted than technical or scientific experts, while a successful test plot—perhaps on a rancher's own land—is the most convincing case of all. To address economic

31.19 GIGATONS REDUCED CO2	$41.6 BILLION NET COST	$699.4 BILLION NET SAVINGS

obstacles, international organizations such as the World Bank and NGOs such as the Nature Conservancy are making loans to enable silvopasture installation—loans a typical bank would not provide. Payments for the ecosystem services silvopasture provides, such as supporting biodiversity, can also make the economics make sense for farmers. As the impacts of global warming progress, silvopasture's appeal will likely grow, as it can help farmers and their livestock adapt to erratic weather and increased drought. Trees create cooler microclimates and more protective environments, and can moderate water availability. Therein lies the climatic win-win of silvopasture: As it averts further greenhouse emissions from one of the world's most polluting sectors, it also protects against changes that are now inevitable.

IMPACT: *We estimate that silvopasture is currently practiced on 351 million acres of land globally. If adoption expands to 554 million acres by 2050—out of the 2.7 billion acres theoretically suitable for silvopasture—carbon dioxide emissions can be reduced by 31.2 gigatons. This reduction is a result of the high annual carbon sequestration rate of 1.95 tons of carbon per acre per year in soil and biomass. Farmers could realize financial gains from revenue diversification of $699 billion, on investment of $42 billion to implement.*

Why Bother?
MICHAEL POLLAN

It can be safely said that no person has had a greater influence on how we choose, consider, cook, and create our food than Michael Pollan. A scholar, gardener, author, and journalist, he has written books that are a constellation of sensible yet highly original insights about our relationship with food and agriculture, and how that relationship became badly skewed by corporate dominance of farming, food science, politics, and advertising. In his best-selling books The Omnivore's Dilemma, The Botany of Desire, *and* In Defense of Food, *he does not advise us what to eat or how to farm, but highlights the fact that foodlike substances are harming our bodies, soil, and country. Pollan restores common sense, as epitomized by his classic maxim: "Eat food, not too much, mostly plants." It could be countered with "Learn about food, as much as possible, mostly from Pollan."* — PH

Why bother? That really is the big question facing us as individuals hoping to do something about climate change, and it's not an easy one to answer. I don't know about you, but for me the most upsetting moment in *An Inconvenient Truth* came long after Al Gore scared the hell out of me, constructing an utterly convincing case that the very survival of life on earth as we know it is threatened by climate change. No, the really dark moment came during the closing credits, when we are asked to . . . change our light bulbs. That's when it got really depressing. The immense disproportion between the magnitude of the problem Gore had described and the puniness of what he was asking us to do about it was enough to sink your heart.

But the drop-in-the-bucket issue is not the only problem lurking behind the "why bother" question. Let's say I do bother, big time. I turn my life upside-down, start biking to work, plant a big garden, turn down the thermostat so low I need the Jimmy Carter signature cardigan, forsake the clothes dryer for a laundry line across the yard, trade in the station wagon for a hybrid, get off the beef, go completely local. I could theoretically do all that, but what would be the point when I know full well that halfway around the world there lives my evil twin, some carbon-footprint doppelgänger who's eager to swallow every bite of meat I forswear and positively itching to replace every last pound of carbon dioxide I'm struggling no longer to emit. So what exactly would I have to show for all my trouble?

A sense of personal virtue, you might suggest, somewhat sheepishly. But what good is that when virtue itself is quickly becoming a term of derision? And not just on the editorial pages of *The Wall Street Journal* or on the lips of the [then] vice president, who famously dismissed energy conservation as a "sign of personal virtue." No, even in the pages of the *New York Times* and *The New Yorker*, it seems the epithet "virtuous," when applied to an act of personal environmental responsibility, may be used only ironically. Tell me: How did it come to pass that virtue—a quality that for most of history has generally been deemed, well, a

virtue—became a mark of liberal softheadedness? How peculiar, that doing the right thing by the environment—buying the hybrid, eating like a locavore—should now set you up for the Ed Begley Jr. treatment.

There are so many stories we can tell ourselves to justify doing nothing, but perhaps the most insidious is that, whatever we do manage to do, it will be too little too late. Climate change is upon us, and it has arrived well ahead of schedule. Scientists' projections that seemed dire a decade ago turn out to have been unduly optimistic: the warming and the melting is occurring much faster than the models predicted. Now truly terrifying feedback loops threaten to boost the rate of change exponentially, as the shift from white ice to blue water in the Arctic absorbs more sunlight and warming soils everywhere become more biologically active, causing them to release their vast stores of carbon into the air. Have you looked into the eyes of a climate scientist recently? They look really scared.

sunlight with fossil fuel fertilizers and pesticides, with a result that the typical calorie of food energy in your diet now requires about 10 calories of fossil fuel energy to produce. It's estimated that the way we feed ourselves (or rather, allow ourselves to be fed) accounts for about a fifth of the greenhouse gas for which each of us is responsible.

Yet the sun still shines down on your yard, and photosynthesis still works so abundantly that in a thoughtfully organized vegetable garden (one planted from seed, nourished by compost from the kitchen and involving not too many drives to the garden center), you can grow the proverbial free lunch—carbon dioxide–free and dollar-free. This is the most-local food you can possibly eat (not to mention the freshest, tastiest and most nutritious), with a carbon footprint faint and slight. And while we're counting carbon, consider too your compost pile, which shrinks the heap of garbage your household needs trucked away even as it feeds your vegetables and sequesters carbon in your soil. What else? Well, you will probably notice that you're getting a pretty good workout there in your garden, burning calories without having to get into the car to drive to the gym.

You begin to see that growing even a little of your own food is, as Wendell Berry pointed out thirty years ago, one of those solutions that, instead of begetting a new set of problems—the way "solutions" like ethanol or nuclear power inevitably do—actually beget other solutions, and not only of the kind that save carbon. Still more valuable are the habits of mind that growing a little of your own food can yield. You quickly learn that you need not be dependent on specialists to provide for yourself—that your body is still good for something and may actually be enlisted in its own support. If the experts are right, if both oil and time are running out, these are skills and habits of mind we're all very soon going to need. We may also need the food. Could gardens provide it? Well, during World War II, victory gardens supplied as much as 40 percent of the produce Americans ate.

But there are sweeter reasons to plant that garden, to bother. At least in this one corner of your yard and life, you will have begun to heal the split between what you think and what you do, to commingle your identities as consumer and producer and citizen. Chances are, your garden will re-engage you with your neighbors, for you will have produce to give away and the need to borrow their tools. You will have reduced the power of the cheap-energy mind by personally overcoming its most debilitating weakness: its helplessness and the fact that it can't do much of anything that doesn't involve division or subtraction. The garden's season-long transit from seed to ripe fruit—will you get a load of that zucchini?!—suggests that the operations of addition and multiplication still obtain, that the abundance of nature is not exhausted. The single greatest lesson the garden teaches is that our relationship to the planet need not be zero-sum, and that as long as the sun still shines and people still can plan and plant, think and do, we can, if we bother to try, find ways to provide for ourselves without diminishing the world. ▣

So do you still want to talk about planting gardens?
I do.

The act I want to talk about is growing some—even just a little—of your own food. Rip out your lawn, if you have one, and if you don't—if you live in a high-rise, or have a yard shrouded in shade—look into getting a plot in a community garden. Measured against the Problem We Face, planting a garden sounds pretty benign, I know, but in fact it's one of the most powerful things an individual can do—to reduce your carbon footprint, sure, but more important, to reduce your sense of dependence and dividedness: to change the cheap-energy mind.

A great many things happen when you plant a vegetable garden, some of them directly related to climate change, others indirect but related nevertheless. Growing food, we forget, comprises the original solar technology: calories produced by means of photosynthesis. Years ago the cheap-energy mind discovered that more food could be produced with less effort by replacing

Excerpted and adapted with permission from Michael Pollan's essay "Why Bother?" in the *New York Times*, April 20, 2008.

REGENERATIVE AGRICULTURE

Regenerative agricultural practices restore degraded land. They include no tillage, diverse cover crops, on-farm fertility (no external nutrient sources required), no or minimal pesticides or synthetic fertilizers, and multiple crop rotations, all of which can be augmented by managed grazing. The purpose of regenerative agriculture is to continually improve and regenerate the health of the soil by restoring its carbon content, which in turn improves plant health, nutrition, and productivity.

As you will see from data at the back of this book, no other mechanism known to humankind is as effective in addressing global warming as capturing carbon dioxide from the air through photosynthesis. When converted to sugars with help from the sun, carbon produces plants and food. It feeds humankind, and, through the use of regenerative agriculture, it feeds the life of the soil. Regenerative agriculture increases organic matter, fertility, texture, water retention, and the existence of trillions of organisms that convey health and protection to the roots and plant itself. Practicing regenerative agriculture addresses all common concerns about fertility, pests, drought, weeds, and yield.

To better appreciate regenerative agriculture, it is helpful to understand what conventional agriculture is, as the dominant farming practice in the world today. It involves photosynthesis too, but does not prioritize capturing soil carbon. Conventional agriculture treats the soil as a medium to which mineral fertilizers and chemicals are added. The soil is plowed, tilled, cultivated, or disked two or more times a year. Herbicides clear the weeds, insect infestation is treated with pesticides, and blight or rust is sprayed with fungicides. Lack of water is compensated for with irrigation, which can cause salinization of the soils. Plowing and tilling release carbon from the soil, and little or none of the carbon from the plants is sequestered.

Looking back not too many years, Americans were (and most still are) eating what author Michael Pollan calls "foodlike substances," highly processed foods with a list of mysterious ingredients longer than this paragraph. A shift started in the 1980s and '90s that is expanding today—the realization that human health depends on real food, not artificial, synthetic, imitation food, and that food quality goes all the way back to the land and farming practices. In conventional agriculture, seeds, synthetic fertilizers, and pesticides go in and food comes out; however, the soil pays a heavy price, as do water, the air, birds, beneficial insects, human health, and the climate. Just as you can manufacture fake food cheaply using fillers, fats, sugars, and starches, conventional industrial agriculture produces food cheaply by not paying the cost of the damage it causes. If you do not provide your body with true nourishment, it becomes obese, diseased, and disabled. If a farmer does not provide nourishment to the soil, it becomes infertile, diseased, and deadened. These are commonsense, simple principles that underlie regenerative practices.

One principle of regenerative agriculture is no tillage. How often do you see bare earth except on a farm, or a road cut? Soil abhors a plant vacuum. Bare land, save for deserts and sand dunes, will naturally revegetate. Plants need a home, and soil

---

The Rodale Institute has been the cornerstone of organic farming in the U.S. since its founding in 1947. Based on the writings and observations of Sir Albert Howard, the godfather of organic agriculture, the Institute publishes, promotes, and conducts extensive ongoing research into organic methods of agriculture. The 333-acre farm in Kutztown, Pennsylvania, shown here, was purchased in 1971 by Robert Rodale, the son of founder J. I. Rodale. The land was played out and exhausted, an inspiration to Rodale to develop regenerative agriculture—systems of farming that are productive but which also increase the capacity for productivity in the future by restoring soil health. He proposed, and the Institute practices, methods of farming that require no external sources of fertility...and certainly no chemicals.

needs a cover. On farms, plows expose the soil and invert it, burying topsoil underneath. When soil is tilled and exposed to the air, the life within it decays quickly and carbon is emitted. Professor Rattan Lal estimates that at least 50 percent of the carbon in the earth's soils has been released into the atmosphere over the past centuries—approximately 80 billion tons. Bringing that carbon back into the soil is a gift to the atmosphere, to be sure, but from a practical agricultural perspective, it is an invitation to farmers to move away from agrochemical farming and bring the carbon back home, where it will help them work with the land more efficiently and productively.

Increasing carbon means increasing the life of the soil. When carbon is stored in soil organic matter, microbial life proliferates, soil texture improves, roots go deeper, worms drag organic matter down their holes and make rich castings of nitrogen, nutrient uptake is enhanced, water retention increases several fold (creating drought tolerance or flood insurance), nourished plants are more pest resistant, and fertility compounds to the point where little or no fertilizers are necessary. This ability to become independent of fertilizers relies upon cover crops. Each additional percent of carbon in the soil is considered equivalent to $300 to $600 of fertilizer stored beneath.

Cover crops sown into harvested plant residues crowd out weeds and provide fertility and tilth to the subsoil. A normal cover crop might be vetch, white clover, or rye, or a combination of them at one time. Experimentation has taught regenerative farmers to plant cover crops containing ten to twenty-five different varieties, each one adding a particular quality or nutrient to the soil. Gabe Brown, a renowned regenerative farmer in North Dakota, once put seventy different varieties in his seed box for pasturage. The possibilities include legumes such as spring peas, clover, vetch, cowpeas, alfalfa, mung beans, lentils, fava beans, sainfoin, and sunn hemp; and brassicas such as kale, mustard, radish, turnips, and collards. Then there are broadleaves such as sunflower, sesame, and chicory; and grasses such as black oats, rye, fescue, teff, brome, and sorghum. Each plant brings distinct additions to the soil, from shading out weeds to fixing nitrogen and making phosphorus, zinc, or calcium bioavailable. When consumed by ruminants, diverse varieties of cover crops afford extraordinary nutrition. This list provides a sense of how regenerative farmers are embracing complex plant communities to grow their crops, soil, and income.

With conventional crop rotation, soy and corn might be planted in alternating years, or wheat may be planted one year and then the field is left fallow the next. That too has changed. Regenerative farms might rotate eight or nine different crops, such as wheat, sunflowers, barley, oats, peas, lentils, alfalfa hay, and flax. Regenerative farmers are creating crop insurance through diversification, which prevents pockets of infestation by pests and fungi. Along with rotation, there is intercropping, in which leguminous companion crops of alfalfa or beans are grown with corn to provide fertility.

Regenerative agriculture is a practical movement, not a purist one. Some regenerative farmers are organic and others drill small amounts of synthetic fertilizer when planting corn, as they make the transition to organic certification. Gabe Brown has applied no fertilizers since 2008, and no pesticides or fungicides for fifteen years. He formerly used herbicides for tough invasive weeds such as Canadian thistle every two years, but has given it up because he no longer needs it.

The impact of regenerative agriculture is hard to measure and model. Individual farms cannot use a cookie-cutter approach. Rates of carbon sequestration will vary considerably in quantity and amount of time required. The results, however, are impressive. Farms are seeing organic matter levels rise from a baseline of 1 to 2 percent up to 5 to 8 percent over ten or more years. Every percent of carbon in the soil represents 8.5 tons per acre. That growth adds up to 25 to 60 tons of carbon per acre.

There has long been a conventional wisdom that the world cannot be fed without chemicals and synthetic fertilizers. However, the U.S. Department of Agriculture is now running trials on farming methodologies that eschew tillage and chemicals. Evidence points to a new wisdom: The world cannot be fed unless the soil is fed. Feeding the soil reduces carbon in the atmosphere. Soil erosion and water depletion cost $37 billion in the United States annually and $400 billion globally. Ninety-six percent of that comes from food production. India and China are losing soil thirty to forty times faster that the U.S. Regenerative agriculture is not the absence of chemicals. It is the presence of observable science—a practice that aligns agriculture with natural principles. It restores, revitalizes, and reinstates healthy agricultural ecosystems. Indeed, regenerative agriculture is one of the greatest opportunities to simultaneously address human, soil, and climate health, along with the financial well-being of farmers. It is about biological alignment—how to live and grow better food in ways that are more productive, safer, and more resilient. ●

IMPACT: *From an estimated 108 million acres of current adoption, we estimate regenerative agriculture to increase to a total of 1 billion acres by 2050. This rapid adoption is based in part on the historic growth rate of organic agriculture, as well as the projected conversion of conservation agriculture to regenerative agriculture over time. This increase could result in a total reduction of 23.2 gigatons of carbon dioxide, from both sequestration and reduced emissions. Regenerative agriculture could provide a $1.9 trillion financial return by 2050 on an investment of $57 billion.*

FOOD
NUTRIENT MANAGEMENT

While nitrogen fertilizers have vastly improved the productive capacity of agricultural systems in the past century, their use also has increased the amount of free, reactive nitrogen in these ecosystems. Some of the synthetic nitrogen is taken up by crops, increasing growth and yield, but nitrogen that is not utilized by plants causes untold problems. Most nitrogen fertilizers are "hot," chemically destroying organic matter in the soil. Nitrogen seeps into groundwater or travels through surface runoff and eventually emerges in streams and rivers, creating algal blooms and oxygen-depleted oceanic dead zones, of which there are five hundred in the world. The elevated levels of nitrogen in aquatic systems have been shown to be responsible for major fish kills. Nitrous oxide, created from nitrate fertilizers by soil bacteria, is 298 times more powerful than carbon dioxide in its atmospheric warming effect.

Proper nutrient management in agricultural systems can improve fertilizer-use efficiency, ensuring that crops take up a greater percentage of the fertilizer applied and reducing the possibility that fertilizer nitrogen in the soil is unused by plants and subsequently converted to nitrous oxide. Effective nutrient management is summarized by the four Rs: right source, right time, right place, and right rate. Collectively, these principles aim to improve nitrogen-use efficiency, which is defined as the ratio of plant productivity to nitrogen applied or residual in the soil.

Right source is primarily about matching fertilizer choices with plant needs or equipment limitations. Fertilizers come in a variety of forms, both dry and liquid, with different nitrogenous compounds that require different delivery mechanisms. Fertilizer manufacturers have begun making slow-release granular products coated with polymers that slow down dissolution after application. The delivery of nitrogen from these products is better synchronized with plant demand, lessening the amount of nitrogen that is lost from the system as nitrous oxide. These products are still relatively new to the market and are not widely employed because of cost. Nonetheless, early studies indicate they are potentially effective at decreasing nitrous oxide emissions.

Right time and *right place* are focused on managing fertilizer applications to deliver nitrogen when and where crop demand is highest. Crop demand for nitrogen is not consistent throughout the growing season. Plants typically require much more as they near growth stages, when their mass increases exponentially or when they are developing fruit or grains. Timing the delivery of nitrogen with these increased periods of demand accelerates the amount that is absorbed by plants and reduces excess. To simplify production and decrease the possibility of equipment damaging plants, producers often will apply fertilizer at planting or just after—times when plant demand for nitrogen is low. Splitting total annual fertilizer into two applications—one at the beginning of the season and one when plants are more mature and their demand for nitrogen is higher—reduces the likelihood that fertilizer will go unused.

Algal bloom off the coast of Sweden in the Baltic Sea.

1.81 GIGATONS
REDUCED CO2

DATA TOO VARIABLE
TO BE DETERMINED

$102.3 BILLION
NET SAVINGS

Arguably the most important decision in addressing nitrous oxide emissions from fertilizer is choosing the *right rate*. Producers often apply more fertilizer than recommended as a buffer against potentially poor growing conditions. As a result, agricultural systems are usually fertilized well beyond the optimum rate, making them more susceptible to nitrous oxide emissions.

Research into how producers make decisions has found that farmers are likely to apply more fertilizer than necessary and prioritize information they receive from fertilizer dealers—even with the knowledge that reducing their rate could lower emissions. Pressures to produce economical yields and to mitigate risk mean that the incentives for farmers to maintain or increase application rates are greater than the incentives to scale back. In addition, nitrogen fertilizers remain relatively cheap in high-production areas and are often subsidized.

Adoption of proper nutrient management requires education and assistance, as well as incentives for farmers and increased regulation that limits the amount farmers can apply. How to balance these tools depends on local context and their political feasibility. In the United States, for example, studies have shown that some farmers are more amenable to incentives and educational programs than they are to regulations. Groups such as the American Carbon Registry have been working with researchers to develop a carbon-offset methodology focused on fertilizer rate reductions that would allow farmers to participate in projects that would ultimately provide them with payments from the carbon offset market.

Regulations concerning fertilizer application and use are highly variable and typically associated with regulatory frameworks that address water quality and pollution. Since nitrogen-fertilizer pollution of water bodies is usually considered nonpoint source pollution (i.e., it cannot be easily linked to a single source), regulations are difficult to create and enforce. Nonetheless, some state agencies, such as in Vermont, have begun to require nutrient-management plans for farms of a certain size to reduce waste and pollution. In the United Kingdom, researchers have identified several Nitrate Vulnerable Zones, where fertilizer application is more regulated. Existing regulatory frameworks such as these may provide a pathway to regulating fertilizer use and reducing associated emissions.

However, government bodies around the world may not adopt or effectively enforce similar regulations. Nations that rely more on domestic production for food security, as well as revenue from export markets, often prioritize production over environmental impact. In China, national objectives of domestic self-reliance and food security undermine public demand for improved environmental quality and related policy and enforcement efforts. Similarly, nations with less productive capacity and greater food insecurity, such as several in sub-Saharan Africa, may require more fertilizer use to close yield gaps and ensure adequate supply for their citizens. In 1991, the European Union created the Nitrates Directive, intended to reduce groundwater and surface water pollution. As of 2017, only two countries had reduced their reliance on synthetic nitrogen fertilizer—Denmark and the Netherlands.

Given the importance of fertilizers to global agricultural production, reductions should be primarily focused on areas where they would have minimal to no impact on agricultural yields. Estimating how many acres of land on which fertilizer use has been reduced would require broad surveys of farmers, which are practically impossible. Furthermore, a farmer can choose to "abandon" nutrient management by simply returning to a higher fertilizer rate, and in practice farmers change rates based on a variety of factors every year.

The Food and Agriculture Organization of the United Nations and the World Bank provide excellent data on fertilizer consumption within every country that clearly demonstrates fertilizer use has been steadily expanding over the past decade in most countries, as have per-acre rates. This data reflects the expansion of agricultural production to meet the food demands of a growing population, and on the surface it seems to indicate that adoption of this solution is very low. The United Nations Environment Programme estimates that a 20 percent improvement in nutrient use would eliminate more than 20 million tons of nitrogen fertilizer and produce potential savings of $50 billion to $400 billion.

Nutrient management is unique among the land-use solutions in this book, in that it is primarily about avoided emissions and not about carbon sequestration. As such, the climate benefits of nutrient management are more continuous and not at risk of saturation; reduction in fertilizer use leads to avoided emissions in perpetuity. Additionally, implementation of this solution is extremely simple, as it only requires farmers to moderately reduce their inputs and not undertake a drastically new practice or install a new technology. That being said, continual application of chemical fertilizers results in loss of fertility, water infiltration, and loss of productivity over time. These impacts can cause farmers to increase fertilization in hopes it will compensate for the overall loss of soil health, which is in actuality a downward spiral. Although this solution focuses on more intelligent nutrient management, the true solution to nutrient management is rotational, regenerative land practices discussed throughout *Drawdown* that eliminate most, if not all, of the need for synthetic nitrogen. ●

IMPACT: *By reducing fertilizer overuse on a total of 2.1 billion acres of farmland by 2050—up from an estimated 177 million acres currently—avoided nitrous oxide emissions could equal 1.8 gigatons of carbon dioxide. No investment is required, and farmers could save $102 billion from reduced fertilizer costs. Our analysis assumes adoption that roughly parallels conservation agriculture, as farmers are likely to be amenable to both practices.*

FOOD
TREE INTERCROPPING

There are two ways to farm. Industrial agriculture sows a single crop over large areas. Regenerative practices such as tree intercropping use diversity to improve soil health and productivity and align with biological principles. Lower inputs, healthier crops, and higher yields are the outcome. Like many solutions in this book, tree intercropping is rarely undertaken to address global warming. Farmers practice it because it works better, though it declined in Europe for most of the twentieth century in the wake of the industrialization of agriculture. Like all regenerative land-use practices, it increases the carbon content of the soil and productivity of the land. Intercropping provides windbreaks that reduce erosion and creates habitat for birds and beneficial insects. Fast-growing annuals, susceptible to being flattened by wind and rain, can be protected. Deep-rooted plants can draw up subsoil minerals and nutrients for shallow-rooted ones. Vines or creepers have a ready trellis. Light-sensitive crops can be protected from excess sunlight.

To top it off, tree intercropping is beautiful—chili peppers and coffee, coconut and marigolds, walnuts and corn, citrus and eggplant, olives and barley, teak and taro, oak and lavender, wild cherry and sunflower, hazel and roses. Triple-cropping is common in tropical areas, with coconut, banana, and ginger grown together. The possible combinations are endless.

To be successful at tree intercropping, a landholder has to carefully assess and know the land, soil type, and climate at hand. Sunlight, nutrient flows, and water availability determine species, density, and spatial overlap of trees and crops. If you drive through the Ardennes in France, you will see poplar trees intercropped with wheat. It may appear that the trees were sown in a row without much thought; however, years of knowledge

17.2 GIGATONS REDUCED CO2	$147 BILLION NET COST	$22.1 BILLION NET SAVINGS

goes into appraising the impacts of wind, light, seasonal changes, and nutrient competition. These in turn determine the configuration and types of plants—in this case, the type of poplar. The arrangement of trees and crops varies with topography, culture, climate, and crop value.

Tree intercropping has many variations. Alley cropping is a system in which trees or hedges are planted in closely spaced rows to fertilize the crops grown between. The small trees or hedges are nitrogen-fixing leguminous species such as riverhemp, gliricidia, and apple-ring acacia. In trials conducted in Malawi over ten years, maize was alley-cropped with gliricidia trees and the yield was compared to unfertilized maize grown in fields without trees. In the alley-cropped field, nitrogen-containing gliricidia prunings were applied to the soil on an annual basis. The result: Alley-cropped maize produced three times as much as the unfertilized maize grown solo. Because of food shortages in Malawi, impoverished smallholders are cropping maize on a continual basis, causing soil degradation and further decline in food security. Though land is "lost" to trees in the alley-cropping system, the increased yield—without chemical inputs—more than makes up for the loss.

Evergreen agriculture, another variation of tree intercropping, calls for a discontinuous cover of scattered trees, such as apple-ring acacia, which provide fodder for livestock. These are planted based on the ecological knowledge of farmers who cultivate crops on lands prone to drought, wind, and erosion. During the rainy growing season, the trees shed their nitrogen-rich leaves, which means maize and other crops do not need to compete for water or light. Yields can increase by a factor of three, without chemical fertilizers or other inputs.

Other variations of intercropping include strip cropping, boundary systems, shade systems, forest farming, forest gardening, mycoforestry, silvopasture, and pasture cropping. Tree intercropping reinforces the idea that human well-being does not depend on an agricultural system that is extractive and hostile to living organisms. Rather, it depends on discovering, innovating, and practicing methods of agriculture that feed a growing population, while providing continuous improvements to soil, fertility, habitat, diversity, and fresh water.

Modern corporations are permeated with the notion of continuous improvement, a concept known as *kaizen* in Japan and based on American quality-engineering principles taught in Japan after World War II. It means getting better at getting better and emphasizes small daily improvements that improve a product and the workplace. As an ancient ecological technique, tree intercropping is the same—a way to both honor and adapt to the land. Displaced and plowed under during the twentieth century to make room for industrialized methods of farming, tree intercropping is one of dozens of techniques that can create an agricultural renaissance, a transformation of food-growing practices that are better at bringing people, regeneration, and abundance back to the land. ●

IMPACT: *Accounting for different sequestration rates across regions and intercropping systems, we estimate total sequestration of 17.2 gigatons of carbon dioxide over thirty years. To achieve that impact, adoption of tree intercropping would need to grow to 571 million acres globally. On an additional investment of $147 billion, savings could be $22 billion over three decades.*

A new freestone peach orchard intercropped with corn in Klickitat County, south central Washington.

FOOD
CONSERVATION AGRICULTURE

Used by hand or drawn by mule, oxen, or tractor, the plow is a standard tool for loosening soil and turning over the top layer before planting crops. Historically viewed as a major advance forward as an agrarian species, plows are absent on farms practicing conservation agriculture, and for good reason. When farmers till their fields to destroy weeds and fold in fertilizer, water in the freshly turned soil evaporates. Soil itself can be blown or washed away and carbon held within it released into the atmosphere. Though intended to prepare a field to be productive, tilling can actually make it nutrient poor and less life giving.

Soil erosion and degradation gave rise to the practice of conservation agriculture in Brazil and Argentina in the 1970s, though in truth most farms were no-till or low-till before the eighteenth century's industrial innovations. Conservation agriculture adheres to three core principles: minimize soil disturbance, maintain soil cover, and manage crop rotation. The Latin root of *conserve* means "to keep together." Conservation agriculture abides by these principles in an effort to keep the soil together as a valuable, living ecosystem that enables food production and can help redress climate change. Conservation agriculture and regenerative agriculture, a separate *Drawdown* solution, both employ a no-till strategy. Most farmers who practice conservation agriculture plant cover crops. Conservation agriculture differs from regenerative practices in its use of synthetic fertilizers and pesticides.

Annual crops, those that are replanted each year, are grown on 89 percent of the world's cropland. Conservation agriculture is practiced on 10 percent of those 3 billion acres. It is prevalent in South America, North America, Australia, and New Zealand, among both large-scale operations and small ones. Absent tilling,

Above: Young, no-till soybeans in central Iowa.

Right: No-till seeder preparing the field and planting soybeans.

RANKING AND RESULTS BY 2050

#16

17.35 GIGATONS
REDUCED CO2

$37.5 BILLION
NET COST

$2.12 TRILLION
NET SAVINGS

farmers seed directly into the soil. They leave crop residues after harvesting or grow cover crops in order to protect the soil. Crop rotation—changing what is grown and where—is almost universally practiced when the crops are grains and legumes.

In part, conservation agriculture is already widespread because farmers can adopt it with relative ease and speed and realize a range of benefits. Water retention makes fields more drought resistant or reduces the need for irrigation. Nutrient retention leads to increased fertility and can lower fertilizer inputs. Most farmers who employ conservation agriculture see costs go down, yields go up, and income rise. Critics point out that modern no-till practices, especially in Western countries, rely heavily on herbicide application and genetically modified crops. Others argue that is not true conservation agriculture. In most of Africa, herbicide is not used in no-till farming.

Conservation agriculture sequesters a relatively small amount of carbon—an average of half a ton per acre. But given the prevalence of annual cropping around the world, those tons could add up and shift this dominant segment of agricultural production from net greenhouse gas emitter to net carbon sink. Because conservation agriculture makes land more resilient to climate-related events such as long droughts and heavy downpours, it is doubly valuable in a warming world.

Conservation agriculture is a well-proven solution. The core challenge to scaling it is the gap between up-front investments and the gains they ultimately bring. This is especially true for smallholder farmers, who may not be able to wait out returns, and farmers who lease rather than own their land, limiting motivation to invest in the long-term health of its soil. With widespread programs to educate, equip, and financially support farmers, millions more could adopt conservation agriculture, reap its benefits, and enhance farmland as a carbon storehouse.

IMPACT: *Based on historic growth on large farming operations, our analysis projects the total area under conservation agriculture will continue growing from 177 million acres to peak at 1 billion acres by 2035. We assume that as regenerative agriculture becomes more widely used, farms that have already adopted conservation agriculture will convert to these more effective soil fertility practices in response to consumer demand for fewer harmful herbicides. The benefits of that conversion are counted by the regenerative agriculture solution. Nonetheless, conservation agriculture offers significant benefits in the interim, reducing carbon dioxide emissions by 17.4 gigatons based on average carbon sequestration rates of .15 to .25 tons of carbon per acre per year, depending on region. Implementation costs are low at $38 billion with a return of $2.1 trillion.*

COMPOSTING

Organic matter *matters*. Sir Albert Howard, English agriculturalist and ardent, prophetic champion of compost, knew this instinctually. Conducting experiments from England to India in the early twentieth century, Howard saw proof in his plants that healthy, living soil was the key to thriving, resilient crops. Though he did not fully understand the web of interactions, he knew that somehow organic matter, soil fertility, and plant health are intrinsically linked. To that end, he orchestrated large-scale composting schemes and probed root structures for answers. Perhaps, Howard thought, compost enhanced the relationship between the roots of plants and mycorrhizal fungi in the soil. Throughout his life, he battled the establishment of the time, which advocated for the use of chemical fertilizers to supply the nutrients plants need. It was the era of the Haber process, the German discovery of how to manufacture affordable nitrogen fertilizers. In its wake, compost and top-dressing fields with organic matter began to be seen as old-fashioned and uneconomic.

The new fertilizer manufacturing processes gained worldwide attention. Fritz Haber and Carl Bosch were awarded separate Nobel Prizes. But Howard was onto something. Human beings have long used compost and manure to feed their crops and gardens, without understanding the mechanics of its benefits. The oldest surviving work of Latin prose, *De Agricultura*, by Cato the Elder, includes guidance on compost—deemed a must for farmers. Shakespeare also knew the power of the true black gold. "Do not spread the compost on the weeds," Hamlet cautions in metaphor. Dutch scientist Antoni van Leeuwenhoek first saw "wee beasties" through a prototypical microscope in the

1670s, but society is only just now coming to understand the power of microbes at the heart of soil ecology.

Fertile soil depends, as was once conjectured, on a mix of weathered rock fragments and decaying organic matter, and there are more microbes in a teaspoon of healthy soil than there are people on the planet. These soil microorganisms play two interlocking roles. They help to break down organic matter from dead plants and animals, putting key nutrients back into circulation within an ecosystem. They also help supply those key nutrients to plants' roots, precisely where they are needed, in exchange for exudates, carbohydrates released by plants—food for bacteria and fungi. From nitrogen to potassium to phosphorus and beyond, microbes keep the plant world thriving, and have their role to play in addressing climate change.

Like all living beings, humans create waste, but that waste can be uniquely problematic. Nearly half of the solid waste produced around the world is organic or biodegradable, meaning it can be decomposed over a few weeks or months. A key contributor to that rubbish flow is food waste, as well as wastelike leaf litter from yards and parks. For millennia this waste made its way back into the natural economy; today, much organic waste ends up in landfills. It decays in the absence of oxygen, producing the potent greenhouse gas methane, which is up to thirty-four times more powerful than carbon dioxide over one hundred years. A quarter of anthropogenic global warming may be due to methane gas alone. While many landfills have some form of methane management, it is far more effective to divert organic waste for composting, both dramatically reducing emissions and putting microbes to work. Composting processes avert methane emissions with proper aeration. Without it, the emissions benefits of composting shrink.

Composting can range in scale from backyard bins to commercial operations. Whatever the scale, the basic process remains the same: ensuring sufficient moisture, air, and heat to cook up an ongoing microbe feast of organic material. Bacteria, protozoa, and fungi chow down on organic matter rich in carbon. It is a process of decomposition that happens constantly, in every single ecosystem. The earth itself has a thin compost heap spread across its various landscapes. Rather than generating methane, as decomposition in a landfill would, the composting process actually converts organic material into stable soil carbon and makes it available to plants. Compost is an incredibly valuable fertilizer, retaining water and nutrients of the original waste matter, and can aid soil carbon sequestration. It is like going from refuse to riches.

Thanks to the work of Howard and others, industrial composting has existed since the early twentieth century. It is especially useful for cities today. With their dense populations, managing urban food waste is no small task. In 2009, San Francisco passed an ordinance that makes composting the city's food waste mandatory. Seattle monitors curbside bins and now tags and fines those who violate its composting requirement. Copenhagen, Denmark, has not sent organic waste to landfill in more than twenty-five years, reaping compost's win-win-win of cost savings, fertilizer production, and carbon mitigation.

Traditionally, landfilling has been cheap and convenient, but that is changing as land-use pressures and landfill regulations grow. These shifts are boosting the appeal of composting, as are its ease and diversity of approaches. Like recycling, successful compost operations require efforts to educate the public about disposal; develop the necessary infrastructure to gather, transport, and process waste; and deploy targeted collection strategies. Compost is nothing new, but needed now are fresh ways to make it a reality at scale. Leonardo da Vinci professed, "We might say that the earth has the spirit of growth; that its flesh is the soil." Composting is a way to both enhance that flesh—its spirit of growth—and keep emissions out of the atmosphere. ●

IMPACT: *In 2015, an estimated 38 percent of food waste was composted in the United States; 57 percent was composted in the European Union. If all lower-income countries reached the U.S. rate and all higher-income countries achieved the E.U. rate, composting could avoid methane emissions from landfills equivalent to 2.3 gigatons of carbon dioxide by 2050. That total excludes additional gains from applying compost to soil. Compost facilities cost less to construct but more to operate, which is reflected in the financial results.*

Large-scale composting of household green waste in the United Kingdom.

FOOD
BIOCHAR

I n ancient Amazonian society, virtually all waste was organic. The disposal method of choice for kitchen crumbs, fish bones, livestock manure, broken pottery, and the like was to bury and burn. Wastes were baked without exposure to air beneath a layer of soil. This process, known as pyrolysis, produced a charcoal soil amendment rich in carbon. The result was *terra preta*, literally "black earth" in Portuguese.

Terra preta contrasts starkly with the yellow, acidic soils typical of the Amazon basin. It was the hallmark of an agricultural system that differs dramatically from pervasive practices today: the wholesale conversion of Amazonian forest to annual crops, such as soybeans for livestock feed. When forest is cleared and vegetation burned, a residual layer of carbon remains, but only for a short period of time. It is difficult to build up organic matter in the tropics. These regions produce the most biomass per acre, but they have the highest rate of decay. Heavy rains leach nutrients more quickly from the thinner soils. The addition of carbon creates a few years of fertility before new plots of land have to be abandoned.

In comparison, *terra preta* agriculture maintained soil fertility for many decades—more than five hundred years in some studies. As with Asia, the Fertile Crescent, and Europe, abundant and dependable long-term agricultural production provided the foundation for cities and urban life. A marginal number of European explorers who ventured deep into the Amazon came back with wondrous reports of large urban settlements. Their accounts were subsequently deemed fantasies, for good reason: The cities had disappeared and could not be found. Smallpox wiped out 90 to 99 percent of the population, and the metropolises were abandoned and quickly covered over by jungle. The surviving residents fled deep into the wilderness to escape both disease and the conquistadores. It is speculated that Amazonian tribes, which have had first contact in the past decades, may be descendants of these fifteenth-century civilizations.

Today, *terra preta* soils cover up to 10 percent of the Amazon basin, retaining extraordinary amounts of carbon. Though soil-supporting charcoal dates back twenty-five hundred years, it was not (re)discovered by modern agronomists until rather recently. Dutch soil researcher Wim Sombroek uncovered this unusual black earth in the Amazon in the 1950s and published his seminal text, *Amazon Soils*, in 1966—a topic he continued to work on throughout his life. Black earth has been found elsewhere in Latin America, as well as Northern Germany and West Africa. These ancient roots of what is now called biochar have modern promise for agriculture and the atmosphere.

Researchers and archeologists from Embrapa, a network of Brazil's agricultural research stations, hover around an excavation showing how deep biochar (*terra preta*) is buried in Amazonian soils. In Manaus, the Embrapa staff has planted annual crops in *terra preta*–laden soils for forty years and has been unable to exhaust or ruin their fertility or productivity. Some scientists call the potential of *terra preta nova* the equivalent of a "black revolution" in agriculture.

.81 GIGATONS
REDUCED CO2

GLOBAL COST AND SAVINGS DATA
TOO VARIABLE TO BE DETERMINED

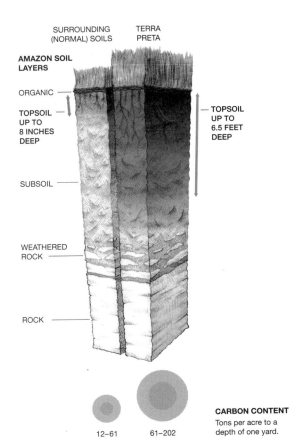

SURROUNDING (NORMAL) SOILS TERRA PRETA

AMAZON SOIL LAYERS

ORGANIC

TOPSOIL UP TO 8 INCHES DEEP

TOPSOIL UP TO 6.5 FEET DEEP

SUBSOIL

WEATHERED ROCK

ROCK

CARBON CONTENT
Tons per acre to a depth of one yard.

12–61   61–202

The pyrolysis process for producing biochar is from the Greek *pyro* for "fire" and *lysis* for "separating." It is the slow baking of biomass in the near or total absence of oxygen. The preferred method is gasification, a higher-temperature pyrolysis that results in more completely carbonized biomass. Biochar is commonly made from waste material ranging from peanut shells to rice straw to wood scraps. As it is heated, gas and oil separate from carbon-rich solids. The output is twofold: fuels that can be used for energy (perhaps for fueling pyrolysis itself) and biochar for soil amendment. Depending on the speed of baking, the ratio of fuel to char can shift. The slower the burn, the more biochar. Pyrolysis is unusual in its versatility. Large, polished industrial systems can produce it, and it can be made in small makeshift kilns. That means biochar is amenable to almost any context in the world . . . and many of the places that need it most.

Why would charred carbon impact soil fertility? When a farmer thinks about increasing yields, he or she thinks in terms of nitrogen, potash, phosphorus, and a few minerals such as calcium and zinc. If you go to buy fertilizers for your farm or garden, carbon will not be among them because it does not fertilize directly; rather, it creates the conditions for greater fertility. Biochar possesses a porous structure, which provides extensive surface area packed into a small space. Think of biochar as a habitat—much like a coral reef, it is riddled with nooks and crannies that catch nutrients, hold on to water, and help vital microorganisms to set up shop. Experts report that just one gram of biochar can have a surface area of twelve hundred to three thousand square yards, thanks to the abundance of tiny pores. It functions as a nutrient magnet, carrying a negative electrical charge that pulls in positively charged elements such as calcium and potassium. This can reduce soil acidity caused by nitrogen fertilizers and increase yield. When tilled into the ground, biochar typically helps plants grow with vigor, but not in all soils. Scientists continue to research where and how biochar is most beneficial to soil and the plants growing in it. Early work indicates that different types of biomass make biochars with different properties; learning how to match soil to an appropriate biochar will help improve its value. Studies show an average crop yield increase of 15 percent, with the greatest impact on soils that are acidic and degraded—the soils often found in areas struggling with food insecurity. What's more, biochar can improve plants' ability to absorb nitrate fertilizers, possibly allowing farmers to get the same effect out of smaller nutrient application, which cuts costs and reduces runoff and damage to aquatic ecosystems.

Pyrolysis produces carbon-dense material from the sugars created by plants during photosynthesis. When biomass decomposes on the surface, carbon and methane escape into the atmosphere. Biochar retains most of the carbon present in biomass feedstock and buries it below. Rendered stable, it can be held for centuries in the soil—a much-delayed return to the atmosphere, effectively interrupting the normal carbon cycle and putting it into slow motion. Theoretically, experts argue, biochar could sequester billions of tons of carbon dioxide every year, in addition to averting emissions from organic waste.

A central issue for biochar is the feedstock employed. When feedstock comes from agricultural or urban waste, converting it to biochar is a means to sequester carbon, increase fertility, and produce energy. Without proper regulation and enforcement, however, stripping the land of biomass or cutting down trees to create biochar damages and degrades the soil.

As interest and activity in biochar grows, the debate around what constitutes sustainable feedstock continues. Biochar manufacturing is a young industry. The science of its use and application is evolving; pyrolysis technologies continue to be developed, though demand is still relatively small. Groups such as the International Biochar Initiative are working to create standards, coherence, and support for the practice, including a certification effort designed to delineate a transparent and sustainable future for biochar. As of 2015, the initiative counted 326 companies, up from 175 in 2013. These are the key players taking biochar from ancient practice to one of the essential solutions to global warming. ●

IMPACT: *Biochar can produce 0.8 gigatons of carbon dioxide emissions reductions by 2050. This analysis draws on total lifecycle assessments of the many ways biochar prevents and sequesters greenhouse gases, while assuming the nascent biochar industry is limited by the availability of global biomass feedstocks.*

FOOD
TROPICAL STAPLE TREES

The marula tree (*Sclerocarya birrea*) ranges from the woodlands of Southern Africa to as far north as the Sahel. It has a wide crown similar to oak trees and is in the same family as mango and cashew. It is a prolific source of food for giraffes, rhinoceros, and elephants, with the latter being the dominant consumer. Marula produces an exceptional and delicious fruit with inner nuts that are a rich source of protein and marula oil. Elephants eat the fruit and branches and will munch on the bark, which is why it is sometimes called the elephant tree. As harsh as their impact is on the tree, elephants also spread the seeds everywhere in the dung to make up for it.

The idea of agriculture conjures up staple crops such as corn, wheat, and rice, pulses such as soy and peanuts, root crops such as potatoes, sweet potatoes, and cassava, and rows of broccoli, tomatoes, and lettuce. These crops have one thing in common: They are annuals—planted, harvested, and then replanted every year. Due to the nature of farming practices, annuals cause a net release of carbon from the soil into the atmosphere every year.

Though it is not widely known, many perennial crops, including trees and other long-lived vines, shrubs, and herbs, produce staple foods as well. Many of these perennial staples have been cultivated and harvested for millennia. A number of them are critical components of the world's food supply, particularly in the tropics, where staples such as bananas and avocados are consumed daily. Staple foods from trees include starchy fruits such as bananas and breadfruit, oil-rich fruits such as avocado, and nuts such as coconut and Brazil. Many legumes are perennial, including the chachafruto tree, pigeon peas, mesquite, and carob. And then there are specialized foods, such as sago, a starchy carbohydrate made from the pulpy pith of the sago palm. Or enset, a banana-like plant in Ethiopia that is fermented in the ground for three to six months to make a traditional staple dish called kocho. Africa abounds with staple tree crops: baobab, mafura, argan, mongongo, marula, dika, monkey orange, moringa, safou, and more.

Today, 89 percent of cultivated land, about 3 billion acres, is devoted to annuals. Of the remaining land in perennial crops, 116 million acres are used for perennial staple crops. Lands converted from annuals to perennial staples sequester, on average, 1.9 tons of carbon per acre every year for decades. In tropical regions, the yield of starches and protein per acre of staple crops matches that of annual crops, and in some cases greatly exceeds it.

Currently, temperate and boreal regions are without candidate crops that can produce yields that compete with those of annual staples. Another challenge facing perennial staple crops is mechanical harvesting. Most of the crops do not lend themselves to being mechanically picked or combined. This disadvantage, however, can be capitalized on by many of the farmers in low-income countries, who cannot compete with commoditized annual crops but can do well with mixed forest-farms of staple crops.

The benefits far outweigh the disadvantages, however. Tropical staple tree crops can take root in forest-farms, multistrata agroforestry, or tree intercropping systems. In each case, they can reverse erosion and runoff and create higher infiltration rates for rainwater. They can be grown on slopes too steep for mechanized annual crop production and are suited to a wider range of soils. Some take a liking to quite arid conditions, where annual crops are marginal or impossible. They require less fuel, fertilizer, and pesticide, if any at all, and there is virtually no tillage after planting.

20.19 GIGATONS REDUCED CO2	$120.1 BILLION NET COST	$627 BILLION NET SAVINGS

Given changes in worldwide weather patterns, perennials are more resilient, providing food where annual crops have failed. Net rainfall is increasing in the world, but not in the way it is wanted or needed. Global warming is creating rainfall patterns that range from prolonged drought conditions to overwhelming rains accompanied by flash floods. Perennial staple tree crops can thrive under conditions that annuals cannot. For example, an enset can go dormant for six to eight years and survive without any rainfall; when rain returns, the ensete returns as well. Annuals are delicate, less durable when compared to a palm or banana tree. Conversion is a wiser use of land and resources, with multilayered benefits to smallholders (who collectively tend about 430 million acres globally, with an average landholding of less than five acres), villages, conservation, and income. •

IMPACT: *Tropical staple crops currently grow on 116 million acres, mostly in the tropics. Their rate of sequestration is high at 1.9 tons per acre per year. Expand this area by another 153 million acres by 2050 and they can sequester 20.2 gigatons of additional carbon dioxide. Our analysis assumes that expansion only occurs on existing cropland, with no forest clearing. Because their yield is 2.4 times higher than annual staples—at 60 percent of the cost—savings are significant, while cost to implement is low.*

FOOD
FARMLAND IRRIGATION

Drip irrigation was invented by Simcha Blass of Israel. His inspiration occurred in the 1930s when a farmer wanted to know why his biggest tree was growing without any water. Blass excavated around the roots and discovered a leaky pipe. It was not until the 1960s and the advent of cheap plastic pipes that his invention could be patented and commercialized, however. It may be that this single invention has saved more water than any other technology.

To irrigate is to supply land with water. The practice dates back to roughly 6000 BC, when waters of the Nile and Tigris-Euphrates were first diverted to feed farmers' fields. Both Egyptians and Mesopotamians used the rise and fall of their rivers to saturate the soil of their croplands. In the wake of the waters, Hapi and Enbilulu emerged as patron gods of flooding and irrigation, so central did that technology become to these ancient societies. Remains of these early water-management systems—canals, embankments, dikes—still exist today.

Eight millennia later, agriculture and irrigation consume 70 percent of the world's freshwater resources, and irrigation is essential for 40 percent of the world's food production. Given its prevalence and scale, irrigation can cause surface and groundwater depletion by tapping rivers and aquifers, and spark competition for water rights between farms, cities, and businesses. Pumping and distributing farm water also requires energy, releasing carbon emissions in the process.

Over the course of human history, the irrigation methods that started in the Nile and Tigris-Euphrates valleys have continued to dominate. Called "flood" or "basin" irrigation, they rely on submerging fields and remain the most common approaches in many parts of the world. But in the mid-twentieth century, a suite of irrigation technologies evolved that help farmers irrigate more precisely and efficiently, thereby saving water and reducing climate impact. Both drip and sprinkler methods make water application more exact, matching as closely as possible the amount crops need to thrive. Drip irrigation achieves 90 percent application efficiency. Sprinkler irrigation reaches 70 percent precision. That means each drop of water generates more value, improving the productivity of irrigation and requiring less water consumption overall.

The benefits of using farm water more efficiently are numerous. In addition to energy demands and associated carbon emissions decreasing, crop yields improve, the costs of cultivation drop, and soil erosion declines. A less humid field environment curtails pests. Surface and groundwater resources are better protected by lowering demand for water use. Conflicts among various stakeholders for water resources may ease. Moreover, drip irrigation can work across a wide range of landscapes. There are disadvantages to navigate, however. More efficient and precise irrigation requires more extensive infrastructure; it is not just a matter of opening the floodgates. That means higher capital costs as well as ongoing upkeep, which may make it untenable for lower-value staple crops.

Crops require varying amounts of water at various stages of growth. With irrigation scheduling, another contemporary efficiency method, farmers can monitor conditions and meet crops' water needs in a timely way. The practice of deficit irrigation is similar in its variable application of water: Crops have more drought-tolerant stages, during which farmers can cut back on irrigation. This strategic water stress can actually improve crop quality. Sensors are also changing the irrigation landscape. They can monitor soil moisture and control irrigation systems automatically, taking out the guesswork—and legwork—for farmers. Where rainwater or runoff can be captured and fed into irrigation systems, farmers have yet another approach for using water efficiently and effectively.

Both drip and sprinkler irrigation are mature technologies. The area of farmland under drip and other "micro" irrigation has increased sixfold in the past twenty years, from roughly 4 million acres to at least 25.5 million. It continues to grow, but amounts to less than 4 percent of the irrigated land worldwide. So far, most of the uptake has been in the United States, New Zealand, and select European countries, leaving low-income regions of the world ripe for greater adoption. Asia is home to the most conventional surface irrigation, and hence the most significant opportunity to improve farm water productivity.

The single greatest barrier to its spread is the cost of purchase and installation, putting drip and sprinkler irrigation beyond the reach of many smallholder farmers. New low-cost drip irrigation technologies are trying to change that. So are targeted loans and subsidies, which are already increasing adoption. Irrigation infrastructure also needs human expertise; education and training can ensure that farmers are equipped with systems *and* the knowledge and skills to optimize them. Where equipment costs come down and the technical capacity of farming communities rises, improved irrigation can be a boon for both cultivation and climate. ●

IMPACT: *At present, use of sprinkler and drip irrigation varies widely around the world, from 42 percent of area in high-income countries to 6 percent in lower-income countries in Asia and Africa. Our analysis assumes the area under improved irrigation grows from 133 million acres in 2020 to 448 million acres in 2050. The highest adoption increases would occur in Asia, where 62 percent of total irrigated area is located and currently only 4 percent of that land is under micro-irrigation. This growth could avoid 1.3 gigatons of carbon dioxide emissions and save 90 billion gallons of water and $430 billion by 2050.*

Joe Del Bosque, president of Del Bosque Farms, Inc., inspects a water hose used for drip irrigation in his almond orchard in Firebaugh, California. In March 2015, California lawmakers approved legislation sought by Governor Jerry Brown that committed $1 billion to addressing the drought gripping the most populous U.S. state for a fourth year.

The Hidden Half of Nature
DAVID R. MONTGOMERY AND ANNE BIKLÉ

The agricultural industry has long argued that the only way we can feed humanity is through the use of chemical fertilizers, pesticides, and, more recently, genetically modified seeds. The conventional wisdom is that biological or organic agricultural methods are incapable of feeding the world—mere specialty practices for smaller farmers that are impractical given the world's food needs. In this excerpt, David Montgomery and Anne Biklé summarize the history of how science "proved" that plants grow best with chemical inputs—the foundation of all industrial agriculture and the prevailing doctrine about how to feed a hungry world.

As Montgomery and Biklé show, the science was incomplete because the role of soil life was unknown at that time. Agronomists and soil scientists of the nineteenth and most of the twentieth century had no inkling of what microbial populations were doing in the soil. In the absence of this knowledge, the chemical fertilizer theory of agricultural productivity was untouchable because it did sustain and increase yields, particularly on degraded soils. However, industrial agriculture came with a heavy price. By the mid- to late twentieth century, chemical-based agricultural practices were causing steady losses of soil carbon, topsoil, and humus, and creating water pollution, crops that were more susceptible to pests, greenhouse gases (nitrous oxide and carbon dioxide), and oceanic dead zones.

What creates soil health, productivity, water infiltration rates, drought tolerance, pest resistance, and water quality are, in large part, the legions of bacteria found in the soil, an unfathomably complex community of life-giving processes. This is the "hidden half of nature" that Montgomery and Biklé write about so eloquently in their book of the same name. All of the land-use practices included in Drawdown deliver enhanced carbon sequestration, productivity, and ecosystem services because they are aligned with life processes. As you will see under "Microbial Farming" in the "Coming Attractions" section, the largest agricultural companies in the world are now racing to understand, patent, and commercialize microbial solutions in order to stem 150 years of land degradation caused by industrial agricultural practices founded on the agrochemical approach.

In 1634, a Flemish chemist and physician, Jan Baptist van Helmont, began looking into the puzzling world of soil fertility and plant growth. This wasn't his first choice for how to spend his time, however. An alchemist by training, he believed that natural objects housed elemental forces that could attract and repel things—and could be understood through observation and experimentation. He ran afoul of the Church in rejecting a role for a divine hand in explaining natural phenomena. The unamused Inquisition condemned van Helmont to house arrest, charging him with impudent arrogance for investigating how God's creation—nature—worked.

Stuck at home for several years, he made the best of it and began thinking about how a tiny seed could turn into a large tree.

How plants grew was far from obvious. Unconvinced by the prevailing idea that plants ate soil, he weighed and then planted a five-pound willow sapling in a pot with two hundred pounds of dried-out soil. Adding only water, he simply let the tree grow, the perfect experiment for someone imprisoned at home. At the end of five years, he reweighed the tree and found it had gained 164 pounds and that the soil had lost just two ounces. He concluded that the tree grew by taking on water.

Spurred on by his findings, van Helmont conducted a wide range of experiments. In one, he burned sixty-two pounds of oak charcoal, carefully collecting and weighing the resulting pound of ash and sixty-one pounds of gas (carbon dioxide). That burning wood produced ash was no surprise. But the production of gas, let alone so much of it, was a new discovery. Before this, the idea that most of a plant was fashioned from an invisible gas would have been laughable.

A century and a half passed before Nicolas-Théodore de Saussure, a Swiss chemist studying plant physiology, put it all together. In 1804, he repeated van Helmont's experiments, carefully weighing and accounting for the water and carbon dioxide that a plant consumed. He demonstrated that plants grow through combining liquid water with carbon dioxide gas in the presence of sunlight—a process we call photosynthesis.

De Saussure's discovery turned the understanding of fertility on its head. Plants did not draw carbon from the humus in the soil; they pulled it out of the air! This reversal challenged the centuries-old notion that plants grew through absorbing humus (decaying organic matter). Still, de Saussure's work remained counterintuitive. After all, generations of farmers knew full well that manure helped their plants grow.

[. . .]

Natural philosophers believed that soil organic matter, or humus—the thin dark layer at the top of soil beneath decomposing plant matter—somehow helped plants grow. The prevailing idea was that this mysterious material directly fed plants. Until, that is, experiments showed that humus would not dissolve in water, thereby discrediting the idea that plants could absorb nutrients directly from rotting organic matter. And if plants couldn't suck humus up through their roots, then how could they use it for growth?

Stumped, scientists of the day cooled on the notion that plants absorbed nutrients directly from humus. German chemist Justus von Liebig picked up the thread and led the charge on discrediting the humus theory of plant nutrition. In 1840, swept up by the Industrial Revolution, he wrote an influential treatise on agricultural chemistry in which he reasoned that carbon in soil organic matter did not fuel plant growth because, as de Saussure had shown, plants obtained the carbon they needed from carbon dioxide in the atmosphere. Using then-standard practices of

analyzing and weighing plant matter before and after burning it, Liebig found that plant ashes were rich in nitrogen and phosphorus. It seemed reasonable to assume that the matter left over in the ash was what nourished plants, and thus crops. This finding, in his view, provided the answer that plant scientists had long sought—soil chemistry held the key to soil fertility.

In short order, Liebig and his disciples identified five key things essential for plants to grow—water (H_2O), carbon dioxide (CO_2), nitrogen (N), and the two rock-derived mineral elements, phosphorus (P) and potassium (K). They then jumped to the conclusion that organic matter played no important role in creating and maintaining soil fertility. By overthrowing the prevailing humus theory, Liebig ushered in a view of soil fertility at the heart of modern agriculture.

The appeal of Liebig's chemical philosophy is easy to understand when you read accounts of the explosive crop growth European farmers realized when they began fertilizing degraded soils with recently imported guano. In 1804, German explorer Alexander von Humboldt brought samples of this magic stuff back to Europe from an island off the coast of Peru. In addition to containing a lot of phosphorus, this white rock held more than thirty times the nitrogen of most manure.

By the time the Peruvian guano islands had been mined to oblivion in the late nineteenth century, the widespread adoption of chemical fertilizers had become firmly entrenched as the guiding philosophy of agricultural production.

[. . .]

[It turns out that] organic matter is the lifeblood of the soil, the currency in the original underground economy. A soil's hunger for organic matter partially explains the mystery of why it disappears so quickly. There beneath your feet, microbes and larger life forms create complex and dynamic communities where everyone has a dual role—eat and be eaten. These microscopic workhorses not only break down organic matter; they also play the role of supplier and distributor of nutrients, trace elements and organic acids that plants need. So, while plants don't directly absorb organic matter, they do absorb the metabolic products of soil organisms that feed on and break down organic matter. For most of his life Liebig was satisfied with the notion that organic matter didn't matter. But now we know otherwise—soil organisms do the heavy lifting that keeps soil fertile and plants fed.

When microorganisms decompose dead plants and animals, they put life's elemental building blocks back into circulation, including the big three—nitrogen, potassium and phosphorus—and all the other major nutrients and assorted micronutrients important for plant health. Moreover, microbes deliver nutrients right back to where they are needed—a plant's roots.

We're only beginning to appreciate the specialized, ancient connections between a plant's roots and soil life. By some estimates, we still only know of about one out of ten soil-dwelling species. Until very recently the field of soil ecology was much like ancient astronomy, when our view was limited to the stars we could see with the naked eye. The hidden half of nature works the skin of the Earth, weaving a carpet of life that ripples up from the soil to the plants and animals that upon death become the foundation for a thriving microbial world. Given the difficulties of observing what goes on in soil, we still have much to learn about below-ground relationships forged on the anvil of deep time. ●

Agronomist Jerry Glover of the Land Institute in Kansas demonstrating the extent of root growth of perennial grasses in native prairie meadows.

Over the long term, grazing animals create extraordinary environments. Study the Serengeti plains of east-central Africa and the tallgrass prairies of the buffalo commons in the United States and this becomes clear. Where original grasslands are still intact, they are abundant lands with carbon-rich soils ten feet deep. When that same land is plowed over and over or grazed by domestic animals, the land degrades over time and loses its soil carbon.

Managed grazing imitates what migratory herds of herbivores do on wildlands. Herbivores cluster to protect themselves and their young from predators; they munch perennial and annual grasses to the crown; they disturb the soil with their hooves, intermixing their urine and feces; and they move on and do not return for a full year. Herbivores such as cattle, sheep, goats, elk, moose, and deer are ruminants, mammals that ferment cellulose in their digestive systems and break it down with methane-emitting microbes. Ruminants cocreated the world's great grasslands, from the pampas in Argentina to the mammoth steppe in Siberia. Put those animals inside a fence, and it is a whole different story. Worse still, if you place cattle in feedlots and measure their impact upon the environment and climate, they rank with coal as being one of the greatest detriments to the planet. What has become apparent, however, is that when cattle and other ruminants are managed on grasslands in a holistic way, it can be the best thing for the land.

French biochemist and farmer André Voisin first put forth a theory of the benefits of managed grazing in 1957. Voisin studied chemistry and physics but was a plant and animal physiologist at heart. When he returned to his farm after World War II, he became intrigued by the relationship between his cows and grass. There is a tendency to take grass for granted: It grows, it is eaten, it dies, and it grows again. Voisin noticed that agronomists paid great attention to which pasture grasses were sown, how pastures were fertilized, and when they were watered, but gave little or no thought to how the animals and the grasses interacted. Was the grass chomped to the crown? Was it grazed once? Was it overgrazed? What was the condition of the grass after multiple feedings? Did it recover? What kind of weight gain was being achieved in differently grazed pastures? Voisin examined the minutiae of grazing. From these observations—and setting aside other variables such as rainfall—he realized that how cows ate grass was the main determinant of a pasture's health and productivity.

When grasses are continuously grazed, nutrient reserves in the roots trail off until they reach a point of exhaustion. As plants go, so goes the soil. This is known as overgrazing; the world is beset with more than a billion acres of land in this condition, according to some estimates. The impact of overgrazing led to the belief that the land would recover if animals were removed. Not so. When herbivores are removed from the land, whether they are wild or domesticated, the land deteriorates. The damage done by overgrazing obscured what happens when grasslands are undergrazed—soil health declines and carbon is lost.

In the course of his studies, Voisin homed in on two key variables: how long an animal grazes on a specific grassland and how long the land rests before animals return. Achieving optimal

16.34 GIGATONS
REDUCED CO2

$50.5 BILLION
NET COST

$735.3 BILLION
NET SAVINGS

results in the cow-grass relationship came to be known as managed grazing. There are three basic managed-grazing techniques that improve soil health, carbon sequestration, water retention, and forage productivity:

1. Improved continuous grazing adjusts standard grazing practices (essentially a pasture free-for-all) and avoids overgrazing by decreasing the number of animals per acre.

2. Rotational grazing systematically moves livestock to fresh paddocks or pastures, allowing those already grazed to recover.

3. Adaptive multipaddock grazing, sometimes known as mob grazing, is the most intensive of the three. It shifts animals to and from smaller paddocks in quick succession, after which the land is given time to recover—a month in a warm, wet climate, or a year in a cooler, drier locale.

Studies report a range of impacts across the three practices. A meta-analysis of research shows that the impacts of grazing depend heavily on local climate, soil coarseness, and the grass species that dominate a landscape. Improved grazing typically sequesters a few hundred pounds of carbon per acre, but in some cases as much as three tons per acre. When methane and nitrous oxide emissions are taken into account, the net sequestration is much lower. However, pastures make up 70 percent of the world's agricultural land, and because managed grazing can be used across geographies, it can have significant impact if scaled.

The changeover from conventional grazing to intensive grazing involves a transitional period from one regime to another. It requires weaning farms off pesticides, herbicides, fungicides, and fertilizers. All of these are conclusions agricultural corporations are unlikely to study and fund. The empirical results achieved by long-term adherents describe a two- to three-year period for the transition—about the same length of time as most of the studies that question the results shown by proponents. What farmers across North America are experiencing is specific to a single farm and thus not included in studies or peer-reviewed papers of managed grazing. Many of the benefits that are reported are consistent across geography, type of ranch or farm, and climes, and paint a different story from conclusions based on short-term observations.

Farmers who use managed grazing report that perennial streams that once went dry have returned. On farms with intensive one- to two-day rotations, the capacity to stock cattle on the land increased by 200 to 300 percent. Native grasses reestablished themselves, crowding out weeds. Not having to sow pastures saved time and diesel fuel. Tillage of pastureland stopped as well, conserving fuel and equipment expenses. The behavior of cattle changed. Rather than lollygagging around a stubbly, overgrazed pasture, they moved quickly and in the process ate weeds (which farmers are discovering are protein rich), thus reducing or eliminating the need for weed control.

Managed-grazing experimentation continues everywhere worldwide, and there are networks of ranchers using social media and face-to-face meetings to share what they are learning. There are no by-the-book techniques. The results seem to improve when grazing is rapid and intense and rest periods are longer. The protein and sugars of the grasses improve, and the more carbon sugars that are fed to the microbes in the soil, the greater the growth in mycorrhizal fungi, which secrete a sticky substance called glomalin. The organic rich soils are clumped together in small granules by the glomalin, which creates crumbly soil with empty spaces in which water can flow. Practitioners report that their soils can soak up eight, ten, and fourteen inches of rain per hour, whereas before the hardened soils would pond and erode with a mere inch of rain. Although rates of carbon sequestration are much discussed by climate activists, the farmers and ranchers who are leading the way are not doing this to sequester carbon or impact climate. They are increasing carbon to create healthy soils

Left: Mob grazing on Brown's Ranch in North Dakota.

Below: Gabe Brown kneeling in a cover crop of plantain, daikon radish, annual ryegrass, triticale, crimson clover, phacelia, and lentils.

and livestock. Many who started at 1 percent organic matter are now at 6 to 8 percent, or more.

Practitioners describe significant increases in income due to higher productivity and reduction in expenditures for herbicides, pesticides, fertilizers, diesel fuel, and veterinary costs. And they describe life returning to their land—flocks of songbirds, native grouse, fox, deer, and pollinating insects such as bees and butterflies. And despite the fact that the methodology is more intense, interviews recount how farmers have more time though they have more animals on the same amount of land. While the U.S. Department of Agriculture tends to waffle on the conservative side, the strongest advocates for moving carbon into pasture soils are farmers themselves.

Will Harris is a fourth-generation farmer operating White Oak Pastures in Clay County, Georgia, one of the poorest counties in the southeastern United States. After half a century of chemical-intensive techniques, out of a "growing sense of heritage and responsibility," Harris began transforming his family's farm into a holistic and humane system. He gave up corn feed, hormone injections, and antibiotics, then pesticides and fertilizers. Now, he says, "What I think about—all day, every day—is how can I make this land better?"

White Oak uses a rotational method modeled after natural grazing patterns in the Serengeti, which specifies that big ruminants be followed by small ruminants and then by birds. That means cows, then sheep, then chickens and turkeys, all given freedom of movement within their pastures. The farm functions more like an ecosystem, with animals expressing what Harris calls their instinctive behaviors, and the White Oak team views the entire operation as a living organism. Instead of using maximum output per acre to measure the farm's success, Harris focuses on health, longevity, and alignment with nature's principles—a profitable business for the long haul. As for sequestration, Harris reports that the carbon-rich organic matter in his 1,250 acres of soil is ten times higher than on conventional farms nearby with the same soil type and rainfall.

Gabe Brown of Brown's Ranch, located east of Bismarck, North Dakota, employs a high-density grazing technique with hundreds of cattle in a single herd moving between his one hundred paddocks, with durations less than a day in some. On one of his plots of land, Brown has taken soil organic matter from 4 percent to 10 percent in six years without using any outside inputs, an increase of fifty tons of carbon per acre. He describes the change in his agricultural practices best: "When I was farming conventionally, I'd wake up and decide what I was going to kill today. Now I wake up and decide what I am going to help live." And he is equally clear where change will come from: "You're not going to change Washington [D.C.]. Consumers are the driving force." ●

IMPACT: *By enhancing carbon sequestration compared to standard grazing practices, this solution can sequester 16.3 gigatons of carbon dioxide by 2050. Note that this does not reduce the 10 gigatons of methane that are emitted on that grazing land today. Growth in adoption of managed grazing practices would need to rise from 195 million acres to 1.1 billion acres over thirty years. Financial returns are $735 billion by 2050, on a $51 billion additional investment.*

White-bearded wildebeest herds mass together during the annual Serengeti migration. This photograph approximates what all herd animals do, which is to stick relatively close together while continuously moving on grassy rangelands. By herding, animals protect the young from hyenas, lions, and other predators that track the migrations. Managed grazing makes use of fences and short rotation times to imitate range behavior in order to optimize the health of the animals and regenerate the land.

WOMEN
AND
GIRLS

This sector is deceptively small in number. The solutions here focus on the majority of humanity, the 51 percent who are female. We call them out specifically because climate change is not gender neutral. Due to existing inequalities, women and girls are disproportionately vulnerable to its impacts, from disease to natural disaster. At the same time, women and girls are pivotal to addressing global warming successfully—and to humanity's overall resilience. As you will see here, suppression and marginalization along gender lines actually hurt everyone, while equity is good for all. These solutions show that enhancing the rights and well-being of women and girls could improve the future of life on this planet.

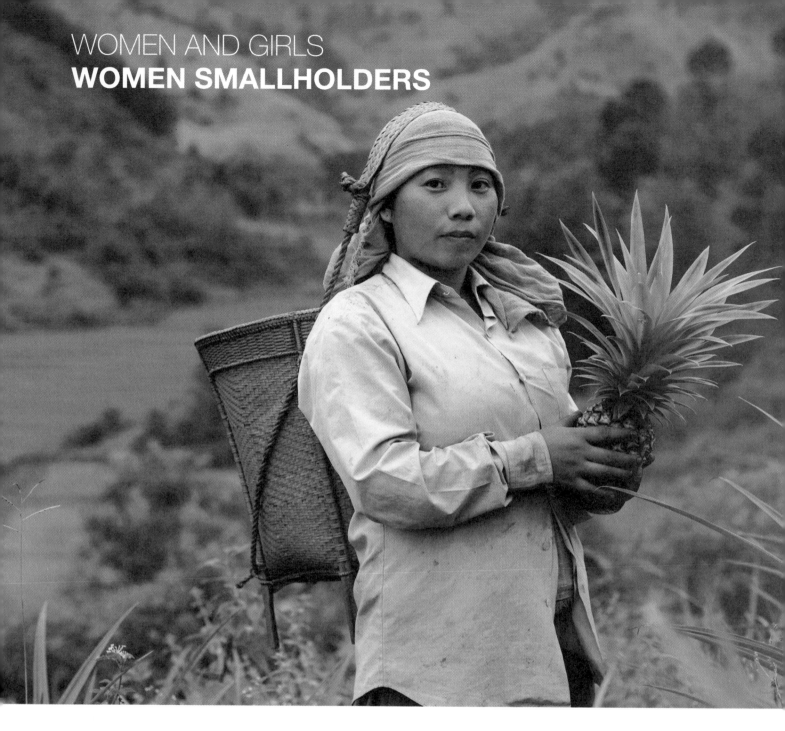

WOMEN AND GIRLS
WOMEN SMALLHOLDERS

There is a gender gap in agriculture in low-income countries—a gap between the resources and rights available to men who work the land and those available to women who do the same. On average, women make up 43 percent of the agricultural labor force and produce 60 to 80 percent of food crops in poorer parts of the world. Often unpaid or low-paid laborers, they cultivate field and tree crops, tend livestock, and grow home gardens. Most of them are part of the 475 million smallholder families who operate on less than 5 acres of land—to some extent for their own subsistence—and are among the world's poorest and most undernourished people. Their stories are diverse but share a key commonality: compared with their male counterparts, women have less access to a range of resources, from land and credit to education and technology.

Even though they farm as capably and efficiently as men, inequality in assets, inputs, and support means women produce less on the same amount of land. Closing this gender gap can improve the lives of women, their families and communities, while addressing global warming.

According to the Food and Agriculture Organization of the United Nations (FAO), if all women smallholders receive equal access to productive resources, their farm yields will rise by 20 to 30 percent, total agricultural output in low-income countries will increase by 2.5 to 4 percent, and the number of undernourished people in the world will drop by 12 to 17 percent. One hundred million to 150 million people will no longer be hungry. A few studies demonstrate that if women have access to the same resources as men—all else being equal—their outputs actually

2.06 GIGATONS	DATA TOO VARIABLE	$87.6 BILLION
REDUCED CO2	TO BE DETERMINED	NET SAVINGS

surpass parity: They exceed men's by 7 to 23 percent. Closing this gender gap can also control emissions. When agricultural plots produce well, there is less pressure to deforest for additional ground, and where regenerative practices replace chemical-intensive ones, soil becomes a carbon storehouse.

Land rights are at the center of the gender gap that women smallholders face. Few countries break down statistics of landownership along the lines of gender, but those that do reveal an underlying inequity: Just 10 to 20 percent of landholders are women, and within that group, insecure land rights are a persistent challenge. Many women are legally prevented from owning or inheriting property in their own right, limiting their decision making and leaving them vulnerable to displacement. In the words of Kindati Lakshmi, of India's Mahabubnagar district, "Owning a piece of land only would enable us to live with dignity and without hunger. We have no other way except to continue our struggle until we get land." Layered onto that reality, women have less access to cash and credit. Lack of capital can mean lack of fertilizer, farm tools, water, and seeds. Their second-class status restricts technical information and support from extension agents, membership in rural cooperatives, and marketing and sales outlets. As more men migrate to cities seeking nonfarm income, women are increasingly central to cultivation in low-income countries. They are hindered, however, from making decisions about and investments in improving the land they farm. Their responsibility grows but their rights and resources may not.

Proven interventions address ways in which current systems fail women, though complexity on the ground defies one-size-fits-all strategies. Bina Agarwal, a professor at the University of Manchester and the author of *A Field of One's Own*, captures the range of measures needed:

- Recognize and affirm women as *farmers* rather than *farm helpers*—a perception that undermines them from the start.
- Increase women's access to land and secure clear, independent tenure—not mediated through and controlled by men.
- Improve women's access to the training and resources they lack, provided with their specific needs in mind—microcredit in particular.
- Focus research and development on crops women cultivate and farming systems they use.
- Foster institutional innovation and collective approaches designed for women smallholders, such as group farming efforts.

Agarwal's last tenet is powerful. When women take part in cooperatives for growing, learning, financing, and selling, they achieve economies of scale in their operations and pool their influence,

know-how, and talent. They also are able to share labor, resources, and risk, such as the uncertain outcomes of trying a new crop or farming technique. Innovation and farm productivity follow. These outcomes are all the more important in a world shifting under global warming, to which farmers must readily adapt.

As with all smallholder farmers, diversity in cultivation helps annual yields to be more resilient and successful over time. For decades, agribusiness and government agencies have promoted techniques that are dependent on synthetic fertilizers, pesticides, and genetically modified seeds, which have left many smallholders at risk of market-commodity collapse, pest infestations, and deteriorated soil. In contrast, diversifying crops through practices such as agroforestry and intercropping does not require the same or, in many cases, any chemical inputs and creates more-resilient landscapes. Women—and men—need support not just in achieving yield gains but in yields gained *sustainably*, in ways that support them in the face of climate change. According to the FAO, "It will be difficult, if not impossible, to eradicate global poverty and end hunger without building resilience to climate change in smallholder agriculture through the widespread adoption of sustainable . . . practices."

As the world's population continues to grow—reaching a projected 9.7 billion by 2050—agricultural production will need to rise (in tandem with reduced food waste and dietary shifts). Given constraints on arable land and the need to protect intact forests, humanity will need to increase the yield of each plot. Growing more food on the same amount of land cannot be done without attending to smallholders, so many of whom are women, whose farming needs have been much overlooked. Countries that have higher levels of gender equality have higher average cereal yields; high levels of inequality correlate with the opposite outcome. If women smallholders get equal rights to land and resources, they will grow more food, feed their families better throughout the year, and gain more household income. When women earn more, they reinvest 90 percent of the money they make into education, health, and nutrition for their families and communities, compared to 30 to 40 percent for men. In Nepal, for example, strengthening women's landownership has a direct link to better health outcomes for children. With this solution, human well-being and climate are tightly linked, and what is good for equity is good for the livelihoods of all genders. ●

IMPACT: *This solution models reduced emissions from avoided deforestation, resulting from increasing the yield of women smallholders. Based on literature in the field, we assume yield per plot can rise by 26 percent, if women's access to finance and resources comes closer to parity with men's. If women managing 98 million acres receive equal assistance and achieve that 26 percent gain, this solution could reduce 2.1 gigatons of carbon dioxide by 2050.*

For women to have children by choice rather than chance and to plan their family size and spacing is a matter of autonomy and dignity. Two hundred and fourteen million women in lower-income countries say they want the ability to choose whether and when to become pregnant but lack the necessary access to contraception—resulting in some 74 million unintended pregnancies each year. The need persists in some high-income countries as well, including the United States, where 45 percent of pregnancies are unintended. Securing the fundamental right to voluntary, high-quality family planning services around the world would have powerful positive impacts on the health, welfare, and life expectancy of both women and their children. The benefits for social and economic development across all genders are myriad and, unto themselves, merit swift and sustained action. Family planning can also have ripple effects on drawing down greenhouse gas emissions.

In the early 1970s, Paul Ehrlich and John Holdren developed the now-famous equation known as "IPAT": *Impact =* *Population* x *Affluence* x *Technology*. In simplified fashion, it argues that the impact human beings have on the environment is a function of number, level of consumption, and the kind of technology used. Much of the work to address global warming has focused on the technology piece of the equation and the shift away from fossil fuels. Some has zeroed in on affluence, aiming to reduce consumer appetite for things, particularly in rich countries. Addressing the third factor, population, remains controversial, despite widespread agreement that greater numbers place more strain on the planet, though not equally so. Each person consumes resources and causes emissions throughout a lifetime; those impacts are much greater for someone in the United States than in Uzbekistan or Uganda. Carbon footprints are a common and comfortable topic. How many feet are leaving their tracks is not, due largely to concerns that linking family planning with environmental health is inherently coercive or cruel—Malthusian in the worst sense. However, when family planning focuses on healthcare provision and meeting women's expressed needs,

59.6 GIGATONS
REDUCED CO2   SEE IMPACT BELOW

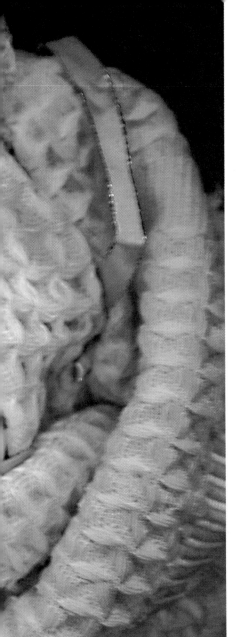

Three-day-old Waleed lies wrapped in blankets at his family's home in the southern Gaza Strip town of Rafah in 2016. Waleed was recognized as the two millionth person born in Gaza, a tiny enclave squeezed between Egypt, Israel, and the Mediterranean Sea. Gaza is just 7.5 miles across at its widest point, and has one of the highest population densities in the world.

across the country: female health workers providing basic care for women and children where they live. These and other success stories show that provision of contraception is rarely sufficient. Family planning requires social reinforcement, for example the radio and television soap operas now used in many places to shift perceptions of what is "normal" or "right."

After being silent on the topic of family planning for more than twenty-five years, the Intergovernmental Panel on Climate Change (IPCC) included access to reproductive health services in its 2014 synthesis report and pointed to population growth as an important factor in greenhouse gas concentrations. Growing evidence suggests that family planning has the additional benefit of building resilience—helping communities and countries better cope with and adapt to inevitable changes brought by global warming. That too has implications for women and girls, who, because of existing inequities, suffer disproportionately when impacts, from disease to natural disaster, hit. Still, this topic continues to be taboo in many countries and institutions, hemmed in by the persistent belief that raising the issue of population, or approaches that reduce it, is inherently draconian and an affront to the worth of human life. It may be the other way around on a warming, crowded planet: To revere human life it is necessary to ensure a viable, vibrant home for all. Honoring the dignity of women and children through family planning is not about centralized governments forcing the birth rate down—or up, through natalist policies. Nor is it about agencies or activists in rich countries, where emissions are highest, telling people elsewhere to stop having children. It is most essentially about freedom and opportunity for women and the recognition of basic human rights. Currently, family planning programs receive just 1 percent of all overseas development assistance. That number could double, with low-income countries aiming to match it—a moral move that happens to have meaning for the planet. ●

empowerment, equality, and well-being are the goal; benefits to the planet are side effects.

Challenges to expanding access to family planning range from basic supply of affordable and culturally appropriate contraception to education about sex and reproduction; from faraway health centers to hostile attitudes of medical providers; from social and religious norms to sexual partners' opposition to using birth control. Currently, the world faces a $5.3 billion funding shortfall for providing the access to reproductive healthcare that women say they want to have.

The success stories in family planning, however, are striking. Iran put a program into place in the early 1990s that has been touted as among the most successful such efforts in history. Completely voluntary, it involved religious leaders, educated the public, and provided free access to contraception. As a result, fertility rates halved in just one decade. In Bangladesh, average birth rates fell from six children in the 1980s to two now, as the door-to-door approach pioneered at the Matlab hospital spread

IMPACT: *Increased adoption of reproductive healthcare and family planning is an essential component to achieve the United Nations' 2015 medium global population projection of 9.7 billion people by 2050. If investment in family planning, particularly in low-income countries, does not materialize, the world's population could come closer to the high projection, adding another 1 billion people to the planet. We model the impact of this solution based on the difference in how much energy, building space, food, waste, and transportation would be used in a world with little to no investment in family planning, compared to one in which the projection of 9.7 billion is realized. The resulting emissions reductions could be 119.2 gigatons of carbon dioxide, at an average annual cost of $10.77 per user in low-income countries. Because educating girls has an important impact on the use of family planning, we allocate 50 percent of the total potential emissions reductions to each solution—59.6 gigatons a piece.*

59.6 GIGATONS
REDUCED CO2   SEE IMPACT BELOW

Girls' education, it turns out, has a dramatic bearing on global warming. Women with more years of education have fewer, healthier children and actively manage their reproductive health. In 2011, the journal *Science* published a demographic analysis of the impact of girls' education on population growth. It details a "fast track" scenario, based on South Korea's actual climb from one of the least to most educated countries in the world. If all nations adopted a similar rate and achieved 100 percent enrollment of girls in primary and secondary school, by 2050 there would be 843 million fewer people worldwide than if current enrollment rates sustain. According to the Brookings Institution, "The difference between a woman with no years of schooling and with 12 years of schooling is almost four to five children per woman. And it is precisely in those areas of the world where girls are having the hardest time getting educated that population growth is the fastest."

In the poorest countries, per capita greenhouse gas emissions are low. People do not have enough energy to properly sanitize their water, read or study at night, or power their small businesses. There are 1.1 billion people who do not have any electricity at all. From one-tenth of a ton of carbon dioxide per person in Madagascar to 1.8 tons in India, per-capita emissions in lower-income countries are a fraction of the U.S. rate of 18 tons per person per year. Nevertheless, changes in fertility rates in these countries would have multiple benefits on virtually every level of global society.

Nobel laureate and girls' education activist Malala Yousafzai has famously said, "One child, one teacher, one book, and one pen, can change the world." An enormous body of evidence supports her conviction: For starters, educated girls realize higher wages and greater upward mobility, contributing to economic growth. Their rates of maternal mortality drop, as do mortality rates of their babies. They are less likely to marry as children or against their will. They have lower incidence of HIV/AIDS and malaria—the "social vaccine" effect. Their agricultural plots are more productive and their families better nourished. They are more empowered at home, at work, and in society. An intrinsic right, education lays a foundation for vibrant lives for girls and women, their families, and their communities. It is the most powerful lever available for breaking the cycle of intergenerational poverty, while mitigating emissions by curbing population growth. A 2010 economic study shows that investment in educating girls is "highly cost-competitive with almost all of the existing options for carbon emissions abatement"—perhaps just $10 per ton of carbon dioxide.

Education also shores up resilience in terms of climate change impacts—something the world needs as warming mounts. Across low-income countries, there is a strong link between women and the natural systems at the heart of family and community life. Women often and increasingly play roles as stewards and managers of food, soil, trees, and water. As educated girls become educated women, they can fuse inherited traditional knowledge with new

Kenya has made significant gains in education, with more than 80 percent of all boys and girls currently enrolled in primary schools. In secondary schools, the rate of enrollment drops to 50 percent for both boys and girls. Poverty is the main cause of low overall enrollment, and given socioeconomic norms, boys receive priority for higher education when there are financial constraints.

Malala Yousafzai is an activist for girls' education who was born in the Swat Valley in northern Pakistan. Largely educated by her father, Yousafzai was recognized early in life by the global community for her commitment to education rights under the specter of the Taliban's growing influence in Swat. In October 2012, a Taliban gunman attempted to assassinate Yousafzai as she was riding a bus home after taking an exam. Malala is the youngest recipient of the Nobel Peace Prize, and continues both her studies and her work through the Malala Fund, which aims to secure 12 years of safe, quality education for girls the world over.

information accessed through the written word. As cycles of change play out in the times to come—new diseases blighting fruit trees, soil composition shifting in garden plots, altered seed-sowing times—educated women can marshal multiple ways of knowing to observe, understand, reevaluate, and take action to sustain themselves and those who depend on them.

Education also equips women to face the most dramatic climatic changes. A 2013 study found that educating girls "is the single most important social and economic factor associated with a reduction in vulnerability to natural disasters." *The single most important.* It is a conclusion drawn from examining the experiences of 125 countries since 1980 and echoes other analyses. Educated girls and women have a better capacity to cope with shocks from natural disasters and extreme weather events and are therefore less likely to be injured, displaced, or killed when one strikes. This decreased vulnerability also extends to their children, families, and the elderly.

In the past twenty-five years, the global community has learned a great deal about educating girls. So many challenges impede girls from realizing their right to education, and yet, around the world, they are striving for a place in the classroom. Economic barriers include lack of family funds for school fees and uniforms, as well as prioritizing the more immediate benefits of having girls fetch water or firewood, or work a market stall or plot of land. Cultural barriers encompass traditional beliefs that girls should tend the home rather than learn to read and write, should be married off at a young age, and, when resources are slim, should be skipped over so boys can be sent to school instead. Barriers are also safety related. Schools that are farther afield put girls at risk of gender-based violence on their way to

and from, not to mention dangers and discomforts at school itself. Disability, pregnancy, childbirth, and female genital mutilation also can be obstacles.

The barriers are real, but so are the solutions. The most effective approaches concurrently tackle access (school affordability, proximity, and suitability for girls) and quality (good teachers and good learning outcomes). Mobilizing communities to support and sustain progress on girls' education is a powerful accelerant. The encyclopedic book *What Works in Girls' Education* maps out seven areas of interconnected interventions:

1. Make school affordable.
   For example, provide family stipends for keeping girls in school.
2. Help girls overcome health barriers.
   For example, offer deworming treatments.
3. Reduce the time and distance to get to school.
   For example, provide girls with bikes.
4. Make schools more girl-friendly.
   For example, offer child-care programs for young mothers.
5. Improve school quality.
   For example, invest in more and better teachers.
6. Increase community engagement.
   For example, train community education activists.
7. Sustain girls' education during emergencies.
   For example, establish schools in refugee camps.

Today, 130 million girls are denied the right to attend school. The situation is most dire in secondary classrooms. In South Asia, less than half of girls—16.3 million—are enrolled in secondary school. In sub-Saharan Africa, fewer than one in three girls attends secondary school, and while 75 percent of all girls start school, just 8 percent finish their secondary education. Currently, international aid for education projects is about $13 billion annually. Given the link between girls' education and climate change, funds for climate mitigation and adaptation could enable the world to scale solutions rapidly. It could be a powerful match between education's need for funds and the world's need for proven climate solutions. Moreover, synchronizing investments in girls' education with those in family planning would be complementary and mutually reinforcing. Education is grounded in the belief that every life bubbles with innate potential. When it comes to climate change, nurturing the promise of each girl can shape the future for all. ●

IMPACT: *Two solutions influence family size and global population: educating girls and family planning. Because the exact dynamic between these solutions is impossible to determine, our models allocate 50 percent of the total potential impact to each. We assume that these impacts result from thirteen years of schooling, including primary through secondary education. According to the United Nations Educational, Scientific, and Cultural Organization, by closing an annual financing gap of $39 billion, universal education in low- and lower-middle-income countries can be achieved. It could result in 59.6 gigatons of emissions reduced by 2050. The return on that investment is incalculable.*

BUILDINGS AND CITIES

Thinking has come full circle on cities, from blaming them for environmental destruction to considering that urban environments, properly designed and managed, can be a kind of biological as well as a cultural ark— places where human beings can have the lowest impact on the planet and be educated, creative, and healthy. This remarkable shift began with the work of writer Jane Jacobs and landscape architect Ian McHarg in the 1960s and spread to architects, mayors, designers, and developers who now help reimagine the life of cities with Mother Nature and human nature as their twofold template. Buildings and broader urban habitats have become a font of innovation with respect to water, energy, lighting, design, and impact. Biologists such as Janine Benyus are now mapping how cities could be more productive in terms of air, water, flora, fauna, pollinators, and carbon sequestration than the original lands they were built upon. Rather than being a source or cause of degeneration, cities are becoming regenerative to the environment and human well-being.

What was once an engineering challenge and architectural oddity has become an alternative construction method available all over the world. A net zero building is one that has zero net energy consumption, producing as much energy as it uses in a year. In some months it may generate excess electricity; at other times it may require electricity. On balance, it is self-supporting. Along with using less energy, net zero buildings are more resilient during disasters and blackouts, are more carefully designed by necessity, and generally have reduced operating costs.

Designing a net zero building means following energy use back to the source. There are multiple ways to reduce energy loads in building. Lighting can be reduced in favor of daylight wherever possible. Spaces are designed to encourage people to walk between floors instead of using elevators. Walls, windows, and ceilings have maximum insulating power (R-value) to retain heat in the winter and maintain coolness in the summer. Window louvers and overhangs are designed to receive sunlight in winter months, when the sun is lower, and to create needed shade in the summer, when the sun is more directly above. Electrochromic glass changes its opacity according to heat, sun, and the difference between indoor and outdoor temperature. If the window is next to you, it can be manually adjusted with an app on your smartphone. Heat exchangers are strategically placed to be sure every stray calorie is put to use. Passive solar gain is achieved by building orientation and artful fenestration. Air-conditioning uses natural ventilation principles from nature, such as termite mounds and belowground thermal mass, to create natural convection and cooling breezes.

When net zero buildings were a novelty, it was rational to be modest in one's construction goals and see net zero as a risky experiment. Today, they are becoming more commonplace, as architects roll out extraordinary buildings across the world. One New England architectural firm refuses commissions for buildings that are not net zero. When queried, the partner said it was in order to preserve their reputation.

Net zero neighborhoods, districts, and communities are being designed and constructed, such as the Kaupuni Village affordable housing project in Hawaii and the Sonnenschiff solar city in Freiburg, Germany, which produces four times the energy it consumes. The city of Cambridge, Massachusetts, has created a plan for all buildings to be net zero by 2040. California is proposing to revise its building code to mandate all new residential construction be net zero by 2020—followed by all new commercial building construction by 2030. There is now a Walgreens drugstore in Chicago that is a net zero building. Newer net zero buildings push the margins further: zero water and zero waste. They harvest rainwater and process sewage on-site into compostable forms.

The conceptual origins of net zero buildings are in organisms. More often than not, buildings have been seen as parts and pieces designed and engineered to fulfill functions—not as the system they are. Engineers in particular had perverse incentives. To avoid future liability, they would calculate the air-conditioning system required for a building and then double the system's capacity, for instance. Compensation for some professionals was based on overall build-out costs—rewarding sufficiency, not efficiency. Once the paradigm shifts, the building, the site, the

RANKING AND RESULTS BY 2050

#79

COST AND SAVINGS MODELED IN RENEWABLE ENERGY,
LED LIGHTING, HEAT PUMPS, INSULATION, ETC.

weather, the arc of the sun, and the building's occupants are all seen as one system. Buildings breathe just like creatures; they inhale and exhale air. They require energy, but as in nature, no waste—the right amount at the right time and place.

When the U.S. Green Building Council first established higher standards for building in 1993, it was arguably the first trade organization in the world calling for higher than government standards. An offshoot organization called the Cascadia Green Building Council, led by architect Jason McLennan, believed that ZEBs (Zero Energy Buildings) could be designed that far surpassed Leadership in Energy and Environmental Design (LEED) standards. They began to promote the net zero concept, which eventually led to the International Living Future Institute and the idea of Living Buildings. In 2005, the same year that McLennan created the Living Building Challenge with architect Bob Berkebile, architect Ed Mazria announced the 2030 Challenge, a staged timeline for all buildings to become carbon neutral by that date. The 2030 Challenge has since been adopted by districts, cities, states, and countries that are using net zero building techniques. Projected U.S. building sector energy consumption in 2030 has declined for eleven successive years since the challenge was issued, a reduction of 18.5 quadrillion British thermal units (BTUs), the equivalent of 1,209 coal-fired 250-megawatt power plants. What was once seen as a marginal if not fanciful notion to construct workplaces and human habitats is now practiced the world over thanks to the concept of zero emission buildings. ●

IMPACT: *There are no numbers at the top of this page because net zero buildings are a mosaic of separate solutions. They draw on smart windows; green roofs; efficient heating, cooling, and water systems; better insulation; distributed energy and storage; and advanced automation. All are treated individually in our analysis. If net zero buildings are calculated as a single solution, assuming 9.7 percent of new buildings will be net zero by 2050, the integrated opportunity is 7.1 gigatons of carbon dioxide.*

The Rocky Mountain Institute Innovation Center is a net zero building on the north shore of the Roaring Fork River in Basalt, Colorado. The two-story, 15,600-square-foot building was constructed using Integrated Project Delivery software and model, a replicable process that can be employed by commercial projects around the country of similar scale. Although located in one of the coldest climate zones in the United States, the insulated building envelope was built with R-50 walls and R-67 roof. It has an 83-kilowatt solar photovoltaic system on the roof that provides more energy than the building is designed to use. The building was designed to use less water than the rain and snow that fall upon the site. Although graywater use is not allowed as yet in Colorado, a graywater system was installed in anticipation of changes in state regulations. To save heating and air-conditioning energy, the Center focused on heating and cooling people, not the space. They addressed the six factors that affect human comfort, which are air temperature, wind speed, humidity, clothing level, activity level, and the temperature of surrounding surfaces. By zeroing in on these factors, the Center has a broader range of comfortable air temperature, from 67 to 82 degrees Fahrenheit compared to the conventional commercial building range of 70 to 76 degrees. This cut energy use by 50 percent, eliminated the air-conditioning system, and requires a small heating system only on the coldest days.

Human beings are walking creatures, made to go about on foot, to amble, to march. For most of history, walking was the primary, if not only, form of transportation. All towns and cities were designed for getting around in a bipedal fashion. Picture Florence or Marrakech. Imagine Dubrovnik or Buenos Aires. Walk around Paris in your mind. That orientation toward walking shifted in the early to mid-twentieth century, with the mass manufacture of automobiles and design (or redesign) of urban and suburban spaces to cater to them. It was a shift that had major implications for health, community, and environment, but need not continue to dominate.

In cities around the world today, *walkability* is again a favored term, thanks in large part to urbanist movements that advocate for well-designed, livable, and sustainable cities. Walkable cities, and walkable streets and neighborhoods within them, prioritize two feet over four wheels through careful planning and design (and typically cater to bikes as well). They minimize the need to use a car and make the choice not to rely on one appealing. This renaissance of pedestrian-oriented urban environments is vital today, because walking can dramatically reduce greenhouse gas emissions from driving. According to the Urban

Land Institute, in more compact developments ripe for walking, people drive 20 to 40 percent less.

Urban planner and author Jeff Speck writes, "The pedestrian is an extremely fragile species, the canary in the coal mine of urban livability. Under the right conditions, this creature thrives and multiplies." Speck's "general theory of walkability" outlines four criteria that must be met for people to opt to walk. A journey on foot must be *useful*, helping an individual meet some need in daily life. It must feel *safe*, including protection from cars and other hazards. It must be *comfortable*, attracting walkers to what Speck calls "outdoor living rooms." And it must be *interesting*, with beauty, liveliness, and variety all around. In other words, walkable trips are not simply those with a manageable distance from point A to point B, perhaps a ten- to fifteen-minute journey on foot. They have "walk appeal," thanks to a density of fellow walkers, a mix of land and real estate uses, and key design elements that create compelling environments for people on foot.

When mulling the idea of walking, focus tends to be on the bipedal trekkers themselves. But a network of infrastructure—what people walk on or in—is needed to make walking trips

2.92 GIGATONS	COST TOO VARIABLE	$3.28 TRILLION
REDUCED CO2	TO BE MODELED	NET SAVINGS

merchants benefit from greater foot traffic. They enable people from all walks of life to get around, regardless of income, thus boosting equity and inclusion. With more people walking, traffic congestion—and associated stress and pollution—declines. There are fewer motor vehicle accidents. The more people walk (and cycle), the safer those modalities become. Increased levels of physical activity boost health and well-being, addressing widespread problems of obesity, heart disease, and diabetes. Social interaction and neighborhood safety rise, as do creativity, civic engagement, and connection to nature and place. Walkable cities are easier and more appealing to live in, making for happier, healthier citizens. Health, prosperity, and sustainability go hand in hand in hand.

As the world's urban populations continue to grow, walkable cityscapes will increase in importance. Urbanites are expected to make up two-thirds of world population in 2050. Construction will rise to accommodate that boom. Today, too many urban spaces remain no- or low-walking ones. Far too many municipal policies still foster low-density, suburban-style development rather than dense, mixed-use neighborhoods—choices communities can get stuck with for a long time to come. And cities continue to invest too few dollars in pedestrian infrastructure. In low-income countries, around 70 percent of urban transportation budgets go toward car-oriented infrastructure, though roughly 70 percent of trips are taken on foot or mass transit. All of these trends run counter to what people want; at present, the demand to live in walkable places far outstrips supply.

To realize the full potential of walkability, real estate practices, zoning ordinances, and municipal policies need to shift. Form-based codes that replace conventional single-use zoning, guidelines such as Leadership in Energy and Environmental Design (LEED) for Neighborhood Development, and walkability indexes such as Walk Score are already making a difference. Practices such as "walking school buses," gathering kids together for treks to school, can establish walking habits early in life. Ultimately, walkable cities will be most successful when they make strolling, striding, and sauntering, once again, the most inviting ways to move around. ●

IMPACT: *The six dimensions of the built environment—demand, density, design, destination, distance, and diversity—are all key drivers of walkability. Our analysis focuses on population density as a proxy for walkable neighborhoods. As cities become denser and city planners, commercial enterprises, and residents invest in the "6Ds," 5 percent of trips currently made by car can be made by foot instead by 2050. That shift could result in 2.9 gigatons of avoided carbon dioxide emissions and reduce costs associated with car ownership by $3.3 trillion.*

The San Telmo barrio in Buenos Aires was always a walkable and intimate neighborhood that gathered people into cafés and shops on its cobbled streets. Today, its old churches, antique shops, alleys, and artists attract tourists from around the world. It is a street experience opposite of that on Avenue 9 de Julio three blocks away — a noisy gouge through Buenos Aires through which traffic pours and where big retail towers indifferently over the human beings.

safe, convenient, and desirable. What does that look like? It is the opposite of sprawl. Homes, cafés, parks, shops, and offices are intermingled at a density that makes them reachable by foot. Sidewalks are wide and protected from motorized traffic whizzing by. Walkways are well lit at night, tree-lined and shaded during the day (vital in hot, humid climates). They connect effectively to one another and perhaps lead to entirely car-free areas. Points of interest across the road, tracks, or waterway are accessible by way of safe and direct pedestrian crossings constructed at regular intervals. At street level, buildings feel abuzz with life, fostering a sense of safety. Beauty invites people outside. Perambulation can easily be combined with cycling or mass transit, with good connectivity between these different modes of mobility. Many such improvements can be achieved at a fraction of the cost of other transportation infrastructure. Walkability also enhances the use, and thus cost-effectiveness, of public transit systems.

Many of the things that make cities more sustainable also make them more livable—perhaps nothing more so than walkability. It is why environmentalists find themselves calling for the same changes as economists and epidemiologists. Walkable urban places attract residents, businesses, and tourists, while local

BUILDINGS AND CITIES
BIKE INFRASTRUCTURE

The bicycle has been an agent of change since it first rolled into nineteenth-century Europe as a leisure item for sporty men. Within a matter of years, cycling became widespread, widely accessible, and widely loved. Bikes allowed adolescents to mix and mingle across neighborhoods and social classes, away from moralizing eyes. They gave women freedom of movement and helped redefine norms of dress and femininity. As suffragist Susan B. Anthony said in 1896, "Let me tell you what I think of bicycling. I think it has done more to emancipate women than anything else in the world."

The arrival of cars in the early twentieth century diverted attention to four wheels, and even Europe's cycling capitals, such as Amsterdam, saw automobiles dominate the middle of the century. But today, bicycles seem to be entering another golden age as cities attempt to untangle traffic and unclog skies, urban dwellers seek affordable transportation, and diseases of inactivity and billowing greenhouse gases become impossible to ignore. As one hub among these interconnected spokes, bikes could be a force for societal change once again.

According to British writer Rob Penn, "The bicycle can be ridden, on a reasonable surface, at four or five times the pace of walking, with the same amount of effort—making it the most efficient, self-powered means of transportation ever invented." At virtually zero emissions, it is exceedingly efficient in a climatic sense as well. But in his praise Penn also identifies a potent obstacle to the bicycle's triumph: "a reasonable surface," aka infrastructure.

Just like pedestrians and cars, bicycles need thoughtfully designed infrastructure. Numerous studies have sought to identify the fundamental elements that support safe and abundant cycling. Time and again they identify a tight link between networks of bike lanes or paths and the prevalence of bikers in a city or town. The more direct, level, and interconnected these tracks are, the better. Thoughtfully designed junctions where bicycles and cars meet—intersections, roundabouts, points of access—are vital to safety and flow. For instance, at red lights cyclists can be funneled ahead of queued cars, so they are fully visible and can proceed first, before any turning motorists. Other critical infrastructure includes secure parking, good lighting, greenery, and connections to desired destinations, including public transport. Equity is essential: Some cities have shown a bias for investing in bicycle infrastructure only in areas of privilege.

The role of bicycle infrastructure is to create safe, pleasant, effective environments in which to ride. Bikers—especially women, research shows—want to be separated from car traffic. But physical infrastructure alone is not enough. In the places where cycling thrives, such as Denmark, Germany, and the Netherlands, programs and policies foster a kind of social infrastructure that complements it. Educational initiatives target cyclists and motorists alike. Stricter liability laws protect those on two wheels. Disincentives for car ownership and use make the bicycle more attractive. Research also shows that city bike-share programs, like Vélib' in Paris, and awareness-raising events, like Ciclovía in Bogotá, increase ridership. Workplace showers can make sweaty commutes viable, and access to affordable parts and maintenance can make bike ownership work. Overarching urban design addresses the density, accessibility, and connectivity within the built environment that are crucial to bicycle friendliness.

In 1967, a Dutch official declared cycling tantamount to suicide. That was about to change. Following World War II, the country had shifted to car-centric development and lifestyles, until a rising number of traffic fatalities, many involving children, sparked a movement that induced government action and a turnaround in the Netherlands' trajectory—within a decade. Amsterdam, Rotterdam, and Utrecht are now some of the world's cycling meccas. In Amsterdam, bikes outnumber cars four to one.

Similarly, Copenhagen's infrastructure investments have made cycling easy and fast. They include innovations such as the "green wave"—traffic lights along main roads synchronized to the pace of bike commuters, so they can maintain their cruising speed for long stretches. Currently, the city is investing in a responsive traffic light system that aims to cut travel time by 10 percent for bicycles and 5 to 20 percent for buses, making both modes more appealing. At the same time, infrastructure for cars is becoming less accommodating, as with the gradual removal of parking spaces.

The numbers speak for themselves: In Denmark, 18 percent of local trips are done on two wheels, and in the Netherlands, 27 percent. In the car-crazy United States, by comparison, just 1 percent of trips are taken by bike. But there is hope: Bike commuting throughout the country grew 60 percent between 2000 and 2012, and in places such as Portland, Oregon, where infrastructure investment is high, it jumped from 1.8 percent to 6.1 percent of commutes during that time. Given that 40 percent of urban car trips are less than two miles in length, many could be made by bike instead.

As Dutch history reminds us, all cities were once bike cities, before we began shaping and reshaping them for the almighty automobile. Hills and heat, storms and arctic chills will always pose their challenges, but most barriers to cycling lie fully within the control of municipalities. This is where the rubber meets the road: The more infrastructure we have, the more cyclists. The more cyclists, the more cultural norms shift—*This is simple, smart, stylish*—and the more society reaps numerous returns on investment, including the health benefits of cleaner air and people going about their days with greater physical activity.

Investment, however, is the key word. In most places bicycle infrastructure continues to receive a small fraction of public funds

spent on transportation—allocations that could shift. Cycling also raises concerns about safety, reasonably so, but a clear correlation exists between high cycling rates, more cycling infrastructure, and reduced risk of fatalities. There is safety in infrastructure and numbers, as people shift out of four wheels and onto two. From Europe's new bike highways to local cycle challenges, the bicycle may reduce emissions while reclaiming its status as economical, salubrious, whimsical, and perhaps game changing.

IMPACT: *In 2014, 5.5 percent of urban trips around the world were completed by bicycle. In some cities, bicycle mode share was over 20 percent. We assume a rise from 5.5 percent to 7.5 percent of urban trips globally by 2050, displacing 2.2 trillion passenger-miles traveled by conventional modes of transportation and avoiding 2.3 gigatons of carbon dioxide emissions. By building bike infrastructure rather than roads, municipal governments and taxpayers can realize $400 billion in savings over thirty years and $2.1 trillion in lifetime savings.*

Copenhagen is considered the most livable city in the world, in no small part because it is the most bike-friendly. Thirty percent of Copenhageners ride to work, school, and market on 18 miles of bike lanes, and along three bicycle superhighways connecting Copenhagen to its outlying suburbs. Twenty-three more such highways are currently in the works. Like virtually all European cities, Copenhagen was bicycle-friendly for much of the twentieth century. After the Second World War and into the 1960s the city became polluted and congested with car traffic. Citizens pushed back and reclaimed the city for biking. Today, the city is a testament to what bicycle infrastructure can do.

From an aerial view, most cities are a patchwork of gray, brown, and black rooftops. But look down over some parts of Stuttgart, Germany, or Linz, Austria, and many rooftops are easily mistaken for small parks or grassy squares. They are affirmation of the modern movement for green or "living" roofs, which has taken off in the past fifty years, especially in Europe. They also evoke a much longer history, back to the heyday of the Viking Age, when such roofs first became popular in Scandinavia. Rewind modern-day Norway to the ninth or tenth century, and you would find a landscape dotted with sod-roofed homes, now called *torvtak*.

Today, the conventional rooftop is a brutal, lifeless terrain, typically serving a sole purpose: protecting the building and inhabitants beneath from the elements. In fulfilling that role, roofs take a beating from sun, wind, rain, and snow. They can endure temperatures up to ninety degrees higher than the air around them on a hot day, making it harder to cool the floors below and contributing to the urban heat island effect. This phenomenon of cities being measurably hotter than nearby rural and suburban areas is particularly harmful for residents who are young, elderly, or ill. Green roofs, on the other hand, are veritable ecosystems in the sky, designed to harness the moderating forces of natural ecosystems and curtail a building's carbon emissions in the process.

Living roofscapes depend on a series of carefully designed layers that ensure the roof itself is protected, rainwater is filtered and drained, and plants can thrive. If aiming for performance with minimal inputs, they may have shallow soil to support a simple carpet of hearty, self-sufficient groundcover such as sedum. Often called stonecrops, these flowering succulents cover more than ten acres atop Ford's truck plant in Dearborn, Michigan. Or green roofs can have intensive systems to sustain full-fledged gardens, parks, or farms—places where people can rest, recreate, and raise flowers or food. That is how once-unused rooftops across Brooklyn have become a mecca of urban agriculture. The intensity of investment, structural requirements, installation, and upkeep depends on the level of greenery chosen.

Though up-front costs for green roofs are higher than those of their conventional cousins, and some maintenance is required, returns are compelling and long-term costs are comparable, sometimes lower. The soil and vegetation function as living insulation, moderating building temperatures year-round—cooler in summer, warmer in winter. Because the energy required for heating and air-conditioning is curbed, greenhouse gas emissions are lower, as are costs. On the floor below a living roof, energy use for cooling can drop by 50 percent. Green roofs also sequester carbon in their soil and biomass, filter air pollutants, reduce

0.77 GIGATONS REDUCED CO2	$1.39 TRILLION NET COST	$988.5 BILLION NET SAVINGS

rainwater runoff, support biodiversity within cityscapes, and address urban heat islands—benefiting not just the floors beneath but nearby buildings as well. Because vegetation protects the roof itself from the elements and UV rays, green roofs have double the life span of conventional ones.

People who live, work, or play near green roofs enjoy more natural beauty and greater well-being—the result of biophilia, humanity's innate affinity for the natural world. At the same time, building developers, owners, and operators enjoy increased property appeal and value. Green roofs bring what people love to encounter on the ground to elevated yet often wasted spaces. Land is generally the most limited urban resource, but green roofs can create acres and acres of opportunity for green space and the climate benefits that come with it. To see the green roof on Chicago's City Hall or Singapore's Nanyang Technological University is to imagine the breadth of the opportunity atop buildings. These signature projects and other demonstration efforts—such as those atop bus stops, visible to pedestrians and passing cars—inspire wider public support.

Hot spots of implementation, such as Germany, offer a key lesson: Construction incentives for green roofs and building policy that encourages or mandates their use are twin drivers of proliferation. They are the stimulus for scaling—from oddity to ordinary. To raise the ratio of green in Singapore, for example, the government covers half the cost of green roof installation. Chicago fast-tracks permits for buildings with green roofs. Regulations around storm water control and retention also can encourage adoption of green roofs. In addition, clear and consistent industry standards and capable architects, engineers, and builders can ensure quality. In October 2016, San Francisco became the first U.S. city to adopt a green roof mandate. As of this year, 15 to 30 percent of roof space on new buildings must be green, use solar power, or both. Other cities should follow suit. By attending to the life both within buildings and on top of them, the world's current patchwork of barren roofs can flower, transforming cities into life-supporting systems.

Cool roofs are kith and kin to green roofs, achieving similar impact but doing so with different methods, hurdles, and boons. *Reflection* is from the Latin for "bending back," and cool roofs do just that. When solar energy hits a conventional dark roof on a 99-degree day, just 5 percent of it is reflected back into space. The rest remains, heating the building and surrounding air. A cool roof, on the other hand, reflects up to 80 percent of that solar energy back into space. Cool roofs take a variety of forms: light-colored metal, shingles, tiles, coatings, membranes, and more being developed. Whatever technology is used, in an increasingly urban and warming world, sending solar energy back to where it came from, rather than absorbing it, is essential. Not only do cool roofs reduce heat taken on by buildings, driving down energy use for cooling, they also reduce the temperature in cities. Recent studies have shown that the capacity of cool roofs to relieve the urban heat island effect is more pronounced during heat waves, when heat islands are particularly intense, sometimes deadly. The growth of cities continues, so making them cleaner, more livable, and better for well-being is essential.

Where green roofs struggle with the high costs and special skills needed for implementation, cool roofs are cheaper, simpler, and more like conventional installs. They are eminently doable. Though regular cleaning is needed to sustain top-notch reflection, maintenance needs are much lower as well. Despite this ease, it is necessary to consider context. Cool roofs can create glare for their neighbors, and their impact depends on local climate. Hotter places benefit more from their cooling effect, while suffering less from their reduced heat retention in cold months. In colder climates, the insulation of green roofs may be more optimal year-round.

Cool roofs are not a new concept but have been slow to take root worldwide. They are on the rise in the United States and European Union, while getting increasing attention, and occasionally official commitment, elsewhere. California has been their greatest champion, integrating cool roofs into the state's building efficiency standards, Title 24, a decade ago. The success there shows the way forward, including the importance of regulations, rebates, and incentive programs. The evolution of cool roof technology is also promising. Traditional building aesthetics have worked against so-called "white roofs," but cool roof materials now come in an array of colors, and adjustable levels of reflection may ultimately address their downside in winter. In the interest of "bending back" not just solar energy and air temperature but also emissions, cool roofs hold considerable promise. ●

IMPACT: *In modeling green and cool roofs, we account for regional applications of each technology. If green roofs cover 30 percent of roof space by 2050 and cool roofs cover 60 percent, a total of 407 billion square feet of efficient roofing would be in place globally. Combined, these technologies could reduce carbon dioxide emissions by 0.8 gigatons at a cost of $1.4 trillion, thirty-year savings of $988 billion, and lifetime savings of $3 trillion.*

Designed by Dr. Stephan Brenneisen, the green roof of the Cantonal Hospital in Basel, Switzerland, overlooks the town and Rhine River. Constructed in 1937, the building welcomed its first green roof in 1990, which mimics the riverbank of the Rhine in design. The vegetated roof features two gravel areas to attract birds, as well as areas of sedum, herbs, moss, and large grass meadows. It is interspersed with big branches and stones to provide cover, and is monitored for birds, spiders, beetles, ladybugs, bumblebees, and more.

BUILDINGS AND CITIES
LED LIGHTING

ike other leading-edge technologies, LEDs (light emitting diodes) have a long lesser-known history. Their origin dates back to the 1874 invention by German physicist Ferdinand Braun of the diode—a crystal semiconductor that conducts electricity in one direction. Since then, the development of diodes has evolved into hundreds of critical applications that make possible much of what is plugged in, turned on, watched, and driven every day. One of the important findings was how, under certain conditions, diodes emit light. Although this was first observed in 1907, scientists did not see any practical use for such a device at that time. That all changed in the 1960s when General Electric, Texas Instruments, and Hewlett-Packard developed, patented, and specialized commercial applications. In 1994 high-brightness LED bulbs were invented by

Commissioned by the U.S. government, the War Industries Board created a number of war-specific agencies during World War I, including the United States Fuel Administration. Headed by Harry Garfield, the son of former president James A. Garfield, the agency's job was to ensure sufficient supply of energy for essential industries. Besides the wonderful art nouveau poster seen here, the agency created daylight savings time, a practice already instituted in Europe.

three Japanese scientists, for which they were awarded the Nobel Prize in Physics in 2014.

There are three main types of lighting, and each employs a different mechanism to create light. Incandescent lightbulbs heat a tungsten filament with an electrical charge in a vacuum. Fluorescent fixtures ionize gases by an arc of electricity. UV light is emitted, which is absorbed by the phosphor coating the tube. The phosphor emits their visible light. LEDs are solid-state; through a process called electroluminescence, they create charged electrons that emit photons—units of light.

Incandescent bulbs are so inefficient they have been likened to space heaters that emit a little light. LED bulbs radiate a lot of light and are more like microcomputers or a solar panel that works in reverse. Solar converts photons to electrons; LEDs convert electrons to photons. Solar and LEDs have the same type of semiconductor but an LED contains a circuit board. A light switch acts as the keyboard. When turned on, an LED uses 90 percent less energy for the same amount of light than an incandescent bulb, and half as much as a compact fluorescent, without toxic mercury. On top of that, an LED bulb will last much longer than either type of bulb—twenty-seven years if turned on five hours a day. This translates into a 10 to 30 percent return on investment if you buy and replace older lighting fixtures with LEDs.

When first commercialized in the 1960s, LEDs were used in electronics, displays, and Christmas lights. Today, they are clustered, grouped, and arranged to make a variety of useful and powerful lamps. Using diffusers, they can illuminate broad areas or focus intensely. They have standard bases and can be screwed into conventional sockets. The diverse range of LED lighting now available means that virtually any type of bulb currently in commercial or residential use can be replaced by an LED bulb. LEDs transfer 80 percent of their energy use into creating light—rather than heat, like older technologies—and reduce air-conditioning loads accordingly.

The question about LEDs is not whether they will become the standard in lighting fixtures; it is when. The price is two to three times that of incandescent and fluorescent fixtures per watt equivalent, but falling rapidly. Current up-front costs remain an obstacle for lower-income households, which end up paying higher energy costs when they use cheaper bulbs. And despite their current cost, LEDs offer an advantage to households that have no access to electricity. Their low energy use makes it possible to turn on the lights with small solar cells, replacing expensive kerosene lamps and their noxious fumes and high greenhouse gas emissions. For families and communities not connected to an electrical grid, solar-LED lights can have a beneficial impact on economic livelihoods. According to the University of California's Lawrence Berkeley National Laboratory, "A sixth of humanity spends upwards of $40 billion per year on

RANKING AND RESULTS BY 2050
(HOUSEHOLD)

#33

7.81 GIGATONS
REDUCED CO2

$323.5 BILLION
NET COST

$1.73 TRILLION
NET SAVINGS

RANKING AND RESULTS BY 2050
(COMMERCIAL)

#44

5.04 GIGATONS
REDUCED CO2

-$205.1 BILLION
NET COST

$1.09 TRILLION
NET SAVINGS

lighting (20 percent of the total energy spend for lighting), yet [receives just] 0.1 percent as much illumination as does the electrified world." Solar-LED products, on the other hand, pay for themselves within a year of purchase. In India alone, nearly 1 million solar lighting systems help students do their homework, birthing clinics operate effectively, and businesses remain open after sunset. Still, when the sun sets, more than a billion people live in the dark. LEDs are as important for addressing light poverty as climate change.

LEDs are also transforming urban spaces with street lighting. LED streetlights can save up to 70 percent of energy and significantly reduce maintenance costs, meaning cities can retrofit old, inefficient streetlights with LEDs that pay for themselves. LEDs can be "tuned" to provide health benefits to humans (greater alertness on highways or sleep-inducement in residential areas) and to protect wildlife (preventing birds and turtles, for example, from being disoriented by artificial light).

The impact solar-LED lights have on human well-being and economic development speaks to the essential role artificial lighting plays in day-to-day life. It extends activity into dark hours and expands the spaces that are useful beyond those that are sunlit. So hardwired into human life is lighting, it accounts for 15 percent of global electricity use—more than that generated by all nuclear plants worldwide. And demand is rising. LEDs will be vital to meeting it, while drawing down energy use and emissions, as well as expense. Countries mandating a shift to this technology are already lighting the way, reaping the rewards, and making the technology more affordable for all. ●

IMPACT: *Our analysis assumes that LEDs will become ubiquitous by 2050, encompassing 90 percent of the household lighting market, and 82 percent of commercial lighting. As LEDs replace less-efficient lighting, 7.8 gigatons of carbon dioxide emissions could be avoided in residences and 5 gigatons in commercial buildings. Additional gains, not counted here, will come from replacing off-grid kerosene lighting with solar-LED technology.*

A Tarahumara woman in Cueche, Mexico, in her home with a single LED lantern. Imagine your house without lighting at night. Now imagine the blessing of even a single LED lantern. If used 5 hours a night, one bulb should last 27 years, making it the least expensive form of illumination on the planet.

HEAT PUMPS

Robert Simmer, director of Stadtwerke Amstetten, a local utility company in Austria, stands in front of a heat pump designed to capture and recycle energy from the sewer.

enjamin Franklin may be the only diplomat to have studied the science of refrigeration. That was in Cambridge, England, in 1758, when he was trying to reduce tensions between King George and the American colony but found time for the lab as well. He and English chemist John Hadley were intrigued by a Scottish scientist's discovery, made a decade earlier, showing how the evaporation of volatile liquids manifested a secondary effect: cooling. The basic principle is that higher-energy (hotter) molecules evaporate first, leaving the lower-energy (cooler) ones behind. In Cambridge, the researchers' equipment consisted of a beaker of ether, a mercury bulb thermometer, and a bellows. After wetting

the thermometer in the ether, they worked the bellows hard to evaporate the liquid as rapidly as possible. The thermometer recorded the temperature, which dropped to seven degrees Fahrenheit in one trial, and the accumulation of ice solidified the experiment. Franklin wrote a friend, "One may see the possibility of freezing a man to death on a warm summer's day." An exaggeration, but the famous polymath was, yet again, on the right track. Could he have foreseen the consequences of his insight?

Gwyn Prins, professor emeritus at the London School of Economics and Political Science, suggests that addiction to air-conditioning (AC) is "the most pervasive and least noticed epidemic" in the United States—where the amount of electricity used to keep buildings cool is equal to what the whole of Africa uses, for everything. It is easy to understand how this happened: Fossil fuels were plentiful and cheap; no one worried about greenhouse gas emissions or global warming; cool air was a welcome relief, at home and at work. Critics argue AC is one road civilization should never have taken and must now exit. Perhaps, but an exit is not likely. A top aspiration of people around the world—many of them living in the hotter climes of Asia and Africa—is the comfort of air-conditioning. Demographics alone dictate that the world is set for a massive increase in AC demand in this century—one study predicts as much as thirty-three-fold by 2100. China's experience foreshadows: In the decade between 1995 and 2007, the percentage of air-conditioned homes in Chinese cities increased from 7 percent to 95 percent. China will soon surpass the United States as the leading consumer of AC.

Air-conditioning grabs most of the headlines when the subject is conservation and efficiency, but heating is just as susceptible to inefficiency and just as prime for improvement. The building sector worldwide uses approximately 32 percent of all energy generated; more than one-third of that is for heating and cooling. Various bodies have analyzed the potential for increased efficiency and projected the results. All agree on two points: Business as usual generates spiraling emissions from heating and cooling; maximum efficiency could cut energy use by 30 to 40 percent.

The means to increase efficiency are at hand and are not necessarily high-tech. For example, smart thermostats that correlate the temperature setting inside the building with the temperature outside and with actual human occupancy make good sense, but are often missing. Fan speeds are surprisingly important and are often set incorrectly. Heat exchangers to recover heat or cold from air ventilated outside are vital. Retrofitting existing structures with these low-tech interventions is more expensive, but in any new building they should be mandatory. They save money, prevent discomfort, and reduce emissions. Combine them with thermostat settings a few degrees warmer in summer and a few degrees cooler in winter and the energy benefits are exponential.

One technology stands out from the rest: Heat pumps could address the world's heating and cooling needs and eliminate almost all emissions if powered by renewable energy. Most people have a variation of a heat pump in their homes already: a refrigerator. The working principle is the same. Both refrigerators and heat pumps have a compressor, condenser, expansion valve, and evaporator, and both transfer heat from a cold space to a hot one. In winter, that means pulling heat from outside and sending it into a building. In summer, heat is pulled from inside and sent out. The source or sink of heat can be the ground, air, or water. Air-source pumps work best in temperate climates, as efficiency drops off when outside temperatures drop below 40 degrees Fahrenheit. However, newer technology is effective down to 5 degrees if a building is well insulated. In areas such as Scandinavia and northern Japan, ground-source heat pumps are the technology of choice, taking advantage of the earth's relatively constant temperature underground.

While cost can be high and efficiency fluctuates depending on local climate, heat pumps are easy to adopt, well understood, and already in use around the world. They can supply indoor heating, cooling, and hot water—all from one integrated unit. When it comes to efficiency, heat pumps have a singular advantage: For every unit of electricity consumed, an equivalent of up to five units of heat energy is transferred. According to the International Energy Agency, a 30 percent penetration of the building sector by appropriate heat pumps could reduce worldwide carbon dioxide emissions by 6 percent. This would be one of the larger contributions of any technology now on the market. When paired with renewable energy sources and building structures designed for efficiency, heat pumps will do more than move warm air; they will move the earth toward drawdown.

IMPACT: *Heating and cooling of residential and commercial building space requires more than 13,000 terawatt-hours of energy and is estimated to increase to more than 18,000 terawatt-hours by 2050. This energy use comes from on-site fuel combustion and electricity-based systems—from gas furnaces to air-conditioning units. High-efficiency heat pumps reduce fuel consumption to zero and use less electricity to generate heating and cooling. Current adoption is low at .02 percent of the market, but we estimate rapid growth as costs continue to decrease by up to 25 percent by 2050. For a cost of $119 billion in addition what would be spent on conventional technologies, operating savings could reach $1.5 trillion over three decades and $3.5 trillion over the technology's lifetime. Emissions reductions in this scenario come to 5.2 gigatons of carbon dioxide.*

BUILDINGS AND CITIES
SMART GLASS

Glass windows were a Roman invention, placed in public baths, important buildings, and homes of great wealth. Although quite opaque, Roman glass was a big step forward from animal hides, cloth, or wood for shutting out the elements. The word *window* itself comes from the Vikings' *vindauga*, meaning "wind eye." A one-time luxury, glass windows are now standard across the world, bringing light and visibility into the built environment without inviting in the weather.

Except windows *do* let in the weather, in the form of heat or cold. They are much less efficient than insulated walls at keeping room temperature in and outside temperature out—by a factor of ten or more, depending on the wall and window. If you take a thermographic image of a typical home in winter, its windows will light up with heat loss. The *U-value* or *U-factor* of a window is a measure of its efficiency, indicating the level of heat flow in or out. A single pane of clear glass may have a U-value of 1.2 to 1.3. With two panes and space between them, the window's U-value drops to 0.5 to 0.7. The lower, the better. (A similar metric, *R-value*, measures *resistance* to heat flow, so, inversely, the higher the better.)

Layering glass is not the only means of improving window efficiency. Low-emissivity (low-e) coatings, virtually invisible reflective surfaces, lower that window's U-value further. So does the injection of insulating gas—often argon or krypton—between panes. Tightly sealed, high-quality frames resist air leakage. Together, these technologies have steadily reduced inefficiency and, thus, the contribution of windows to a building's heating and cooling load. Based on window ratings by the U.S. Energy Star program, the most efficient windows clock in around 0.15 to 0.2 U-value.

More adaptive technologies, dubbed "smart glass," make windows responsive in real time to the weather. In chemistry, chromism is any process that causes material to change color. Electricity triggers the process in electrochromism; heat, in thermochromism; and light, in photochromism. Electrochromic glass was developed in the 1970s and '80s by researchers at the National Renewable Energy Laboratory near Denver, the Lawrence Berkeley National Laboratory in California, and other institutes. What makes it electrochromic is a thin layer of nanoscale metal oxides—one-fiftieth the thickness of a human hair—the exact recipe for which varies by manufacturer and continues to evolve through research. When exposed to a brief burst of voltage, ions move into another layer and the tint and reflectiveness of the glass change. Tuned by smartphone or tablet, electrochromic glass is as switchable as indoor lighting.

The most advanced electrochromic windows disaggregate light and heat for optimal performance. On a cold winter day, both visible light from the sun and its thermal radiation can penetrate. In summer, the glass can be activated to admit visible light while blocking heat. Or, at a slightly different voltage, both are reflected, darkening the room—no need to close the blinds, or even have them. (The Boeing 787-9 Dreamliner uses electrochromic glass in lieu of window shades.)

A kindred technology, thermochromic glass requires no jolt of electricity. Based on outside temperature, it transitions automatically from transparent to opaque and back again. It is the mood ring of windows. Photochromic windows operate similarly, on the basis of light exposure. Certain eyeglass lenses use the same chemistry. In both cases, the clear advantage is that there is no action required, but thermochromic and photochromic windows lack the adaptability and control of electrochromic options. On-demand smart windows have the added benefit of reducing energy load for lighting, along with improving heating and cooling efficiency.

In Japan, tests of electrochromic glass have shown that cooling loads can drop by more than 30 percent on hot days. According to the California-based company View, its electrochromic line reduces energy use by 20 percent compared with traditional windows. They are also 50 percent more expensive, which is the fundamental drawback of smart glass. Some of that cost may be made up elsewhere, if the need for curtains and blinds is eliminated and smaller, more efficient air-conditioning units are used. Cost-effectiveness may be greatest in hot climates or on facades with high sun exposure. Price declines should continue as the market grows. Once a futuristic technology featured in movies such as *Blade Runner* (1982), switchable smart glass will become a common tool for increasing building efficiency in the years to come. •

IMPACT: *Smart glass is an up-and-coming solution with a current adoption in only .004 percent of commercial building space. We assume that growth will occur primarily in the commercial sector in high-income countries and can reach 29 percent of new commercial building space by 2050. The potential energy efficiency from cooling is estimated at 23 percent and lighting at 35 percent. Both will vary depending on local climate and building location. Adopting smart glass can result in 2.2 gigatons of emissions reductions from decreased energy use. The financial cost is high at $932 billion, yielding thirty-year operational savings of $325 billion and lifetime savings of $3.6 trillion.*

RANKING AND RESULTS BY 2050

2.19 GIGATONS
REDUCED CO2

$74.2 BILLION
NET COST

$325.1 BILLION
NET SAVINGS

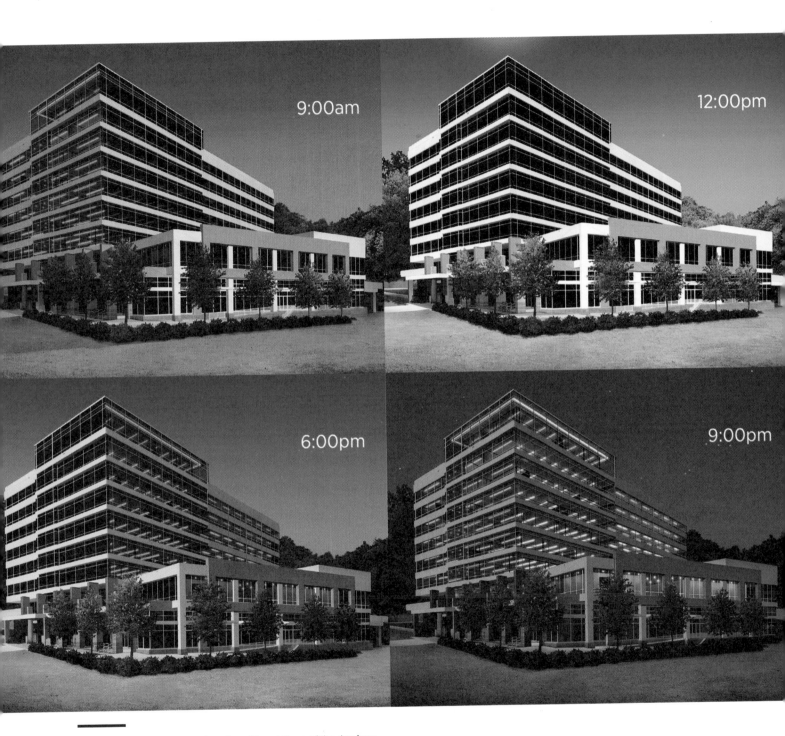

9:00am

12:00pm

6:00pm

9:00pm

Electrochromic glass responds to four different times of the day from two facings of a building. When tinted, the glass is reducing solar radiation and workplace glare, as well as the air-conditioning load, while maintaining daylight illumination inside. Sensors and even real-time weather data will override settings for daytime and allow more incoming light. The building is programmed by algorithms to respond to seasonal shifts in temperature and light; however, single panes of glass can be controlled from a smartphone at the user's desk to adjust glare, light, and tint.

BUILDINGS AND CITIES
SMART THERMOSTATS

2.62 GIGATONS REDUCED CO2	-$74.2 BILLION NET COST	$640.1 BILLION NET SAVINGS

An inconspicuous box or orb on the wall, the thermostat is easy to underestimate, yet in many buildings it is mission control for heating and cooling energy. According to the European Commission, maintaining temperate residential, commercial, and industrial buildings accounts for half the European Union's energy use. Residential thermostats alone control 9 percent of U.S. energy consumption. Smarter, programmable, sensor-connected thermostats that give real-time feedback to homeowners, tenants, and building managers are becoming integral to the management of that energy use. At present, the majority of thermostats require manual operation or preset programming, and studies show people are notoriously unreliable in doing either efficiently. Imagine if homes were only heated and cooled when, where, and to the extent needed, without any heavy lifting. That is the power of smart thermostats such as the Nest Learning Thermostat and the Ecobee. They are "smart" in the sense of being able to learn and take independent action, thereby eliminating the capriciousness of human behavior and driving more predictable energy savings.

Despite nearly two centuries of existence, thermostat technology saw minimal innovation until the past decade. The Nest came to market in 2011, developed by a team of former iPhone engineers who saw an opportunity to bring smartphone thinking to the antiquated temperature controls in homes. Thanks to algorithms and sensors, next-generation thermostats learn over time by gathering and analyzing data. You can still turn the temperature up and down, but these devices will remember your choices

and memorize your routines. Easy to install and simple to operate, they adapt to the dynamic nature of day-to-day living in a way programmable thermostats cannot. People do not always follow predictable schedules—some days departing for work early, and some evenings staying out late. Smart thermostats detect occupancy, learn inhabitants' preferences, and nudge users toward more efficient behavior. The newest technologies also integrate demand response; they can reduce consumption at times of peak energy use, peak prices, and peak emissions. More comprehensive home management systems also control hot water. The net effect: Residences are more energy efficient, more comfortable, and less costly to operate.

Where homes have HVAC systems and broadband, and residents have smartphones, smart thermostats can be highly effective interconnecting devices. Over two years, Nest Labs studied the impacts of its thermostats on energy use and cost savings. According to a company white paper, three separate studies produced similar results: energy savings of 10 to 12 percent on heating and 15 percent on central air-conditioning. Exact savings depend on individual thermostat use before upgrading to smart technology. Many industry estimates hover around 20 percent. Where homes are grouped in buildings or districts or connected to microgrids, individual thermostats can provide data to make the whole system more efficient.

Originating in North America and migrating to Europe, smart thermostats occupy a mere fraction of the addressable market at present. Realizing their room for growth hinges on one key factor: cost. People already own thermostats, so they need good reason and low barriers to elect to purchase and install new ones. Lower prices and incentive programs can encourage homeowners to replace existing thermostats. Price should also drop as the technology evolves and competition grows, and some utility companies are already offering incentives. (Even at their current prices, smart thermostats achieve payback in less than two years.) Amended building codes will help expand adoption, and thermostats that also monitor carbon monoxide and smoke may increase consumer appeal. ●

IMPACT: *We project that smart thermostats could grow from .4 percent to 46 percent of households with Internet access by 2050. In this scenario, 704 million homes would have them. Reduced energy use could avoid 2.6 gigatons of carbon dioxide emissions. Return on investment is high: smart thermostats can save their owners $640 billion on utility bills by 2050.*

BUILDINGS AND CITIES
DISTRICT HEATING

9.38 GIGATONS REDUCED CO2	$457.1 BILLION NET COST	$3.54 TRILLION NET SAVINGS

Density is a defining characteristic of cities. Compact urban spaces allow us to move about on foot and by bicycle, intermingle people and ideas, and create rich cultural mosaics. That density can also enable efficient heating and cooling of a city's buildings. In district heating and cooling (DHC) systems, a central plant channels hot and/or cool water via a network of underground pipes to many buildings. Heat exchangers and heat pumps separate buildings from the distribution network, so that heating and cooling are centralized while thermostats remain independent. Rather than having small boilers and chilling units whir away at each structure, DHC provides thermal energy collectively—and more efficiently.

The earliest examples of district heating are Roman. Hot water was used to warm temples, baths, even greenhouses. Its modern incarnations date back to 1882, when the New York Steam Company began pumping steam under Manhattan's busy streets to serve customers with district heating. Engineer Birdsill Holly first tested the invention at his own property in Lockport, New York, and it quickly spread to many U.S. cities. Canada began implementing district heating around the same time, with the University of Toronto installing its system in 1911. (Campuses continue to be popular locations for DHC.) By the 1930s, the Soviets were constructing networks to send heat from industrial processes into homes. Nordic cities began investing in district heating during the 1970s fuel crisis.

Copenhagen, Denmark, has become the global standout in DHC. It now meets 98 percent of heating demand with the world's largest district system, fueled with waste heat from coal-fired power plants and waste-to-energy plants. (In the coming years, biomass will replace all coal use.) Since 2010, Copenhagen has also tapped the chilly waters of the Øresund Strait for district cooling, sent through pipes that run parallel to thermal ones. Both sources are examples of how DHC can leverage innovative resources and turn waste streams into revenue streams.

Copenhagen's ongoing shift in fuel sources highlights a major advantage of DHC: Once a distribution network is in place, what powers it can morph and evolve. Coal can give way to geothermal, solar water heating, or sustainable biomass. A city's wasted heat—from industrial facilities to data centers to in-household wastewater—can be captured and repurposed. Indeed, DHC comes to life around the world in varied and increasingly clean ways. Renewable sources that might not be cost-effective at the scale of a building can become viable at the municipal level. DHC's collective supply creates economies of scale that save money. In parallel, improved building efficiency reduces heating and cooling needs over time.

Compared to individual heating and cooling systems, Tokyo's district system cuts energy use and carbon dioxide emissions in half—a powerful example of DHC's potential. Although

Dutch king Willem-Alexander attends the opening of BioWarmteCentrale (bioheating station) in Purmerend, the Netherlands. It provides 80 percent green energy for 25,000 people, powered by 110,000 tons of biomass per year.

it is a tried and tested technology, especially in Northern Europe, it is still new and unfamiliar in many parts of the world, and high up-front costs and system complexity continue to be obstacles. To date, district cooling is much less prevalent than heating, though it is becoming more relevant as cities in hot parts of the world grow—and as the world grows hotter. One of the world's largest systems is in Paris, keeping art lovers comfortable at the Louvre and Musée d'Orsay and preserving their masterpieces.

Whether they deploy it for heating, cooling, or both, municipal governments play the most essential role in taking this solution to scale. They are involved in planning, regulation, financing, and infrastructure, as well as setting aspirations around energy and emissions—all of which impact the viability of district systems. Urban decision makers can be, and in some places already are, the essential catalysts for collectively and efficiently heating and cooling the world's cities.

IMPACT: *By replacing existing stand-alone water- and space-heating systems, district heating can reduce carbon dioxide emissions by 9.4 gigatons by 2050 and save $3.5 trillion in energy costs. Our analysis estimates current adoption at .01 percent of heating demand, growing to 10 percent over the next thirty years. While natural gas is currently the most prevalent fuel source for district heating facilities, we model the impact only of alternative sources such as geothermal and solar thermal energy that will become more prevalent over time.*

BUILDINGS AND CITIES
LANDFILL METHANE

2.5 GIGATONS
REDUCED CO2

-$1.8 BILLION
NET COST

$67.6 BILLION
NET SAVINGS

Methane is a mighty molecule. Over the course of a century, it has up to thirty-four times the greenhouse effect of carbon dioxide. Landfills are a top source of methane emissions, releasing 12 percent of the world's total—equivalent to 800 million tons of carbon dioxide. But methane is also a fuel. Landfill methane can be tapped for capture and use as a fairly clean energy source for generating electricity or heat, rather than leaking into the air or being dispersed as waste. The climate benefit is twofold: prevent landfill emissions and displace coal, oil, or natural gas that might otherwise be used.

The world's cities create 1.4 billion tons of solid waste each year; that total may reach 2.4 billion tons by 2025. Globally, we send at least 375 million tons of solid waste to landfills, predominantly in developed countries. That outcome is far inferior to more sustainable waste-diversion approaches: reduction, reuse, recycling, and recovery. Nonetheless, sending waste to a well-engineered sanitary landfill is far better than disposing of it in an open dump, where it discharges pollution, contaminates water, and diminishes health. That remains the prevailing approach in lower-income countries—as it was in most places until the twentieth century.

Most landfill content is organic matter: food scraps, yard trimmings, junk wood, wastepaper. At first, aerobic bacteria decompose those materials, but as layers of garbage get compacted and covered—and ultimately sealed beneath a landfill cap—oxygen is depleted. In its absence, anaerobic bacteria take over, and decomposition produces biogas, a roughly equal blend of carbon dioxide and methane accompanied by a smattering of other gases. Carbon dioxide would be part of nature's cycles, but the methane is anthropogenic, created because we dump organic waste into sanitary landfills. Ideally, we would do it differently. Paper would be diverted for recycling and food scraps sent to composting or run through methane digesters. When they are not entombed, those wastes can create real value. But as long as landfills are piling up, we must manage the methane coming out of them. Even if we stopped landfilling immediately, existing sites would continue polluting for decades to come.

The technology to manage biogas is relatively simple. Dispersed, perforated tubes are sent down into a landfill's depths to collect gas, which is piped to a central collection area where it can be vented or flared. Better still, it can be compressed and purified for use as fuel—in generators, garbage trucks, or mixed into natural gas supply. Generating electricity from landfill gas is not without drawbacks: Pollutants from the combustion process diminish local air quality—a real concern for cities struggling with smog. Nonetheless, it is better than using raw fossil fuels, and has the additional benefits of reducing both odor and the risk of explosion or fire. (Totally clean renewables win the day.)

The amount of methane produced varies from landfill to landfill, as does the amount that can be captured. The more contained the site, the easier and more effective capture can be. According to a study of U.S. landfills, methane collection at closed sites was 17 percent more efficient than at sites actively receiving waste, but open landfills—which have the most active decomposition due to fresh deposits—were responsible for more than 90 percent of methane emissions. So while extraction wells can more thoroughly siphon landfill gas that is sealed within a closed and capped landfill, the biggest culprits, most in need of our attention, are those where rubbish continues to collect.

Landfills need not be hotbeds of emissions. As part of a comprehensive strategy to decrease and divert trash into higher uses, landfills should be—and increasingly are—designed, managed, and regulated with methane recovery in mind. A concentrated problem presents a concentrated opportunity to deliver real results. ●

IMPACT: *This solution sits at the bottom of the waste hierarchy. Landfill waste will decline as diets change, waste is reduced, and recycling and composting grow. What cannot, or should not, be combusted in waste-to-energy facilities will reach landfills as a last resort. These solutions will not be adopted globally overnight, so we assume landfill methane capture will continue to play a role. Combusting landfill methane for electricity production can result in emissions reductions equivalent to 2.5 gigatons of carbon dioxide.*

Wellhead for methane capture at a landfill in Michigan.

BUILDINGS AND CITIES
INSULATION

RANKING AND RESULTS BY 2050

#31

8.27 GIGATONS
REDUCED CO2

$3.66 TRILLION
NET COST

$2.51 TRILLION
NET SAVINGS

The word "insulation" comes from the Latin root *insula* for "island." In terms of thermal flow, making buildings into islands is exactly what insulation aims to do. Heat always moves from warmer areas to cooler areas, until a temperature equilibrium is reached. For keeping buildings within a desirable range of 67 to 78 degrees Fahrenheit, this heat flow presents a central challenge. During summer, hot air infiltrates indoor spaces, causing air conditioners to work overtime. During winter, warm air seeps out, finding its way to unheated attics and basements, up chimneys, and through gaps around windows and doors, so heating systems must work harder. To close the gap on unwanted heat gain or loss and maintain comfortable room temperature, we use more energy, whether fuel like natural gas or electricity. According to the U.S. Green Building Council, air infiltration accounts for 25 to 60 percent of energy used to heat and cool a home—energy that is simply wasted. By better insulating a building envelope, heat exchange can be reduced, energy saved, and emissions avoided.

What makes insulation effective is its capacity for thermal resistance: how effectively it resists heat flow through conduction (direct heat exchange through materials), convection (circulation of heat through air or fluids), and radiation (transfer of heat by electromagnetic waves). R-value is the system of measurement for thermal resistance. The higher the R-value, the more effective the insulation, which varies by type, thickness, and density, as well as where it is installed in a building and how. Ideally, a building's thermal layer should cover all sides—bottom floor, exterior walls, and roof—and be continuous to prevent an effect known as thermal bridging—the higher transfer of heat through other building materials like studs and joists. Air leaks and drafts impact insulation performance as well, which is why sealing gaps and cracks is critical to a more effective building envelope.

Insulation is one of the most practical and cost-effective ways to make buildings more energy efficient—both in new construction and through retrofitting older buildings that often are not well encased. At relatively low cost, insulation results in lower utility bills, while keeping out moisture and improving air quality. The range of insulating materials is wide. Fiberglass is among the most common, either in blanket-like batts or loose fill; plastics fibers can be made into similar products. Mineral wool is not wool at all, but manufactured material from basalt or blast furnace slag. Recycled newsprint finds its way into cellulose insulation, packed densely into cavities. Polystyrene insulation spans from rigid boards to sprayed foam. Also used are natural fibers, such as hemp, sheep's wool, and straw. Reflective barriers are designed to address radiant heat. Innovation in insulation materials continues with the aim of improving their performance and producing them more sustainably, for example capitalizing on the air-trapping power of waste poultry feathers.

Insulation is not new. Villagers and farmers in the north have been using turf roofs for a thousand years. This one is located in the Gjógv Village in the Faroe Islands, a small archipelago located in the Atlantic Ocean between Iceland and Norway with an average temperature of 53 degrees Fahrenheit during the "warm season."

The power of insulation is taken to the extreme with Passivhaus, or Passive House in English, a rigorous building method and standard created in Germany in the early 1990s and intensely focused on saving energy—by as much as 90 percent over conventional comparisons. This approach zealously focuses on creating an airtight envelope for a building, to separate inside from outside below, above, and around all sides. The result is a structure so hermetically sealed that warm air cannot leak out when snow is on the ground and cool air cannot escape when the dog days arrive. Some Passive House dwellings are so efficient they can be heated with the equivalent of a hairdryer. A thermos-like building envelope relies on thick, super-insulated foundation, walls, and roof; sealing all cracks, joints, and seams; addressing conductive thermal bridges; and using high-performance, triple-pane windows. Aggressively reducing energy needed for heating and cooling lays the foundation for meeting energy demand with on-site renewables and ultimately achieving net zero energy use. Passive House sets a high bar for insulation, and most buildings will not reach it in the near-term. But encouraged by financial incentives, building efficiency requirements, and enlightened self-interest, insulation can play a key role in lightening the load buildings place on the planet. ●

IMPACT: *Retrofitting buildings with insulation is a cost-effective solution for reducing energy required for heating and cooling. If 54 percent of existing residential and commercial buildings install insulation, 8.3 gigatons of emissions can be avoided at an implementation cost of $3.7 trillion. Over thirty years, net savings could be $2.5 trillion. However, insulation measures can last one hundred years or more, realizing lifetime savings in excess of $4.2 trillion.*

A receptionist sits behind the information booth during the $530 million retrofit of the Empire State Building, originally constructed in 1931. The retrofit to the Art Deco icon saw the replacement of all 6,500 windows and all heating, cooling, and lighting systems—an energy reduction of 38 percent.

The Empire State Building was never intended to be green. It was intended to be tall. Born out of competition between titans of industry to construct "the world's tallest building," it went up in just over a year and officially opened on May 1, 1931, when President Herbert Hoover ceremonially flicked on its lights from Washington, D.C. The building held its title of tallest until 1972. Once the poster child for bravado and might rendered in steel, limestone, and granite, the Empire State Building is now the poster child for retrofitting to achieve energy efficiency in the built environment—that is, addressing how much heat and cold are escaping or entering the building, what internal systems cool or warm inhabitants, and how the building is illuminated.

Global warming will not be addressed without attending to the buildings that house human-kind day and night. Worldwide, buildings account for 32 percent of energy use and 19 percent of energy-related greenhouse emissions. In the United States, buildings' energy consumption is more than 40 percent of the nation's total. They pull from the electric grid or natural gas lines to heat, cool, and light the spaces within them and to power all manner of appliances and machinery.

COST AND SAVINGS MODELED IN RENEWABLE ENERGY,
LED LIGHTING, HEAT PUMPS, INSULATION, ETC.

As much as 80 percent of the energy consumed is wasted—lights and electronics are left on unnecessarily and gaps in the building's envelope allow air to seep in and out, for example.

Much of the attention paid to green buildings is in new design construction. Various standards—Leadership in Energy and Environmental Design (LEED), Net Zero from the International Living Futures Institute, Passivhaus from the German institute of the same name, and R-2000 developed by Natural Resources Canada, to name a few—specify how to build well from the start, so that wasteful energy use is designed out of the building before it goes from the drafting table to real life. While it is important to look forward and shape the structures to come, it is equally critical to modify existing buildings—and not just commercial buildings. There are 140 million buildings in the United States and 5.6 million are commercial. These structures hold the greatest potential for energy reduction. Because old buildings are replaced by new at a rate of 1 to 3 percent per year, most of the existing building stock will still be here fifteen to twenty years from now.

Ramping up retrofitting was a central impetus for the Empire State Building endeavor. New York City has pledged to cut greenhouse gas emissions by 80 percent by 2050. To meet its goal, buildings need to be retrofitted. In the early years of the twenty-first century, the Empire State Building used as much energy in a single day as forty thousand single-family homes. The retrofit project—a collaboration between private, philanthropic, and nonprofit entities—set out to cut that usage by 40 percent.

The Empire State Building will save $4.4 million in energy costs and avert more than 100,000 tons of greenhouse gas emissions. The building's 6,514 windows were key for advancing efficiency. To save waste and money—more than $15 million worth—they were rebuilt on-site, with a layer of insulating film placed between the existing panes. Although the Empire State Building is a splendid example because of its art-deco heritage and cultural cachet, the 38 percent energy reduction it will achieve is just the beginning. The Willis Tower in Chicago, built in 1970, saved 70 percent of its energy use through a retrofit. Net zero retrofits now exist for older buildings. In the United States, there are 8,000 buildings over 500,000 square feet like Empire State and Willis. They should not avert focus from the other 139.5 million buildings that need retrofitting and for which energy savings, payback, and job creation would be extraordinary.

Retrofitting is a well-understood practice, and good building performance data is making it increasingly effective. The payback on retrofits, depending on the building, is five to seven years on average. Lenders such as Fannie Mae will increase commercial mortgages by 5 percent if the loan is used for greening a building. Yet existing commercial buildings are being upgraded at a rate of just 2.2 percent per year. This being real estate, the common obstacle is money. However, money can be found because the payback is there. There are now consultants in every city who will guide clients through any kind of retrofit desired and help arrange financing. Most utilities will consult as well and specify a wide range of appliances, lighting choices, variable speed pumps, and heating and cooling alternatives that can keep energy in the ground and put money in your pocket. Another payback is rarely mentioned: Retrofitted buildings have higher occupancy rates.

Tenants want healthy green spaces and will pay more for them in most cities today. Studies show people to be more creative, productive, and happy in well-designed green workplaces, and employers find it easier to recruit and retain talent. Developers such as Jonathan Rose Companies seek out and buy older office buildings in downtown areas from New York to Portland, Oregon, retrofit them, and rent them out again. The retrofit raises the quality and desirability of the workspace, which increases demand. Retrofitting extends the life of the building and increases its value. Green buildings, new or old, are better places to live and work—and to own.

For those who can see it and crack it, the business opportunity in retrofits is substantial. According to market sizing and analysis done by the Rockefeller Foundation and Deutsche Bank's climate change shop, $279 billion could be invested in the United States in retrofitting residential, commercial, and institutional buildings, yielding more than $1 trillion in energy savings over ten years—equal to 30 percent of the country's annual spending on electricity. In the process, more than 3.3 million cumulative job years of employment would be generated across all parts of the country, and U.S. emissions would be cut by almost 10 percent.

To realize the massive financial and emissions savings that are possible, a building-by-building approach to the world's 1.6 trillion square feet of building stock (99 percent of which is not green) is probably not the way to go. The Rocky Mountain Institute is piloting a more industrialized strategy in Chicago: Limit the scope of retrofitting to a set of highly effective, broadly applicable measures; pursue additional measures on the basis of impeccable analysis; and undertake multiple buildings simultaneously to gain economies of scale. Early results show it can reduce retrofit costs by more than 30 percent and achieve payback within four years. It is this sort of effort that is needed to connect the dots between people and energy, well-being and economics, and the future of the atmosphere. ●

IMPACT: *As with net zero buildings, there are no results presented from our models here. Building owners who retrofit existing residential and commercial building space install better insulation, improved heating and cooling equipment, upgraded management systems, etc. These solutions are accounted for individually. No retrofit will look exactly the same, making forecasting costs and savings nearly impossible.*

Water is heavy. Pumping it from source to treatment plant to storage and distribution requires enormous amounts of energy. In fact, electricity is the major cost driver of processing and distributing water within cities, underlying the sums on water bills. But those bills do not account for all of the water flowing through municipal systems. Utilities use the phrase "non-revenue water" to describe the gap between what goes in and what ultimately comes out the tap. The World Bank calculates that 8.6 trillion gallons are lost each year through leaks, split roughly in half between high- and low-income countries.

That the gallons lost during distribution are dubbed "non-revenue water" reveals what is at stake for utilities and municipalities: a sinking bottom line. Also at stake are emissions from needlessly producing billions of kilowatt-hours of electricity to pump water not into homes or businesses but through breaks in the world's water-distribution networks. Minimizing those leaks and losses means using energy more sparingly, while conserving water as a scarce resource.

In many places, aging water infrastructure and its deteriorating pipes and valves are a challenge. But their wholesale replacement is neither financially tenable nor necessary outside of extreme cases or whenever public health is at risk. Instead, improving the efficiency of water distribution largely depends on management practices. Those at the tap end of a water system know that pressure matters. It is just as fundamental for the system's health overall. To borrow a description from the *New York Times*: "A steady, moderately low level of pressure is best—just as [with blood flow] in the human body." Too much pressure and water looks for ways to escape; too little and water lines can suck in liquids and contaminants that surround them. Water utilities face a quest for pressure that is "just right." One of their common approaches is creating contained "district metered areas" within the larger system, each with a special valve that acts as a gatekeeper.

Even under conditions of first-rate pressure management, leaks can and will happen. The torrential bursts that cut off service and submerge streets are not actually the worst from a waste perspective: They demand attention and immediate remediation. The bigger problem is with smaller, long-running leaks that are less detectable. Vigilant, thorough detection and speed to resolution are key. A range of tools and techniques can aid in scanning for and pinpointing leaks, a process most effective at night, when the system is relatively quiet. Ongoing evolution of sensors and software is aiding both leak detection and pressure management. In fact, an entire industry has emerged to address water loss, growing out of groundbreaking work by what the *New York Times* called "a bunch of brilliant, obsessive, far-thinking engineers in Britain who started something called the National Leakage Initiative in the early 1990s." Their methodologies and techniques are now in use far beyond the British Isles.

The issue of water loss exists around the world. In the United States, an estimated one-sixth of distributed water escapes the system. Losses are typically much higher in low-income regions—sometimes 50 percent of total volume. If those losses alone were halved, that water could supply some 90 million people. Manila, the capital city of the Philippines, did just that. By successfully cutting its losses in half, the water utility was able to serve an additional 1.3 million people and achieve twenty-four-hour supply for almost everyone.

To date, success stories like Manila's are few and far between, even in high-income countries. Too often utilities fail to tackle the issue of water loss because their institutional or technical capacity is weak, they are not incented or required to act, or even because building new treatment facilities is easier and more exciting, if costly. Because acknowledging leakage problems also means acknowledging management problems—and potentially provoking the ire of customers and politicians—utilities are loath to do so, yet pressure is growing to insist they must. Given the financial investments and engineering excellence that can be required, global enabling efforts such as the World Bank–International Water Association partnership are essential.

The high-water mark for municipalities is this: In addition to increasing a utility's efficiency and improving customer experience, addressing leaks is the cheapest way to source new supply and serve a growing population. Those same practices make municipal water systems more resilient to water shortages, increasingly common events on a warming planet. Water distribution efficiency can be put to work to address climate change and to cope with its effects—a solution that is proactive and protective at the same time. ●

IMPACT: *Modeling only the impact of pressure management and active leakage control, we estimate that water losses can be reduced by an additional 20 percent globally by 2050. The resulting emissions reduction from pumped distribution could be .9 gigatons of carbon dioxide. Total installation cost is $137 billion and operating savings for utilities could be $903 billion by 2050. Implementing this simple solution could save 215 quadrillion gallons of water over thirty years.*

4.62 GIGATONS REDUCED CO2	$68.1 BILLION NET COST	$880.6 BILLION NET SAVINGS

BUILDINGS AND CITIES
BUILDING AUTOMATION

Buildings are complex systems in the guise of static structures. Energy courses through them—in heating and air-conditioning systems, electrical wiring, water heating, lighting, information and communications systems, security and access systems, fire alarms, elevators, appliances, and indirectly through plumbing. Most large commercial buildings have some form of centralized, computer-based building management system, which makes it possible to monitor, evaluate, and control those systems and seize opportunities for raising their energy efficiency, while improving the experience of occupants. But building management systems are manual and susceptible to human error. Adopting automated systems will secure efficiency gains otherwise left on the table, reducing energy consumption by 10 to 20 percent in an average building.

A building automation system (BAS) is a building's brain. Equipped with sensors, BAS buildings are constantly scanning and rebalancing for greatest efficiency and effectiveness. Lights switch off when no one's around, for example, and windows vent to improve air quality and temperature. A conventional system tells building managers what action to take, like a car's dashboard; buildings with automated systems take action themselves, like a self-driving car. New buildings can be equipped with BAS from the start; older ones can be retrofitted to incorporate it and reap its benefits.

The market for BAS is expanding. It is fueled by growing appreciation of the impact automated systems have on occupants' well-being and productivity, as well as energy savings and reduced operations and maintenance costs. Automation systems can help to improve thermal and lighting comfort and indoor air quality, which directly impact occupant satisfaction. According to the World Green Building Council, indoor air quality can contribute to increases in productivity of 8 to 11 percent. For building operators, BAS makes it easier to see when something is going wrong and to fix it fast. Less work is required when management of all systems is centralized and simplified through automation. For green buildings in particular, BAS can measure and verify key building metrics to ensure and maintain efficiency, which can be compromised by human and other factors. Green buildings can have high efficiency ratings, but they are only efficient if ratings match their actual operation.

Barriers to adoption exist. Energy expenditures are typically a small cost driver for businesses, not a place to seek significant savings. For BAS to be worthwhile, it must yield a high return on high up-front cost, and quickly. If projected returns fail to materialize, as they have in some instances, the broader credibility of BAS suffers. Landlord-tenant arrangements are another challenge. When a building's owner and its occupants are distinct actors, the incentive to maximize efficiency is muted: The former makes decisions about the building's systems, while the latter bears the cost of energy use. Occupant comfort is an aspiration they are more likely to share, given its impact on tenant satisfaction and thus retention.

The static structure of buildings makes it easy to forget their contribution to climate change. According to the Intergovernmental Panel on Climate Change, buildings are responsible for roughly one-third of global energy use and one-fifth of global greenhouse gas emissions. Building automation systems are one powerful solution for reining in that energy use. Critically, they circumvent individual behaviors such as adjusting the thermostat, making a step change in efficiency possible. BAS is becoming increasingly necessary to meet local and national building-efficiency requirements, and as buildings themselves become more complex—with distributed energy generation, exterior shading, switchable glass, and the like—BAS sophistication must continue to grow. These systems are the "neural networks" buildings need. ●

IMPACT: *BAS can result in up to 20 percent more efficient heating and cooling and 11.5 percent more efficient energy use for lighting, appliances, etc. Expanding these systems from 34 percent of commercial floor space in 2014 to 50 percent by mid-century—at an added cost of $68 billion—building owners could save $881 billion in operating costs. 4.6 gigatons of carbon dioxide emissions could be avoided.*

LAND
USE

The word *drawdown* describes the reduction of greenhouse gas concentrations in the atmosphere. There are two means by which to achieve it: a radical decrease in human-caused emissions and widespread adoption of proven land and ocean practices that sequester carbon from the air and store it for decades and even centuries. In order to properly measure the impact of land-based practices that would actually affect drawdown, we broke them up into discrete solutions. Thirteen are included under Food because they relate to food production, and nine are detailed here. We first assessed how land was being used the world over; then we calculated what would happen if the use were different, or if the techniques specifically being employed to graze or grow were altered. Although not included in the calculations, the research vividly shows how all twenty-two are no-regrets solutions. Implementation increases soil moisture, cloud cover, crop yields, biodiversity, employment, human health, income and resilience, while dramatically reducing the need for synthetic fertilizers and pesticides on farmland.

6.2 GIGATONS
REDUCED CO2

GLOBAL COST AND SAVINGS DATA
TOO VARIABLE TO BE DETERMINED

896.29 GIGATONS
CO2 PROTECTED

The most critical of all forest types is primary forest, known as old-growth or virgin forest. Examples include the Great Bear Rainforest of British Columbia and those of the Amazon and the Congo. These are forests that have achieved great age with mature canopy trees and complex understories, making them the greatest repositories of biodiversity on the planet. Forests contain 300 billion tons of carbon yet they are still being logged, sometimes under the guise of harvest being "sustainable." Research shows that once an intact primary forest begins to be cut, even under sustainable forest-management systems, it leads to biological degradation.

At one time, the planet's forests covered vast tracts of land and human incursions were relatively negligible. Stone axes were felling trees ten thousand years ago, but hunter-gatherers did not need significant amounts of wood. That began to shift as agriculture took root and communities remained in place. By 5500 BC, civilization and nation states began to bloom in what was known as the Fertile Crescent, nurtured by agricultural bounty. The first iron tools, writing systems, and crops were developed by the ancient Iraqis and other peoples of the Middle East. Populations swelled, fed by wild wheat, peas, fruits, sheep, pigs, goats, and cows. Abundant food surpluses supported art, politics, governance, laws, mathematics, science, and education.

What happened? Forests were cut. Soil erosion accelerated. Rain no longer fed the forest soil but removed it. Subsequent irrigation produced salinization; deadened salt pans emerged where crops once flourished. Overgrazing on drying soils caused them to blow away. The story of ancient Iraq and its environs is playing out across the world. Many of the conflict zones in today's world have been deforested: Syria, South Sudan, Libya, Yemen, Nigeria, Somalia, Rwanda, Pakistan, Nepal, the Philippines, Haiti, and Afghanistan. All suffer from deforestation, uncontrolled cutting of fuelwood, overgrazing, soil erosion, and desertification. The following areas have lost 90 percent or more of their original forest habitat: Burma, Thailand, India, Borneo, Sumatra, the Philippines, the Mata Atlântica forest of Brazil, Somalia, Kenya, Madagascar, and Saudi Arabia.

A 2015 estimate of the world's tree population: three trillion. That count is substantially higher than previously thought, but more than 15 billion trees are cut down each year. Since humans began farming, the number of trees on earth has fallen by 46 percent. (Today, forests cover 15.4 million square miles of the earth's surface — or roughly 30 percent of its land area.) The color of China's Yellow River is caused by soil eroding off the Loess Plateau, the result of centuries of deforestation and overgrazing. European forests were cleared from the seventeenth to twentieth centuries. America did the same in the nineteenth and twentieth centuries. Logging, slash-and-burn removal for pasture, and clearing of forests for palm oil wreaked havoc in Central and South America, Southeast Asia, and Africa in the twentieth century. According to the World Wildlife Fund, the world continues to lose forty-eight football fields' worth of forest every minute.

The Kermode bear is known as the spirit bear by the Tsimshian people of the Great Bear Rainforest—a 250-mile coastal temperate rainforest in British Columbia (BC). The Kermode bear is rarely seen but more easily spotted during salmon season when they feast near streams and falls as you see in this photograph. The forest is largely intact today due to the Great Bear Rainforest Campaign, one of the most successful campaigns ever undertaken to stop clear-cutting and logging. Beginning in 1984 in Clayoquot Sound, First Nations peoples and environmental NGOs set up blockades to protest logging rights that had been granted to Macmillan Bloedel. After 22 years of unrelenting work by campaigners, BC premier Christy Clark announced in February 2016 that an agreement between First Nations, timber companies, and environmental organizations would protect 85 percent of the 15.8 million acres.

Henri des Roziers, a French Catholic priest who doubles as a human rights lawyer, has emerged in Brazil as the next likely target of big landowners bent on turning parts of the rainforest into grazing land for cattle. The price for killing him is estimated around $38,000.

Below Right: Malaysia's tropical hardwoods have been in demand for centuries, intensively so in the last twenty years. During that time, timber companies have not only profited from the sale of timber, they compounded their gains by installing palm oil plantations. Much of the logging was illegal, as was the appropriation of the land. The effects have been devastating. Logging has degraded or destroyed the vast majority of Malaysian rainforests, and the deforestation rate is faster there than in any other tropical country. Home to one of the most intelligent primates, the critically endangered orangutan, it is estimated that only 20 percent of Borneo's rainforests remain. This photo shows the silt-laden waters of the Miri River, colored orange by runoff from upstream logging, and the herringbone tethering of smaller-diameter trees, which indicate that forests are not being allowed to recover before being logged again.

Carbon emissions from deforestation and associated land use change are estimated to be 10–15 percent of the world's total. In gigatons, these emissions dropped by 25 percent from 2001 to 2015, but deforestation rates may climb again in order for food production to increase by 2050. Either more food will have to be grown on existing crop- and pastureland, or more forests and other ecosystems will need to be converted for food production.

In addition to the loss of aboveground biomass carbon held in trees, significant losses of belowground carbon held in soil can accompany deforestation processes. This is particularly true when fire is employed as the land-clearing technique and in peatland areas where there are dense underground stocks of soil carbon. Conversion of forest to agricultural fields or pasture has been estimated to result in a 20 to 40 percent decrease in soil carbon.

Stopping all deforestation and restoring forest resources could offset up to one-third of all carbon emissions worldwide. A number of governmental and private initiatives have this outcome as their goal and are implementing a combination of approaches to some degree around the world. These strategies include public policy and the enforcement of existing anti-logging laws; the protection of indigenous lands; truly sustainable timber and agricultural practices; and numerous programs that enable wealthy nations and corporations to make payments to countries for maintaining their tropical forests.

The most prominent pay-for-performance program is the United Nations Reducing Emissions from Deforestation and Forest Degradation (REDD+) program, which began to take shape in 2005. Funding programs are emerging out of the 2014 New York Declaration on Forests, endorsed by forty countries and nearly sixty multinational corporations, among others. The Forest Carbon Partnership Facility, a multi-sector effort meant to assist REDD+ efforts, has established two funds of nearly $1.1 billion total to reward forested nations for conserving and increasing forest carbon stocks and for reducing deforestation and degradation. The rewards to landowners, forest dwellers, and other constituencies are intended to make conserving forests more economically advantageous than clearing them.

The benefits of forest conservation are many and various: nontimber products (bush meat, wild food, forage and fodder); erosion control; free pollination and pest and mosquito control provided by birds, bats, and bees; and other ecosystem services. However, the benefits of forest conservation are elusive for marginalized people who eke out a living on previously forested land. The people who live at the edge of forests are key actors. There needs to be some form of compensation and livelihoods for them that extract value from standing forests.

Tropical forests are home to two-thirds of all terrestrial plants and animals, an irreplaceable stock of biodiversity. They are the source of genetic material for new pharmaceuticals, of which one-fourth are derived either directly or indirectly from medicinal plants or from synthesizing new compounds based on traditional uses of plants. These values are difficult to quantify or envision, and their benefits may not be immediate.

An effective agenda to save the forests requires a collective understanding of ecology, the danger posed by global warming, political will, local buy-in, and noncorrupt governance. In this regard, no nation has matched Brazil, where slashing and burning peaked between 1998 and 2004, taking out 120,000 square miles of forest—an area the size of Poland. In the following decade, this loss was cut by 80 percent when the country aggressively pursued a multipronged strategy. Brazil enacted strong enforcement policies and engaged world-class scientific monitoring (in conjunction with Germany), including satellite photos that triggered alerts about new deforestation. It revised ownership codes that allowed settlers to claim ownership without clearing the land and established land registry programs. In the state of Pará, ground zero for deforestation, the registry expanded from 500 properties in 2009 to more than 112,000 today, covering 62 percent of the private land in the state. Additionally, Brazil withheld credit from government entities with high deforestation rates, financed projects devoted to sustainable development and reduction of deforestation, and increased productivity of the land already devoted to agriculture.

Also important was the voluntary agreement from soy traders to embargo products from recently deforested land and the 2009 agreement between the three largest Amazon meatpackers and Greenpeace, which sought to ban purchases from suppliers who deforested. Compliance from suppliers hit 93 percent in 2013. Sixty-five of 95 slaughterhouses signed zero-deforestation commitments. All the while, production of cattle and soy increased.

In 2015, Brazil earned the final $100 million payment of a $1 billion grant from Norway, which had set up the fund in 2008 to reward countries that achieved targeted goals of reducing their rate of deforestation. Achim Steiner, former executive director of the United Nations Environment Programme, said, "There is no question that Brazil has made a fundamental departure from the past, and it has given credence to the notion that forest conservation may be an important mechanism for international cooperation on climate." In 2016, however, the number of forested acres cleared for agriculture ticked back up, despite tight enforcement still in place. No one can quite explain the backsliding, but the message is clear: The cattle "launderers," as they are called, are

resourceful too, and the key to the conservation campaign is an unwavering will and commitment.

Without question, the Amazon is the greatest single natural resource in the world. Rainforests are being cut down at a rate that will eliminate them in forty years. Norway's lead in financing forest protection is a model of what could be done. It is difficult to estimate what it would "cost" to save it all. One study asserts that for $50 billion per year—about 3 percent of the world's military spending—tropical deforestation could be reduced by two-thirds. The opening image of the spirit bear gnawing on fresh salmon in the Great Bear Rainforest is talismanic of what escapes pricing, calculation, or monetary value because it inevitably exceeds all of them. When you add up the impact on carbon sequestration and storage, forest protection and tropical and temperate forest restoration together are one of the most powerful solutions available to address global warming. ●

IMPACT: *For each acre of forest protected, the threat of deforestation and degradation is removed. By protecting an additional 687 million acres of forest, this solution could avoid carbon dioxide emissions totaling 6.2 gigatons by 2050. Perhaps more importantly, this solution could bring the total protected forest area to almost 2.3 billion acres, securing an estimated protected stock of 245 gigatons of carbon, roughly equivalent to over 895 gigatons of carbon dioxide if released into the atmosphere. Financials are not projected, as they are not incurred at the landholder level.*

LAND USE
COASTAL WETLANDS

Along the fringes of coasts, where land and ocean meet among shallow, brackish waters, lie the world's salt marshes, mangroves, and sea grasses. These coastal wetland ecosystems are found on every continent except Antarctica. They provide nurseries for fish, feeding grounds for migratory birds, a first line of defense against storm surges and floodwaters, and natural filtration systems that boost water quality and recharge aquifers. Relative to their land area, they also sequester huge amounts of carbon in plants aboveground and in roots and soils below.

Absorbed over centuries, maybe millennia, this "blue carbon"—so called because of its seaside location—was overlooked for years, although coastal wetlands can store five times as much carbon as tropical forests over the long term, mostly in deep wetland soils. According to the journal *Nature*, the soil of mangrove forests alone may hold the equivalent of more than two years of global emissions—22 billion tons of carbon, much of which would escape if these ecosystems were lost. Thanks to research and advocacy efforts, things are changing. The international community has a growing appreciation for these unsung carbon sinks, as well as the pressures they face.

Often in human history, "wetland" has meant "wasteland"—a place to dike, dredge, and drain for purposes ranging from farming to homesteading. These coastal ecosystems have suffered from mosquito spraying, pollution and sediment runoff, timber extraction, invasive species, and operations of the fossil fuel industry. They have been cleared to make way for shrimp farms, palm plantations, condo developments, and golf courses. Over the past few decades, more than one-third of the world's mangroves have been lost. As the intersecting trends of global population growth and demand for food continue to rise, the pressures on wetlands will mount accordingly.

Whether or not coastal wetlands succumb to them will influence climate change, for better or for worse. When they are intact and healthy, marshes, mangroves, and meadows of sea grass absorb and hold on to carbon. Thanks to rapid plant growth and a paucity of oxygen, the bodies of dead plants build up fast and break down slowly in soggy and anaerobic conditions, producing carbon-rich soil. To quote the journal *Nature*, "Some 2.4–4.6 percent of the world's carbon emissions are captured and sequestered by living organisms in the oceans, and the United Nations estimates that at least half of that sequestration takes place in 'blue-carbon' wetlands." When these ecosystems are degraded or destroyed, this process of taking in carbon does not simply halt; coastal wetlands then become a potent source of emissions, releasing volumes of carbon long sequestered.

3.19 GIGATONS
REDUCED CO2

GLOBAL COST AND SAVINGS DATA
TOO VARIABLE TO BE DETERMINED

53.34 GIGATONS
CO2 PROTECTED

Mudflats and marsh grasses along Alaska's Turnagain Arm waterway—so named because it proved a dead end ("turn again") during British explorer James Cook's search for the Northwest Passage. The area is known for its huge tidal range, exposing a wide mudflat at low tide and submerging it when waters rise. The Chugach Mountains form a backdrop for this intertidal wetland.

As awareness grows about the role blue carbon plays in curbing (or contributing to) climate change, it is also becoming apparent that wetlands are critical to coping with its impacts. Sea level rise due to melting ice and thermal expansion and increased storm activity threaten coastal communities, and shoreline ecosystems are vital protection from battering waves and rushing waters. That is especially true as man-made barriers—levees, dams, embankments—prove increasingly inadequate. The shielding and buffering function of wetlands makes it even more crucial to ensure that they are healthy today and resilient for the future.

The optimal scenario, of course, is to safeguard coastal wetlands before they can be damaged and keep a lid on the carbon they contain. Accelerated by the 1971 Ramsar Convention on Wetlands, government regulation and nonprofit programs are helping to protect wetlands of critical importance such as Wasur National Park in Indonesia and Florida's Everglades National Park. Designating protected areas will continue to be important, but preserving large swaths of land can be challenging—and costly—when it means reducing the area available for agriculture or development, often a hot-button issue. Groups such as the Smithsonian Environmental Research Center on the Chesapeake Bay are building a body of science about ways to maximize sequestration.

Alongside the designation of protected areas, it is possible to rehabilitate and restore coastal wetlands that already have been degraded, although their effectiveness as carbon sinks cannot compare to those that are unscathed. Restoration efforts range from simply allowing the ecosystem's processes to play out to redressing the legacy of dikes, ditches, drainage, and development. Passive restoration tends to be less expensive and more effective in the long run. But when wetlands are severely degraded, intensive efforts may be required to help tidal waters to flow freely and natural habitat to flourish. From the Delaware Bay to the coasts of the Netherlands, "living shorelines" are bringing back unfettered tidal zones. In addition to removing infrastructure, such as roads, to nurture living shorelines, it is also helpful to give coastal wetlands room to roam. As sea levels continue to rise, these ecosystems will need to migrate inland toward higher ground, and human settlement could impede that shift.

In contrast to carbon sequestration efforts on land, those along coasts are nascent. Since 2008, a group of European companies has been active in Senegal, spending millions on mangrove restoration and receiving carbon credits to offset emissions back home. Local people—mostly women—have planted tens of millions of trees on land traditionally held in common as a resource for firewood, fish, and mollusks. They discovered later that carbon credits would be sold and companies would profit from their low-paid work. They also found, to their dismay, that they could no longer access those key resources in the replanted coastal areas, lest gathering cockles and wood disturb the new trees and the carbon-sinking process. At the same time, villagers are now experiencing the layered benefits of rebuilding ocean buffers, protecting land from waves and wind, and restoring habitat for birds, monkeys, and mongoose and an important nursery for fish.

What is true in Senegal is true around the world: Human livelihoods and coastal ecosystems intertwine in complex ways that beg greater understanding. Equity in the effort to address global warming—with blue carbon or otherwise—requires discipline by practitioners and vigilance by observers. When coastal wetland investment is done well, returns can be manifold, locally and globally. Conserving coastal ecosystems can benefit the atmosphere, enhance biodiversity, water quality, and storm protection, and respect the rights and well-being of local communities, all at the same time.

IMPACT: *Of the 121 million acres of coastal wetlands globally, 18 million acres are protected today. If an additional 57 million acres are protected by 2050, the resulting avoided emissions and continued sequestration could total 3.2 gigatons of carbon dioxide. While limited in area, coastal wetlands contain large carbon sinks; protecting them would secure an estimated 15 gigatons of carbon, equivalent to over 53 gigatons of carbon dioxide if released into the atmosphere.*

LAND USE
TROPICAL FORESTS

In recent decades, tropical forests—those located within 23.5 degrees north or south of the equator—have suffered extensive clearing, fragmentation, degradation, and depletion of flora and fauna. Once blanketing 12 percent of the world's landmasses, they now cover just 5 percent. In many places, the destruction continues. However, restoration, both passive and intentional, is now a growing trend. According to a 2011 study measuring the global carbon sink that forests represent, "the tropics have the world's largest forest area, the most intense contemporary land-use change, and the highest carbon uptake, but also the greatest uncertainty." Yet even as deforestation persists, the regrowth of tropical forests sequesters as much as six gigatons of carbon dioxide per year. That is equivalent to 11 percent of annual greenhouse gas emissions worldwide or all those emanating from the United States.

When we lose forests, primarily to agricultural expansion or human settlement, carbon dioxide discharges into the atmosphere. Tropical forest loss alone is responsible for 16 to 19 percent of greenhouse gas emissions caused by human activity. Restoring forests does just the opposite. As forest ecosystems come back to life, trees, soil, leaf litter, and other vegetation absorb and hold carbon, taking it out of global warming rotation. Though not immediately equal in their diversity to old-growth landscapes, restored forests support the water cycle, conserve soil, protect habitat and pollinators, provide food, medicine, and fiber, and give people places to live, adventure, and worship. Particularly important for rural, often-marginalized forest-fringe dwellers, these ecosystem goods and services will become more important as climate change persists and communities are faced with adapting to its impacts.

According to the World Resources Institute (WRI), 30 percent of the world's forestland has been cleared completely. Another 20 percent has been degraded. "More than 2 billion hectares [4.9 billion acres] worldwide offer opportunities for restoration—an area larger than South America," a team of WRI researchers reports. Three-quarters of that land would be best

61.23 GIGATONS
REDUCED CO2

GLOBAL COST AND SAVINGS DATA
TOO VARIABLE TO BE DETERMINED

suited to a "mosaic" forest restoration approach, blending forests, trees, and agricultural land uses. Up to 1.2 billion acres are ripe for full restoration of large forests with dense canopy cover, in areas where human residents are sparser. The opportunity is enormous, and the majority of it lies in tropical regions.

Restoration means taking action to help a damaged forest ecosystem recuperate to its original form and function. Flora and fauna return. Interactions between organisms and species revive. The forest regains its multidimensional roles. As Bill McKibben wrote in 1995, chronicling the resurgence of forests along the U.S. East Coast, "what matters is not simply the number of trunks but the quality of the forest." In general, the more harm an ecosystem has sustained, the more complex and expensive restoration will be. Recent research has upended long-standing assumptions about the immutability of razed tropical forest: They are, in fact, much more resilient than we previously thought. In a median time of sixty-six years, tropical forests can recover 90 percent of the biomass that old-growth landscapes contain.

The specific mechanics vary for restoring or rehabilitating a tropical forest. The simplest scenario is to release land from nonforest use, such as growing crops or damming a valley, and let a young forest rise up on its own, following a course of natural regeneration and succession. Protective measures can keep pressures such as fire, erosion, or grazing at bay. Other techniques are

---

Above Left: Burning continues to be the preferred means of clearing land in the Amazon to make way for cattle. It is a delusional act because the thin acid soils quickly degrade and fail. This picture was taken in Rondonia State, just northeast of Bolivia.

Above: The Monteverde Cloud Forest Reserve in Costa Rica comprises 26,000 acres of virgin forest, possibly containing the most diverse biome in the world. It was named by Quaker farmers who moved from Alabama to Costa Rica to avoid being drafted for the Korean War. (Costa Rica had just abolished its armies, which was the driver for choosing Costa Rica.) It was Green Mountain to them; Monteverde ever since.

more intensive, such as cultivating and planting native seedlings and removing invasives. These techniques give vital species an opportunity to thrive and provide an accelerant to natural ecological processes. They are critical in places where soil has been severely degraded and natural seed banks—such as nearby forests or seeds remaining underground—are not present. As seedlings grow, they enhance soil health, shade out grasses, and attract birds and other seed dispersers, which further aid revival and subsequent natural regeneration and succession.

While restoration homes in on the forest ecosystem, human systems are critical to its success. The days of sweeping, untouched landscapes are largely gone. Because forests and people rarely exist in isolation in today's heavily populated world, restoring a forest means more than making it ecologically robust again. It needs to be socially and economically viable or, better yet, valuable—a source of pride and profit, play and provisions for local communities at large. From a climate perspective specifically, the global benefits of carbon mitigation should meet local benefits of adaptation to global warming and its impacts. Without achieving these interwoven benefits, restoration may simply never get off the ground, or, worse, could see its investment reversed by subsequent damage. Local communities will have a stake in what is growing if it is to sustain.

Given the interconnectedness of people and forests, a particular framework for restoration has emerged: forest landscape restoration (FLR). This approach, proposed by the Food and Agriculture Organization (FAO) of the United Nations, means "regarding the landscape as an integrated whole . . . looking at different land uses together, their connections, interactions, and a mosaic of [restoration] interventions." It means there is no single formula for forest restoration. Growing trees is an essential intervention, of course, but FLR insists human stakeholders and their participation are equally crucial. (Of the ten guiding principles for restoration developed by the FAO, just one of them is "planting trees.") Making restoration a collaborative process can ensure it is done with and for local communities, and that root causes of forest damage are addressed, a suite of sometimes competing objectives can be met, and the revitalized forest has champions, not challengers. Restoration cannot be done in the halls of power alone. It starts and ends on the ground.

Today, we can point to a veritable global movement for forest restoration. A critical year in its evolution was 2011, when the Bonn Challenge set an ambitious target of restoring 370 million acres (150 million hectares) of forest worldwide by 2020. The 2014 New York Declaration on Forests affirmed that aim and added a cumulative target of 865 million acres (350 million hectares) restored globally by 2030. (These goals accompany others focused on halting deforestation in the first place.) Should the world restore 865 million acres of forest by 2030, a total of 12 to 33 gigatons of carbon dioxide would be removed from the atmosphere and become terrestrial once again, alongside provision of myriad other goods and services.

According to recent analysis, active forest restoration, which is not always required, typically costs $400 to $1,200 per acre. Those numbers do not include land costs, and they vary according to the species planted, methods used, starting conditions, and project scale. Restoring 865 million acres of forest between now and 2030 could cost $350 billion and as much as $1 trillion. The return on investment would be larger. According to estimates from the International Union for Conservation of Nature, "Achieving the 350 million-hectare [865 million–acre] goal could generate $170 billion per year in net benefits from watershed protection, improved crop yields, and forest products, [while sequestering] up to 1.7 gigatons of carbon dioxide equivalent annually."

The bulk of restoration opportunities lies primarily within low-income countries in tropical regions. Those countries cannot manage the level of investment required, nor should they, since the benefits of restoration provide value and a service to all. The relevant stakeholders are the entire human race, and some bear greater responsibility for the problem of climate change than others.

Tropical forest restoration is vital for development. Forests are a source of income, from timber to tourism; food security, from bush meat to crop pollination; energy, from firewood to in-stream hydropower; health, from clean water to mosquito control; and safety, from landslide prevention to flood control. They are dynamic engines of human sustenance and well-being. These layers of benefits have sparked powerful regional and national commitments to tropical forest restoration. AFR100, the African Forest Landscape Restoration Initiative, is committed to restoring 247 million acres of degraded land on the continent by 2030—an area three times the size of Germany. Having cut Amazonian deforestation rates by 80 percent from 2005 to 2015—a feat that once seemed impossible—Brazil is restoring more than 29 million acres of forest. Restoration is a means of both reaping national development rewards and receiving international compensation for carbon sinks.

Because forest restoration is such a potent solution, commitments and funding need to be a global priority. And because restoration efforts have ranged from success to failure, it is important to analyze why, scale best practices, and eliminate those that do not work. Initiatives need to respect land rights and tenure, especially those of indigenous people, be well equipped and technically adept, and ensure effective enforcement of strong policies. Success depends on changing land-use practices and reducing meat consumption, so we can feed a growing global population without expanding agricultural acreage. One of the dominant storylines of the nineteenth and twentieth centuries was the vast loss of forestland. Its restoration and re-wilding could be the twenty-first-century story. ●

IMPACT: *In theory, 751 million acres of degraded land in the tropics could be restored to continuous, intact forest. Using current and estimated commitments from the Bonn Challenge and New York Declaration on Forests, our model assumes that restoration could occur on 435 million acres. Through natural regrowth, committed land could sequester 1.4 tons of carbon dioxide per acre annually, for a total of 61.2 gigatons of carbon dioxide by 2050.*

LAND USE
BAMBOO

In the Philippine creation story, the first man, *Malakas* (Strong One), and the first woman, *Maganda* (Beautiful One), emerged from the two halves of a bamboo tree. It is one of many Asian origin myths that features bamboo—a plant that human beings have cultivated for more than a thousand uses. Addressing global warming is another way it could be put to use; bamboo rapidly sequesters carbon in biomass and soil, taking it out of the air faster than almost any other plant, and can thrive on inhospitable degraded lands. Some species, in the right environment, are capable of sequestering seventy-five to three hundred tons of carbon per acre over a lifetime.

Bamboo is not a plant that needs encouragement. There is a top-ten list of the fastest-growing plants in the world, and duckweed, algae, and kudzu did not have a chance of making number one. You can sit by timber bamboo in the spring and watch it grow more than one inch an hour. Bamboo reaches its full height in one growing season, at which time it can be harvested for pulp or allowed to grow to maturity over four to eight years. After being cut, bamboo re-sprouts and grows again. Managed bamboo is cultivated on more than 57 million acres worldwide.

Just a grass, bamboo has the compressive strength of concrete and the tensile strength of steel. It is used in almost every aspect of buildings from frame to floor to shingles, as well as food, paper, furniture, bicycles, boats, baskets, fabric, charcoal, biofuels, animal feed, and even plumbing. Although bamboo's value is well understood in Asia (called the "friend of the people" in China), it is still considered a weed in much of the world. But its versatile uses, including carbon sequestration, place it among the world's most useful plants.

Because bamboo is a grass, it contains minute silica structures called plant stones, or phytoliths. Composed of minerals, phytoliths resist degradation longer than other plant material. The carbon they store can remain sequestered in the soil for hundreds or thousands of years. The combination of phytoliths and bamboo's rapid growth rate make it a prolific means to sequester carbon. The carbon impact of bamboo is even greater, due to its ability to replace high-emissions materials such as cotton, plastics, steel, aluminum, and concrete. As a replacement for pulp used for paper, bamboo can produce six times as much pulp as a conventional pine plantation.

Bamboo can pose ecological problems. An invasive species in many places, it can spread with detrimental effects to native ecosystems. Care should be taken to select appropriate locations and manage its growth. Bamboo can also have some of the same drawbacks as monoculture tree plantations used for afforestation. By focusing on commercial use on degraded lands, especially those with steep slopes or significant erosion, it is possible to maximize the positive impacts of bamboo—useful products, carbon sequestration, and avoided emissions from alternative materials—while minimizing the negatives. ●

IMPACT: *Bamboo is planted on 77 million acres today. We assume that it will be grown on an additional 37 million acres of degraded or abandoned lands. Our carbon sequestration calculations include both living biomass and long-lived bamboo products, with an annual rate of 2.9 tons of carbon per acre. Where bamboo is substituted for aluminum, concrete, plastic, or steel, there can be avoided emissions as well, which are not included in the total of 7.2 gigatons of carbon dioxide sequestered by 2050. An initial investment of $24 billion could yield a thirty-year financial return of $265 billion.*

The Man Who Stopped the Desert
MARK HERTSGAARD

Studies have shown that 98 percent of the news published or broadcast about climate change is negative and essentially gloom inducing. In this excerpt from Mark Hertsgaard's book Hot: Living Through the Next Fifty Years on Earth, *the news is different—it is a story of desertification being reversed in the face of more challenging rainfall conditions. The hero in this story is Yacouba Sawadogo, known in Burkina Faso, Africa, as "the man who stopped the desert." This is a story about how solutions arise from practice and place, from people who know the land—farmers who have made important discoveries about what is known as tree intercropping. Tree intercropping is not a new discovery; it has been around for millennia. One of the gifts global warming bestows to the world is the impetus to find our way back to practices once known and understood. In the West, there has been a long-standing premise that it had to help Africa "develop." The Western aid and development model for addressing poverty has been dismantled by both Africans and many studies, yet it persists. In Mark's work, people are growing three things: trees, crops, and wisdom. Foreign aid, sacks of genetically modified corn, and handouts come and go, but if we are to successfully address global warming, we should learn to trust the capacity of people everywhere to understand the consequences and imagine place-based solutions on a collaborative basis, and not force solutions upon them, however well intentioned.*
—PH

Yacouba Sawadogo was not sure how old he was. With a hatchet slung over his shoulder, he strode through the woods and fields of his farm with an easy grace. But up close his beard was gray, and it turned out he had great-grandchildren, so he had to be at least sixty and perhaps closer to seventy years old. That means he was born well before 1960, the year the country now known as Burkina Faso gained independence from France, which explains why he was never taught to read and write.

Nor did he learn French. He spoke his tribal language, Mòoré, in a deep, unhurried rumble, occasionally punctuating sentences with a brief grunt. Yet despite his illiteracy, Yacouba Sawadogo is a pioneer of the tree-based approach to farming that has transformed the western Sahel over the last twenty years.

"Climate change is a subject I have something to say about," said Sawadogo, who unlike most local farmers had some understanding of the term. Wearing a brown cotton gown, he sat beneath acacia and zizyphus trees that shaded a pen holding guinea fowl. Two cows dozed at his feet; bleats of goats floated through the still late-afternoon air. His farm in northern Burkina Faso was large by local standards—fifty acres—and had been in his family for generations. The rest of his family abandoned it after the terrible droughts of the 1980s, when a 20 percent decline in annual rainfall slashed food production throughout the Sahel, turned vast stretches of savanna into desert, and caused millions of deaths by hunger. For Sawadogo, leaving the farm

was unthinkable. "My father is buried here," he said simply. In his mind, the droughts of the 1980s marked the beginning of climate change, and he may be right: scientists are still analyzing when man-made climate change began, some dating its onset to the mid-twentieth century. In any case, Sawadogo said he had been adapting to a hotter, drier climate for twenty years now.

"In the drought years, people found themselves in such a terrible situation they had to think in new ways," said Sawadogo, who prided himself on being an innovator. For example, it was a long-standing practice among local farmers to dig what they called zai—shallow pits that collected and concentrated scarce rainfall onto the roots of crops. Sawadogo increased the size of his zai in hopes of capturing more rainfall. But his most important innovation, he said, was to add manure to the zai during the dry season, a practice his peers derided as wasteful.

Sawadogo's experiments proved out: crop yields duly increased. But the most important result was one he hadn't anticipated: trees began to sprout amid his rows of millet and sorghum, thanks to seeds contained in the manure. As one growing season followed another, it became apparent that the trees—now a few feet high—were further increasing his yields of millet and sorghum while also restoring the degraded soil's vitality. "Since I began this technique of rehabilitating degraded land, my family has enjoyed food security in good years and bad," Sawadogo told me.

Farmers in the western Sahel have achieved a remarkable success by deploying a secret weapon often overlooked in wealthier places: trees. Not planting trees. Growing them. Chris Reij, a Dutch environmental specialist at VU University Amsterdam who has worked on agricultural issues in the Sahel for thirty years, and other scientists who have studied the technique say that mixing trees and crops—a practice they have named "farmer-managed natural regeneration," or what is known as agro-forestry—brings a range of benefits. For example, the trees' shade and bulk offer crops relief from the overwhelming heat and gusting winds. "In the past, farmers sometimes had to sow their fields three, four, or five times because wind-blown sand would cover or destroy seedlings," said Reij, a silver-haired Dutchman with the zeal of a missionary. "With trees to buffer the wind and anchor the soil, farmers need sow only once."

Leaves serve other purposes. After they fall to the ground, they act as mulch, boosting soil fertility; they also provide fodder for livestock in a season when little other food is available. In emergencies, people too can eat the leaves to avoid starvation.

The improved planting pits developed by Sawadogo and other simple water-harvesting techniques have enabled more water to infiltrate the soil. Amazingly, underground water tables that plummeted after the droughts of the 1980s had now begun recharging. "In the 1980s, water tables on the Central Plateau of

Yacouba Sawadogo

Burkina Faso were falling by an average of one meter a year," Reij said. "Since FMNR and the water-harvesting techniques began to take hold in the late 1980s, water tables in many villages have risen by at least five meters, despite a growing population."

Some analysts attributed the rise in water tables to an increase in rainfall that occurred beginning in 1994, Reij added, "but that doesn't make sense—the water tables began rising well before that." Studies have documented the same phenomenon in some villages in Niger, where extensive water-harvesting measures helped raise water tables by fifteen meters between the early 1990s and 2005.

Over time, Sawadogo grew more and more enamored of trees, until now his land looked less like a farm than a forest, albeit a forest composed of trees that, to my California eyes, often looked rather thin and patchy. Trees can be harvested—their branches pruned and sold—and then they grow back, and their benefits for the soil make it easier for additional trees to grow. "The more trees you have, the more you get," Sawadogo explained. Wood is the main energy source in rural Africa, and as his tree cover expanded, Sawadogo sold wood for cooking, furniture making, and construction, thus increasing and diversifying his income—a key adaptation tactic. Trees, he says, are also a source of natural medicines, no small advantage in an area where modern health care is scarce and expensive.

"I think trees are at least a partial answer to climate change, and I've tried to share this information with others," Sawadogo added. "My conviction, based on personal experience, is that trees are like lungs. If we do not protect them, and increase their numbers, it will be the end of the world."

Sawadogo was not an anomaly. In Mali, the practice of growing trees amid rows of cropland seemed to be everywhere. As word of such successes travels, agro-forestry has spread throughout the region, according to Salif Ali, a neighboring farmer. "Twenty years ago, after the drought, our situation here was quite desperate, but now we live much better," he said. "Before, most families had only one granary each. Now, they have three or four, though the land they cultivate has not increased. And we have more livestock as well." After extolling the many benefits trees have provided—shade, livestock fodder, drought protection, firewood, even the return of hares and other small wildlife—Salif was asked by one member of our group, almost in disbelief, "Can we find anyone around here who doesn't practice this type of agro-forestry?"

"Good luck," he replied. "Nowadays, everyone does it this way."

According to Tony Rinaudo, an Australian missionary and development worker who was one of the original champions of what is called farmer managed natural regeneration, "The great thing about agro-forestry is that it's free. They stop seeing trees as weeds and start seeing them as assets." But only if they're not penalized for doing so.

Agro-forestry has spread largely by itself, from farmer to farmer and village to village, as people see the results with their own eyes and move to adopt the practice. Not until Gray Tappan of the U.S. Geological Survey compared aerial photos from 1975 with satellite images of the same region in 2005 was it apparent just how widespread agro-forestry had become. Reij, Rinaudo, and other advocates were surprised by the satellite evidence; they had had no idea so many farmers in so many places had grown so many trees.

"This is probably the largest positive environmental transformation in the Sahel and perhaps in all of Africa," said Reij. Combining the satellite evidence with ground surveys and anecdotal evidence, Reij estimated that in Niger alone farmers had grown 200 million trees and rehabilitated 12.5 million acres of land. "Many people believe the Sahel is nothing but doom and gloom, and I could tell lots of doom-and-gloom stories myself," he said. "But many farmers in the Sahel are better off now than they were thirty years ago because of the agro-forestry innovations they have made."

What makes agro-forestry so empowering—and sustainable—Reij added, is that Africans themselves own the technology, which is simply the knowledge that nurturing trees alongside one's crops brings many benefits. "Before this trip, I always thought about what external inputs were required to increase food production," Gabriel Coulibaly said at a debriefing session after our fact-finding expedition. Coulibaly, a Malian who worked as a consultant to the European Union and other international organizations, added, "But now I see that farmers can create solutions themselves, and that is what will make those solutions sustainable. Farmers manage this technology, so no one can take it away from them."

And agro-forestry's success does not depend on large donations from foreign governments or humanitarian groups—donations that often do not materialize or can be withdrawn when money gets tight. This is one reason Reij sees agro-forestry as superior to the Millennium Villages model promoted by Jeffrey Sachs, the economist who directs Columbia University's Earth Institute. The Millennium Villages program focuses on twelve villages in various parts of Africa, providing them free of charge with what are said to be the building blocks of development: modern seeds and fertilizer, boreholes for clean water, health clinics. "If you read their website, tears come to your eyes," said Reij. "It's beautiful, their vision of ending hunger in Africa. The problem is, it can only work temporarily for a small number of selected villages. Millennium Villages require continuing external inputs—not just fertilizer and other technology, but the money to pay for them—and that is not a sustainable solution. It's hard to imagine the outside world providing free or subsidized fertilizer and boreholes to every African village that needs them."

Outsiders do have a role to play, however. Overseas governments and NGOs can encourage the necessary policy changes by African governments, such as granting farmers ownership of trees. And they can fund, at very low cost, the grassroots information sharing that has spread agro-forestry so effectively in the western Sahel. Although farmers have done the most to alert peers to agro-forestry's benefits, crucial assistance has come from a handful of activists like Reij and Rinaudo and NGOs such as Sahel-Eco and World Vision Australia. These advocates now hope to encourage the adoption of agro-forestry in other African countries through an initiative called "Re-greening the Sahel," said Reij.

If humanity is to avoid the unmanageable and manage the unavoidable of climate change, we must pursue the best options available. Agro-forestry certainly seems to be one of them, at least for the poorest members of the human family. "Let's look at what's already been achieved in Africa and build on that," urged Reij. "In the end, what happens in Africa will depend on what Africans do, so they must own the process. For our part, we must realize that farmers in Africa know a lot, so there are things we can learn from them as well." ●

LAND USE
PERENNIAL BIOMASS

RANKING AND RESULTS BY 2050

#51

3.33 GIGATONS
REDUCED CO2

$77.9 BILLION
NET COST

$541.9 BILLION
NET SAVINGS

Plant in the spring. Grow through the summer. Harvest in the fall. This rhythm has existed for ten thousand years of humanity's agricultural history. It is how we think about the cycle of production but does not apply to all crops. Gardeners know well the difference between perennials and annuals: Daffodils bloom season after season, while dahlias require yearly effort. At that scale, it is a matter of taste and time. At the scale of a farmer's field, more critical dynamics are at play: Compared to annuals, perennials have the potential to avoid leaching nutrients, eroding soil, spraying synthetic fertilizers, and running diesel-swigging equipment as often. Bioenergy crops present an opportunity to swap annuals for perennials, and draw down carbon in the process.

Plant material is used in a variety of ways to create energy: combusted to produce heat or electricity; anaerobically digested to produce methane; and converted to ethanol, biodiesel, or hydrogenated vegetable oil for fuel. Within transportation, bioenergy makes up 2.8 percent of fuel consumed. Within the power sector, it comprises 2 percent of the total. The whole bioenergy lineup is projected to grow.

Whether plant material used for bioenergy is annual or perennial (or waste content) makes all the difference. The United States leads the world in the production of liquid biofuels. Forty percent of the corn grown nationally becomes ethanol. Huge subsidies go into this annual crop, often for little or no benefit to the climate because energy inputs are so high. Producing corn ethanol can threaten water supplies and raise food prices without making any progress on cutting emissions.

Perennial bioenergy crops can be different. Cultivated appropriately, they can reduce emissions by 85 percent compared to corn ethanol. Switchgrass, fountain grasses, and silver grass (Miscanthus giganteus) are robust herbaceous plants that require less water and nutrients than food crops and can be harvested year after year without sowing. Short-rotation woody crops such as poplar, willow, eucalyptus, and locust have a twenty- to thirty-year lifetime. They can be harvested through a process called coppicing: cutting close to the ground, followed by rapid and repeated regrowth. Most important, the impact of perennials on soil carbon is dramatically different from that of annuals. If existing annual bioenergy crops are replaced with perennials, they can make a net-positive contribution through sequestration. In addition, many are prime candidates to grow on degraded land not suited to food production. Compared to corn and other annuals, total production of plant material can be lower with perennials, and they prevent erosion, produce more stable yields, are less vulnerable to pests, and support pollinators and biodiversity.

Heated debate about bioenergy continues—whether and to what extent it can benefit the climate, without endangering food supply or encroaching on forests. The story of bioenergy is

Miscanthus is sometimes called elephant grass because of its height, growing to ten feet tall in a single season. The farmer in his field at harvest time.

not singular, and perennials, though seldom discussed, are pivotal for its outcome. That does not mean they are a silver bullet. Given the amount of energy we use and food we need to produce, there is simply not enough land to meet all of our needs with plant-based fuels. But it is not an either-or proposition: We need a host of solutions to reverse global warming. Where more efficient renewables such as solar and wind can replace fossil fuels, they should. When it comes to uses such as airplane fuel that are more stubborn, bioenergy can provide a vital substitute. Executed thoughtfully and well, perennial bioenergy crops are a solution, among many, that merits attention. ●

IMPACT: *Perennial biomass crops provide the feedstock for biomass energy generation, making those emissions reductions possible. They also can generate their own climate impact of 3.3 gigatons of carbon dioxide by 2050, as they replace annual feedstocks and sequester more soil carbon. Our analysis assumes a rise from .5 million acres currently to 143 million acres by 2050. The cultivation of perennials is costlier than annuals, but returns over thirty years could be $542 billion.*

LAND USE
PEATLANDS

"The ground itself is kind, black butter," wrote Seamus Heaney in his 1969 poem "Bogland." Though Heaney had Ireland in mind, his is a vivid metaphor for peatlands—also known as bogs or mires— around the world. They are neither solid ground nor water, but something in between. Peat is a thick, mucky, waterlogged substance made up of dead and decomposing plant matter. It develops over hundreds, even thousands of years, as a soupy mix of wetland moss, grass, and other vegetation slowly decays beneath a living layer of flora in the near absence of oxygen. That acidic, anaerobic environment has preserved human remains, so-called "bog bodies" from the Iron Age and earlier. Given enough time, pressure, and heat, peat would become coal.

Ranging in depth from two feet to more than sixty, layers of peat contain enormous amounts of carbon. Their typical carbon content is over 50 percent. For that very reason, as well as its accessibility, peat was the first fossil fuel widely used. From Ireland to Finland to Russia, burning dried bricks of peat for heat, cooking, and eventually electricity is an age-old custom, still practiced in some places. Peat was key to the Dutch Golden Age of the seventeenth century. An abundant, cheap, and easily transported

energy source, it enabled Dutch industry and production of goods for the international market to flourish. Today, though these unique ecosystems cover just 3 percent of the earth's land area, they are second only to oceans in the amount of carbon they store—*twice* that held by the world's forests, at an estimated five hundred to six hundred gigatons. Though forests have gotten more attention in recent decades, society is waking up to the invaluable role of peatlands as a carbon storehouse . . . so long as they stay wet.

For peatlands to stockpile carbon effectively, they must have plants to absorb and store it through photosynthesis, and water to create anaerobic conditions that keep carbon from escaping back into the atmosphere. Eighty-five percent of the world's peatlands have the water retention that is crucial. As intact ancient ecosystems they can effectively collect carbon, while absorbing and purifying water, protecting against floods, and supporting biodiversity from foxes to orangutans. Safeguarding them, through land preservation and fire prevention, is a prime opportunity to manage global greenhouse gases, and a cost-effective one by comparison. (While unspoiled peatlands do emit some methane, the carbon they sequester vastly outweighs the methane they release.)

The ability to siphon and hold carbon has a flip side, of course. Holding up to ten times more carbon per acre than other ecosystems, these wetlands can become powerful greenhouse chimneys if disrupted. Fifteen percent already have been. When peat is exposed to the air, the carbon it contains gets oxidized into carbon dioxide. It can take thousands of years to build up peat, but a matter of only a few to release its greenhouse cache once it is degraded. Drained peatlands make up 0.3 percent of the world's land area, yet they produce 5 percent of all carbon dioxide emissions caused by human beings.

The causes of peatland degradation are varied. These boggy ecosystems are found predominantly in temperate-cold climates across the northern hemisphere, covering large swaths of North America, Northern Europe, and Russia, as well as in tropical-subtropical climates, such as Indonesia and Malaysia. In Southeast Asia, forest fires and clearing for palm oil and pulpwood plantations are major drivers of peatland damage—and on the increase. Indeed, it is why Indonesia's greenhouse gas emissions are so high. When emissions from land-use change and forestry are included in country totals, Indonesia consistently ranks in the top five emitters in the world, in tandem with India and Russia. As global warming grows, so does the risk of peatland fires. In more temperate parts of the world, mining peat for fuel, extracting peat moss as a horticultural commodity, and draining peatlands for timber production and grazing are the main culprits.

Though not as effective as halting degradation before it starts, restoring drained and damaged peatlands is an essential

This diagram shows some of the plants that have adapted to peatlands. They include sedges, mosses, the carnivorous sundews, orchids, bog myrtle, and many others that thrive in a waterlogged environment where nutrients are scarce.

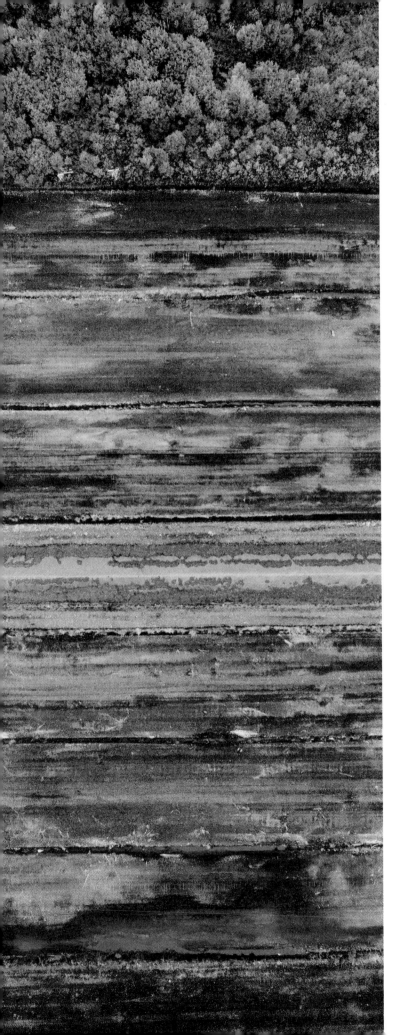

| 21.57 GIGATONS REDUCED CO2 | GLOBAL COST AND SAVINGS DATA TOO VARIABLE TO BE DETERMINED |

1,230.38 GIGATONS
CO2 PROTECTED

strategy. Rewetting is the chief priority—an aptly named process that aims to saturate an expanse of peat by retaining water and raising the water table. In other words, stop water from escaping and reflood the soils. Once the peatland is wet again, oxidation and carbon release are curbed. Paludiculture, from the Latin *palus* for "marsh" and *cultura* for "growing," can build on the success of rewetting by cultivating biomass to protect and regenerate peat. It is the artful creation of vegetation decay that can renew peat layers over time, and can accommodate certain crops such as oranges and tea trees. Taken together, restoration practices should help the ecosystem become whole again.

Protection of peatlands is still in its infancy. Mapping and monitoring them is crucial: knowing where they are and what is happening to them, so that knowledge can guide action. But scientists still have a lot to learn; indeed, a team discovered a bog the size of England in a remote part of Congo-Brazzaville in 2014. How peatlands will respond to a warming climate remains unclear. Developing incentives to maintain or restore their ecological integrity is key, especially if that means forgoing other economic gains from growing food or timber. From Sweden to Sumatra, a variety of national and cross-border initiatives have cropped up to protect and restore peatlands. They range from outright preservation of intact peatlands and bans on further drainage to rewetting schemes, public awareness campaigns, and training in responsible management practices. For millennia, peatlands have been sacred, ritual spaces—sometimes viewed as a gateway to the gods. A similar reverence today could ensure that peat's layers of death and decomposition can continue to be a life-giving force. ●

IMPACT: *If the total protected area of peatlands increases from 7.9 million acres to 608 million acres by 2050, or 67 percent of all currently intact peatlands, 21.6 gigatons of carbon dioxide emissions can be avoided. At 608 million acres, peatlands would hold a protected stock of 336 gigatons of carbon, or roughly 1,230 gigtons of carbon dioxide if released into the atmosphere. Though peatlands comprise only 3 percent of global land area, they are the most organic-rich soils; their degradation would release an enormous amount of carbon. Financials are not projected as they are not incurred at the landholder level.*

Harvested peatlands in Ireland as seen from a drone. Peatland ecosystems cover 17 percent of the Irish Republic and have been hand cut—what was known as "working in the moss"—for fuel and winter warmth since Roman times. Today, machines employed by the state-owned company Bord na Móna have replaced people, leaving boglands irreparably damaged. In 2015, the company announced it would phase out all peat cutting by 2030 and make a transition to sustainable biomass, wind, and solar power.

LAND USE
INDIGENOUS PEOPLES' LAND MANAGEMENT

ndigenous communities are among those most dramatically impacted by climate change, despite contributing the least to its causes. They are particularly vulnerable to the negative effects of environmental change because of their land-based livelihoods, histories of colonization, and social marginalization. Their homelands may be in more fragile locations such as primary forests, small islands, high altitudes, and desert margins. As their ecosystems transform, these communities are responding—drawing on local knowledge, traditional practices, and scientific technologies to adapt their livelihoods and management of local resources. Beyond adapting to their specific circumstances, they are mitigating global warming to a degree that benefits everyone.

Indigenous communities have long been the front line of resistance against deforestation; mineral, oil, and gas extraction; and the expansion of monocrop plantations. Their resistance prevents land-based carbon emissions and maintains or increases carbon sequestration. Traditional indigenous practices and land management conserve biodiversity, maintain a range of ecosystems services, and safeguard rich cultures and traditional ways of life. Indigenous and community-owned lands represent 18 percent of all land area, including at least 1.2 billion acres of forest (approximately 14 percent of global forestlands). These forests in turn contain 37.7 billion tons of carbon stock.

For indigenous communities, climate change affects more than their physical landscapes; it challenges their human rights,

6.19 GIGATONS
REDUCED CO2

GLOBAL COST AND SAVINGS DATA
TOO VARIABLE TO BE DETERMINED

849.37 GIGATONS
CO2 PROTECTED

culture, stores of knowledge, and customary governance. The Intergovernmental Panel on Climate Change has recognized the unique impacts climate change has on these communities, as well as the important contributions that traditional knowledge and science can have when developing strategies for adapting to and curbing it. Many initiatives all over the world are working to support the effective participation of indigenous and local communities, so that traditional knowledge and practices become global warming solutions—solutions that are relevant to local contexts and responsive to the needs of the most vulnerable.

Traditional systems have the potential for increasing above- and below-ground carbon stocks and reducing greenhouse gas emissions via a range of practices. Local indigenous communities practice many different ways of living within ecosystem boundaries through swidden or shifting agriculture, agroforestry, pastoralism, fishing, hunting and gathering, and traditional forest management. Many of these cultures have coexisted with nature's cycles and resources over long durations of time, without depleting them, in places inhabited over millennia in some cases.

Home Gardens. Often occurring in communities that live close to forests, home gardens represent a form of small-scale agriculture that has been practiced in many parts of the world since time immemorial. South and Southeast Asian home gardens constitute a substantial portion of cropped lands, with about 12.7 million acres in Indonesia, 1.3 million in Bangladesh, and 2.6 million in Sri Lanka. Home gardening systems facilitate multiple advantages for practitioners and the landscape, such as efficient nutrient cycling, high productivity, diverse species composition, and maintenance of social and cultural values. These varied systems aid in conserving biodiversity, meeting local food security, and conserving soil and water resources. Home gardens hold higher carbon sequestration potential compared to monocrop production systems, with sequestration rates comparable to those of mature forest stands.

Agroforestry. A significant amount of carbon is sequestered by agroforestry systems, which integrate trees and crop production. Agroforestry systems are well studied and known for their capacity to protect the land from soil erosion, recycle organic matter and soil nutrients, hedge smallholder income against market and weather events that impact single crops, and maintain high species diversity.

Swidden. Another indigenous practice is swidden cultivation, where cultivated land continues to shift from year to year. The term *swidden* refers to the burning and clearing of forestland for annual cultivation and the subsequent fallowing of the land over some period to allow regeneration. Governments have attempted to eliminate shifting cultivation, considering it inefficient and destructive to forests and soils. However, studies show that shifting cultivation is not a major cause of deforestation relative to land conversion, and that more carbon is being sequestered under shifting cultivation than under annual cropping or plantations.

Pastoralism. Indigenous pastoralists around the world manage the vast and often harsh terrain of rangelands, making productive use of these systems to meet their subsistence needs and maintaining ecosystems that sequester considerable amounts of carbon. Rangelands are approximately 40 percent of global land area and comprise the largest single land use in the world. Much of these lands have historically been utilized and managed by indigenous groups for hunting, gathering, grazing, and seasonal agriculture. The indigenous communities who engage in pastoral management are characteristically nomadic, living in low-density, highly mobile populations. Rangelands continue to support the livelihoods of 100 million to 200 million pastoralists, who steward more than 1.2 billion acres of rangelands globally. These systems are biologically diverse and highly productive, and conserve large stores of carbon. Literature suggests that these lands store up to 30 percent of the world's soil carbon and have the potential to sequester significantly more carbon by 2030 under improved rangeland management practices. Further, pastoralism has been shown to be more productive per acre than commercial ranching or sedentary livestock in similar environments. The temporary grazing of livestock helps secure carbon that may otherwise be released into the atmosphere, compared to other land-use systems such as annual crop production and bioenergy crop production.

Traditional pastoral systems are currently under duress due to climate change and pressures on pastoral communities to modernize. Pastoralists contribute significantly to local, regional, and national economies yet face historic and ongoing negative treatment. Their subsistence practices and cultures are perceived as inefficient, irrational, low tech, primitive, and environmentally destructive. These entrenched views undergird policies that attempt to dispossess pastoralists of their land and traditional practices—for example, through efforts to nationalize traditional

This image and the one on the succeeding page were taken on behalf of the International Conservation Fund of Canada (ICFC), which has worked with the Kayapo people to protect their 26-million-acre landholdings from encroaching loggers, miners, and Brazilian frontier society. From satellite photographs, the Kayapo traditional lands in Mato Grosso and Pará are a jewel of the Amazon, emerald and unblemished. On the margins of their land are the roads, clearings, frontier towns, and billowing smoke from fires clearing the land for cattle and farming. The efforts of the Kayapo have not always succeeded: Construction on the hugely destructive Belo Monte Dam on the Xingu River in Pará commenced in March 2011 after decades of legal and political resistance. It continues to be contested on legal fronts.

rangelands. At worst, these stereotypes of pastoralists can breed ethnic intolerance, which can lead to forced evictions and human rights violations. Nomadic pastoralism and traditional rangeland management practices persist in much of the world's rangeland in the face of social and political pressure on pastoralists to settle and modernize. Modern arrangements such as communal-area conservation agreements and the granting of landownership or return of native lands to indigenous communities are helping to secure rights for continued use of rangelands by pastoral populations.

Fire Management. Across the world human populations have practiced fire ecology historically and up to the present for a multitude of reasons. Native Americans throughout North America engaged in a wide range of land-management practices that utilized burning, which can be documented in historical and archaeological evidence. Sophisticated burning techniques were applied to produce favorable environments for certain food sources, game, and plant materials across a broad landscape. In the Pacific Northwest, indigenous populations used fire management to influence a range of ecosystems—from forest clearings to prairies—in order to create habitats and increase yields for beneficial plant and animal species. In northern Australia, aboriginal peoples have practiced techniques to regulate seasonal fires. Fires have been used to keep forest and countryside open, control vegetation growth, flush game, and comply with cultural obligations. Traditional fire management employs low-intensity, early dry season burns to clear vegetation, which can decrease intensity of naturally or human-caused fires.

Community Managed Forests. Indigenous and community management of forestlands has taken place for centuries. These lands may or may not be recognized formally by states as owned or managed by indigenous peoples or local communities. However, much indigenous or community-managed forestland is under traditional practices and customary law. An estimated 400 million to 500 million people globally depend on forests for their livelihoods; among these are 60 million forest-dependent indigenous peoples in Latin America, West Africa, and Southeast Asia. Estimates of total forestlands under commons management, regardless of ownership, reach 8 billion acres.

A wide range of practices can be considered indigenous or community forest management, including fallows management, forest groves with domesticated species, sacred groves, selective cultivation of forest species and trees, and intensive forest management. Indigenous management includes individual forest management practices and involves the process of collective decision making about the use and conservation of forest in communities. Loss of forest tenure and insecure land rights play a significant role in deforestation and degradation of indigenous or community-managed forestlands. Numerous studies have demonstrated that tenure-secure community forests exhibit lower deforestation rates and produce healthier ecosystem outcomes compared to similar forests without tenure security. Community management helps lower rates of degradation, enhance biomass growth, increase sequestration rates, and reduce emissions from forests. In a review of 118 cases that assess associations between tenure and forest change, it was found that tenure security is associated with positive forest outcomes and less deforestation. In another study it was shown that community-managed forests on average increased carbon storage by 2 tons per acre per year compared to nonmanaged forests.

Despite trends of declining forestlands, the global area of forest designated for or owned by indigenous peoples and communities has increased from 951 million acres in 2002 to 1.2 billion acres in 2013. The percentage these lands represent as a proportion of all forestlands increased from 10.8 percent to 15.4 percent in the same period. While the global trend appears positive, country-level proportions of indigenous and community forests vary widely.

Despite widely varying rates, given continued trends of policy supporting indigenous and community forest designation and ownership, both the global gross area and proportion of all total forest area under these designations can be expected to expand. Beyond legal recognition of forest rights, government actions are needed to ensure provision of technical assistance, indigenous peoples' engagement in decision-making processes, community mapping, expulsion of illegal settlers, and promotion of community forest management in order to enhance forest security. To increase indigenous land management requires supportive tenure policy environments and government collaboration to protect land rights. ●

IMPACT: *Indigenous peoples have secure land tenure on 1.3 billion acres globally, though they live on and manage much more. Our analysis assumes higher rates of carbon sequestration and lower rates of deforestation on lands managed by indigenous peoples. If forestland under secure tenure grows by 909 million acres by 2050, reduced deforestation could result in 6.1 gigatons of carbon dioxide emissions avoided. This solution could bring the total forest area under indigenous management to 2.2 billion acres, securing an estimated protected stock of 232 gigatons of carbon, roughly equivalent to over 850 gigatons of carbon dioxide if released into the atmosphere.*

LAND USE
TEMPERATE FORESTS

A quarter of the world's forests lie in the temperate zone, between 30 and 50 to 55 degrees latitude, mostly in the Northern Hemisphere. Some are deciduous, dropping their leaves in the winter months; others are evergreen. Until the late nineteenth century, temperate forests were the epicenter of deforestation. Over the course of history, 99 percent of temperate forests have been altered in some way—timbered, converted to agriculture, disrupted by development. However, forests are resilient. They are dynamic systems that constantly recover from impacts of either natural or human origin, even if regaining their full ecological integrity may require centuries.

Today, forests are on the rise across large swaths of the temperate world, due to reliance on timber imports, improved agricultural productivity resulting in the abandonment of once cleared land, improved forest management practices, and deliberate conservation efforts. These trends have relieved some degraded and deforested lands from other land uses and made recovery possible, whether passively allowed or actively aided. The world's 1.9 billion acres of temperate forests are now a net-carbon sink. Rising biomass density and overall increase in area mean these ecosystems absorb roughly 0.8 gigatons of carbon each year. There is an opportunity for more sequestration through restoration. According to the World Resources Institute (WRI), more than 1.4 billion additional acres are candidates for restoration—either large-scale, closed forest or mixed mosaics of forests, more sparsely growing trees, and land uses such as agriculture.

A collaboration between WRI, the International Union for Conservation of Nature, and South Dakota State University produced the global Atlas of Forest and Landscape Restoration Opportunities, which both quantifies and visualizes the prospect before us. Toggle between map layers of current and potential forest coverage, and the eastern half of the United States and continental Europe transform from speckled to dark green. The atlas classifies 84 percent of Ireland as opportunity area for either wide-scale or mosaic restoration. The Emerald Isle was once almost entirely forested, though by the eighteenth century most of its woodlands had been converted to pasture. The United States has substantial opportunities for restoration, building on trends

already in motion. From the 1990s to 2000s, the carbon sink provided by U.S. forestland rose 33 percent. The country's East Coast is home to a renaissance, as forests along the ancient Appalachian Mountains, running from Georgia to Maine, continue to grow in size and improve in health. Abandoned farmland has been the main impetus, with forests slowly rising up where fields were before—an example of passive restoration.

While temperate forests are not threatened by the same large-scale deforestation that afflicts the tropics, they continue to be fragmented by development. A warming world poses new challenges that restoration efforts will have to contend with going forward. Some suggest this is an era of "megadisturbance," given the mounting pressures on temperate forests. They are experiencing hotter and more frequent droughts, longer heat waves, and more severe wildfires, as well as worsening insect and pathogen outbreaks. These disturbances can compound, pushing temperate forests beyond their capacity for resilience, and have replaced overexploitation as the major threat to their persistence and health. Restoration efforts will need to continue evolving in response.

Preventing loss of forest is always better than trying to bring forest back and cure razed land. Because a restored forest never fully recovers its original biodiversity, structure, and complexity, and because it takes decades to sequester the amount of carbon lost in one fell swoop of deforestation, restoration is no replacement for protection. ●

IMPACT: *We project that temperate forest restoration will expand to an additional 235 million acres through natural regeneration. Though this is much lower than the available area for tropical forest restoration, it still sequesters 22.6 gigatons of carbon dioxide by 2050.*

Moss, fern, and southern beech trees in Fiordland National Park on the South Island of New Zealand. The 3-million-acre forested landscape traverses the mountaintops to the seas, with lakes and rainforests in between. It is said that rainfall in Fiordland is measured in meters. The steep inclines, deep ravines, and nonstop moisture kept all but the hardiest from trying to inhabit the land until it became a park in 1952.

The Hidden Life of Trees
PETER WOHLLEBEN

Years ago, I stumbled across a patch of strange-looking mossy stones in one of the preserves of old beech trees that grows in the forest I manage. Casting my mind back, I realized I had passed by them many times before without paying them any heed. But that day, I stopped and bent down to take a good look. The stones were an unusual shape: they were gently curved with hollowed-out areas. Carefully, I lifted the moss on one of the stones. What I found underneath was tree bark. So, these were not stones, after all, but old wood. I was surprised at how hard the "stone" was, because it usually takes only a few years for beechwood lying on damp ground to decompose. But what surprised me most was that I couldn't lift the wood. It was obviously attached to the ground in some way.

I took out my pocketknife and carefully scraped away some of the bark until I got down to a greenish layer. Green? This color is found only in chlorophyll, which makes new leaves green; reserves of chlorophyll are also stored in the trunks of living trees. That could mean only one thing: this piece of wood was still alive! I suddenly noticed that the remaining "stones" formed a distinct pattern: they were arranged in a circle with a diameter of about five feet. What I had stumbled upon were the gnarled remains of an enormous ancient tree stump. All that was left were vestiges of the outermost edge. The interior had completely rotted into humus long ago—a clear indication that the tree must have been felled at least four or five hundred years earlier. But how could the remains have clung onto life for so long?

Living cells must have food in the form of sugar, they must breathe, and they must grow, at least a little. But without leaves—and therefore without photosynthesis—that's impossible. No being on our planet can maintain a centuries-long fast, not even the remains of a tree, and certainly not a stump that has had to survive on its own. It was clear that something else was happening with this stump. It must be getting assistance from neighboring trees, specifically from their roots. Scientists investigating similar situations have discovered that assistance may either be delivered remotely by fungal networks around the root tips—which facilitate nutrient exchange between trees—or the roots themselves may be interconnected. In the case of the stump I had stumbled upon, I couldn't find out what was going on, because I didn't want to injure the old stump by digging around it, but one thing was clear: the surrounding beeches were pumping sugar to the stump to keep it alive.

If you look at roadside embankments, you might be able to see how trees connect with each other through their root systems. On these slopes, rain often washes away the soil, leaving the underground networks exposed. Scientists in the Harz Mountains in Germany have discovered that this really is a case of interdependence, and most individual trees of the same species growing in the same stand are connected to each other through their root systems. It appears that nutrient exchange and helping neighbors in times of need is the rule, and this leads to the conclusion that forests are superorganisms with interconnections much like ant colonies.

Of course, it makes sense to ask whether tree roots are simply wandering around aimlessly underground and connecting up when they happen to bump into roots of their own kind. Once connected, they have no choice but to exchange nutrients. They create what looks like a social network, but what they are experiencing is nothing more than a purely accidental give and take. In this scenario, chance encounters replace the more emotionally charged image of active support, though even chance encounters offer benefits for the forest ecosystem. But Nature is more complicated than that. According to Massimo Maffei from the University of Turin, plants—and that includes trees—are perfectly capable of distinguishing their own roots from the roots of other species and even from the roots of related individuals.

But why are trees such social beings? Why do they share food with their own species and sometimes even go so far as to nourish their competitors? The reasons are the same as for human communities: there are advantages to working together. A tree is not a forest. On its own, a tree cannot establish a consistent local climate. It is at the mercy of wind and weather. But together, many trees create an ecosystem that moderates extremes of heat and cold, stores a great deal of water, and generates a great deal of humidity. And in this protected environment, trees can live to be very old. To get to this point, the community must remain intact no matter what. If every tree were looking out only for itself, then quite a few of them would never reach old age. Regular fatalities would result in many large gaps in the tree canopy, which would make it easier for storms to get inside the forest and uproot more trees. The heat of summer would reach the forest floor and dry it out. Every tree would suffer.

Every tree, therefore, is valuable to the community and worth keeping around for as long as possible. And that is why even sick individuals are supported and nourished until they recover. Next time, perhaps it will be the other way round, and the supporting tree might be the one in need of assistance. When thick silver-gray beeches behave like this, they remind me of a herd of elephants. Like the herd, they, too, look after their own, and they help their sick and weak back up onto their feet. They are even reluctant to abandon their dead.

Every tree is a member of this community, but there are different levels of membership. For example, most stumps rot away into humus and disappear within a couple of hundred years (which is not very long for a tree). Only a few individuals are kept alive over the centuries, like the mossy "stones" I've just described. What's the difference? Do tree societies have second-class citizens just like human societies? It seems they do,

though the idea of "class" doesn't quite fit. It is rather the degree of connection—or maybe even affection—that decides how helpful a tree's colleagues will be.

You can check this out for yourself simply by looking up into the forest canopy. The average tree grows its branches out until it encounters the branch tips of a neighboring tree of the same height. It doesn't grow any wider because the air and better light in this space are already taken. However, it heavily reinforces the branches it has extended, so you get the impression that there's quite a shoving match going on up there. But a pair of true friends is careful right from the outset not to grow overly thick branches in each other's direction. The trees don't want to take anything away from each other, and so they develop sturdy branches only at the outer edges of their crowns, that is to say, only in the direction of "non-friends." Such partners are often so tightly connected at the roots that sometimes they even die together.

Tree roots extend a long way, more than twice the spread of the crown. So the root systems of neighboring trees inevitably intersect and grow into one another—though there are always some exceptions. Even in a forest, there are loners, would-be hermits who want little to do with others. Can such antisocial trees block alarm calls simply by not participating? Luckily, they can't. For usually there are fungi present that act as intermediaries to guarantee quick dissemination of news. These fungi operate like fiber-optic Internet cables. Their thin filaments penetrate the ground, weaving through it in almost unbelievable density.

One teaspoon of forest soil contains many miles of these "hyphae." Over centuries, a single fungus can cover many square miles and network an entire forest. The fungal connections transmit signals from one tree to the next, helping the trees exchange news about insects, drought, and other dangers. Science has adopted a term first coined by the journal *Nature* for Dr. Simard's discovery of the "wood wide web" pervading our forests. What and how much information is exchanged are subjects we have only just begun to research. For instance, Simard discovered that different tree species are in contact with one another, even when they regard each other as competitors. And the fungi are pursuing their own agendas and appear to be very much in favor of conciliation and equitable distribution of information and resources.

Under the canopy of the trees, daily dramas and moving love stories are played out. Here is the last remaining piece of Nature, right on our doorstep, where adventures are to be experienced and secrets discovered. And who knows, perhaps one day the language of trees will eventually be deciphered, giving us the raw material for further amazing stories. Until then, when you take your next walk in the forest, give free rein to your imagination—in many cases, what you imagine is not so far removed from reality, after all. ◉

Excerpted from *The Hidden Life of Trees: What They Feel, How They Communicate, Discoveries from a Secret World*, by Peter Wohlleben, 2016 (Greystone Books).

LAND USE
AFFORESTATION

The capacity of trees to synthesize and sequester carbon through photosynthesis as they grow has made afforestation an important practice in the age of warming. Creating new forests where there were none before in areas that have been treeless for at least fifty years is the aim of afforestation. Degraded pasture and agricultural lands, or other lands severely corrupted from uses such as mining, are ripe for strategic planting of trees and perennial biomass. So are eroding slopes, industrial properties, abandoned lots, highway medians, and wastelands of all stripes—almost any space that is unattended or forgotten can help draw down atmospheric carbon.

The most successful afforestation projects are those that plant native trees. Replanting, however, can take a variety of forms—from seeding dense plots of diverse indigenous species to introducing a single exotic as a plantation crop, such as the fast-growing Monterey pine, the most widely planted tree in the world. Whatever the structure, they all function as carbon sinks, drawing in and holding on to carbon, and distributing carbon into the soil. How much carbon is sequestered annually depends on the details of species, site, soil conditions, and structure.

A recent paper out of the University of Oxford makes a conservative estimate that afforestation could draw down one to three gigatons of carbon dioxide per year in 2030. Global availability of land is a key variable and one that is difficult to predict, affected by factors ranging from population and diet to crop yields and bioenergy demands. While afforestation projects have significant carbon sequestration potential, forests, new or old, are vulnerable to fire, drought, pests, and the ax or saw.

To date, plantations comprise the majority of afforestation projects and are on the rise globally, planting trees for timber and fiber and, increasingly, selling carbon offsets as well. (While plantation forestry makes up just 7 percent of total forest cover, it generates roughly 60 percent of commercial wood.) Plantations have been and remain controversial because they are often created with purely economic motives and little regard for the long-term well-being of the land, environment, or surrounding

18.06 GIGATONS REDUCED CO2	$29.4 BILLION NET COST	$392.3 BILLION NET SAVINGS

communities. Some displace natural forests or other vital ecosystems and then support much lower levels of fauna, from songbirds to snails. They are susceptible to disease, often requiring chemicals to control infestation, and can sap groundwater, as has been the case with China's Three-North Shelter Program—the "Great Green Wall." In their wake, the rights and interests of local and indigenous communities can be disregarded or deliberately transgressed, particularly in low-income countries where foreign interests acquire land to establish them. It has led to strong pushback against how afforestation is being implemented and to concerns about a profit-fueled land rush following the Paris Agreement, potentially accompanied by forced relocation, cultural deracination, and violation of human rights.

These issues are part of the reason efforts have cropped up to make plantation forestry more sustainable, efforts such as third-party certification schemes that disallow conversion of natural forest. But there is no denying the benefits plantations provide. Beyond their usefulness in wood production and carbon

Far Left: This is a typical single-story tree plantation in Umatilla, Oregon, consisting of cottonwood trees planted with eight-foot spacing in order to force upward, knot-free growth.

Below: Single-story afforestation consists of monocultures of pines, poplars, and other fast-growing trees, some of which are genetically modified to speed up their growth. Although single-story plantations sequester carbon in significant quantities, they are the equivalent of arboreal deserts due to their lack of biodiversity and the speed with which they exhaust and acidify the soil. Below is the Miyawaki method or what is called analog forestry, an afforestation technique that mimics natural forest formation. It creates a multistory forest consisting of diverse upper, middle, and lower canopy trees, shrubs, and plants—an ecosystem sustainable for a hundred years or more. This method of afforestation has a higher ratio of biodiversity to biomass, is more productive, and sequesters far more carbon. However, it is unsuitable for the harvesting methods employed in even-aged, industrial tree farms where all trees are cut at the same time.

sequestration, tree farms have a "plantation conservation benefit": They can actually reduce logging of natural forest. A 2014 study calculates a 26 percent reduction in natural forest harvest around the world thanks to planted forests. Initiatives such as New Generation Plantations, out of the World Wide Fund for Nature (WWF), are working to ensure that well-designed plantations and inclusive management practices become mainstream, so that the good (and the goods) of plantations can be optimized, while ensuring the integrity of ecosystems and communities. Because plantations are here to stay, groups such as WWF know it is critically important to engage key actors such as companies and governments, and to identify degraded lands ideal for afforestation. Multipurpose plantations can meet a variety of social, economic, and environmental aims (including providing jobs in places where few exist), but they have to be conceived and implemented with those aims in mind.

Plantations are far from the only option. To counter the ecological deserts of monoculture tree farms, which often introduce invasive species with their potential negative impact, an extraordinary Japanese botanist, Akira Miyawaki, devised a completely different method of afforestation. In the 1970s and '80s, Miyawaki studied the temples and shrines of Japan to better understand the country's original forests. Over decades, perhaps centuries, indigenous oaks, chestnuts, and laurels had been replaced almost completely by pine, cypress, and cedar introduced for timber. These ersatz native forests, he realized, were not resilient or adaptable to climate change. Drawing on a German technique called potential natural vegetation, Miyawaki became a passionate champion of creating indigenous, authentic forests; he now has been part of planting more than 40 million trees around the world.

The Miyawaki method calls for dozens of native tree species and other indigenous flora to be planted close together, often on degraded land devoid of organic matter. As these saplings grow, natural selection plays out and a richly biodiverse, resilient forest results. Miyawaki's forests are completely self-sustaining after the first two years, when weeding and watering are required, and mature in just ten to twenty years—rather than the centuries nature requires to regrow a forest. In the same amount of space, they are one hundred times more biodiverse and thirty times denser than a conventional plantation, while sequestering more carbon. They provide beauty, habitat, food, and tsunami protection.

We think of afforestation as something happening on large tracts of land, but individuals can do this everywhere. Inspired by Miyawaki's approach and drawing on Toyota's assembly line process, entrepreneur Shubhendu Sharma's company Afforestt is developing an open-source methodology to enable anyone to create forest ecosystems on any patch of land. In an area the size of six parking spaces, a three-hundred-tree forest can come to life—for as little as the cost of an iPhone.

Jadav Payeng, the "forest-man of India," single-handedly afforested a 1,300-acre area on Majuli, the world's largest river island. Jadav, without any subsidy or financial support, tilled and sowed native species based on traditional knowledge, on the completely denuded sandbars of the Brahmaputra River, paving the way for natural regeneration. Today, Jadav's forest is home to an astounding array of floral and faunal biodiversity, at the same time serving as a natural erosion control method for the island.

Many of the places prime for afforestation are located in low-income countries, where there is often a multifaceted opportunity for impact. Creating new forest can sink carbon and support biodiversity, address human needs for firewood, food, and medicine, and provide ecosystem services such as flood and drought protection. Engaging local communities in afforestation projects by making them aware of the socioeconomic and environmental benefits of forests is the key to success. Because afforestation is a multidecade endeavor, what properly enables it are provisions for up-front costs, developing markets for forest products, and ensuring clear land rights in order to maintain continuity between planting and eventual harvest. Emerging geospatial and remote sensing technologies along with mobile-based ground validation can serve as powerful monitoring tools to ensure healthy plantations. Applying these approaches can do more than draw down atmospheric carbon; it can create new forests in ways that are ecologically sound, socially just, and economically beneficial. ●

IMPACT: *As of 2014, 709 million acres of land were used for afforestation. Establishing timber plantations on an additional 204 million acres of marginal lands can sequester 18.1 gigatons of carbon dioxide by 2050. The use of marginal lands for afforestation also indirectly avoids deforestation that otherwise would be done in the conventional system. At a cost of $29 billion to implement, this additional area of timber plantations could produce a net profit for landowners of over $392 billion by 2050.*

TRANSPORT

The transport sector is double-edged. You will find here solutions that significantly improve the fuel efficiency of the planes, trains, ships, cars, and trucks that continue to rely on fossil fuels. However, unless use of these modes of transport is curtailed, the efficiency improvements will be devoured by increased consumption. Also included are solutions that can move transport beyond fossil fuels. Electric vehicles are four times as efficient as gas-powered ones, and when powered by wind turbines at today's prices, the electrical equivalent of gasoline is thirty to fifty cents per gallon. Bicycles also offer mobility without fuel. The use and sustainability of transportation cannot be separated from how and where people live, work, and play; two major influences going forward will be the design of the urban environment and reduction of excess consumption.

MASS TRANSIT

Curitiba, Brazil, did not have climate change in mind when it established its bus network. In 1971, a young architect named Jaime Lerner became mayor of the city, appointed by Brazil's then-dictatorship under the mistaken assumption that he would toe the authoritarian line. Of course, creatives rarely do. Subways and light rail were popular among city planners at the time, but Lerner saw that implementing any system based on rail would be too expensive and slow. (He is famous for saying, "If you want creativity, cut one zero from the budget. If you want sustainability, cut two zeros!")

Lerner devised an alternative that focused on something wholly unfashionable—buses—but he gave them the advantages of rail. The main advantage was dedicated lanes along main thoroughfares—separate corridors allowing buses to avoid entanglement with automobiles—at installation costs fifty times less than that of rail. Then, in the early 1990s, Curitiba's bus stops were redesigned to be more like metro stations, facilitating passenger flow. Instead of paying onboard, riders pay at the station; instead of a single point of entry, there are multiple. These signature tubelike stations now pepper the city's terrain (and anchor its brand), and 2 million passengers move through them every day. (By comparison, London's Tube has 3 million riders on the average day.)

Curitiba pioneered what is known as bus rapid transit (BRT), a model replicated across Latin America (e.g., Bogotá's famously successful TransMilenio) and in more than two hundred cities worldwide. BRT is one of the modes of mass transit currently vying with cars for passengers and their miles. Whatever its form, public transportation uses scale to its emissions advantage. When someone opts to ride a streetcar or bus rather than driving a car or hailing a cab, greenhouse gases are averted. To use technical parlance, it is all about *modal shift*.

The transport sector is responsible for 23 percent of global emissions. Urban transport is the single greatest source and growing—largely because the use of cars is on the rise. Of course, most transit was mass transit until World War II, when the automobile became affordable to the masses in high-income countries. Freedom from fixed routes and schedules had—and continues to have—strong allure, while urban and suburban spaces designed around cars made them increasingly essential. Cars and sprawl became coconspirators, especially in the United States. Across the U.S. metropolitan areas that do have transit, less than 5 percent of daily commuters use it. By contrast, in Singapore and London, half of trips are made via public transportation. In low-income countries, mass transit remains a primary form of urban mobility, though car use is on the rise in emerging economies (even in Curitiba). Buses, whether part of BRT systems or intermingled with other vehicles, are the most common mode of public transportation globally.

There are many benefits of mass transit beyond emissions reduction. Perhaps the most obvious is relieving traffic congestion: Mass transit can move greater numbers of people in smaller footprints than cars, and some forms, like subways, shift volumes of travelers off roads entirely and onto separate tracks. The London Underground and Bangkok's Skytrain both name the second advantage outright. With fewer people in cars, fewer accidents and fatalities take place; drivers, riders, and pedestrians are safer. Because public transportation is spatially leaner than car-centric systems (think of the parking spaces alone), it preserves more of a city's land for other and higher uses—green space, housing, places of business—and more economic activity. Overall, air pollution drops. Buses have historically been powered by polluting diesel engines; newer buses are cleaner, with some fueled by electricity or natural gas instead.

Mass transit also has a crucial social advantage: It makes cities more equitable by serving those who cannot drive—the young and the old, those with physical limitations, and those unable to afford car ownership. They are far from its sole users, but

6.57 GIGATONS REDUCED CO2	DATA TOO VARIABLE TO BE DETERMINED	$2.38 TRILLION NET SAVINGS

they might otherwise be excluded from accessing mobility. Mass transit is one manifestation of the public square, in which people of many stripes encounter and share space with one another. As Adam Gopnik put it in *The New Yorker*, "A train is a small society, headed somewhere more or less on time, more or less together, more or less sharing the same window, with a common view and a singular destination"—a unique civic experience, as well as a means of conveyance.

Despite its advantages, mass transit has faced—and continues to reckon with—a variety of challenges. The appeal of cars is strong and culturally entrenched in many places (less so among younger generations), and shifting habits is difficult, especially if behavior change requires more effort, more time, or more money. Public transportation is most successful where it is not just viable but efficient and attractive. One key piece is making the use of multiple modes more seamless, such as a single card to pay for metro, bus, bike share, and rideshare, or a single smartphone app to plan trips that use more than one. Beyond appealing to passengers, mass transit relies on overall urban design. A city's density is the pivotal factor, necessary for ensuring people live and work close enough to transit to use it (what is known as the first-mile/last-mile problem) and for achieving the high-occupancy rates that make transit profitable and efficient. An empty bus is not a solution. Achieving that density may present some cities with the need for fundamental reorganization and "redensification," and those still growing with an opportunity to plan ahead. Compact urban spaces can readily become connected urban spaces, at lower cost.

Even in ideal conditions, investing in transit infrastructure can be a challenge fiscally or politically, but those investments pay dividends. The benefits of mass transit accrue to all city dwellers, not just those who use it. (And its absence places burdens none can escape.) Without putting money where buses, subways, and streetcars are, or could be, modal shift may go towards private cars and their attendant congestion and pollution, rather than lower-emissions transit options. In tandem with biking and walking—and infrastructure to enable them—mass transit can embed mobility, livability, and equity in cities. Movement is a fundamental part of being human, going from here to there for reasons of necessity, pleasure, or curiosity. Mobility brings vitality to individual lives and cities overall; the atmosphere need not be forfeited to achieve it. ●

IMPACT: *Use of mass transit is projected to decline from 37 percent of urban travel to 21 percent as the low-income world gains wealth. If use grows instead to 40 percent of urban travel by 2050, this solution can save 6.6 gigatons of carbon dioxide emissions from cars. Our analysis includes diverse mass transit options (bus, metro, tram, and commuter rail) and examines the costs that travelers pay (car purchase and use compared to buying transit tickets).*

Upper Left: Evening rush hour in the Garden Ring, Moscow, Russia.

Below: An eastbound Metropolitan Area Express light rail train stops at Yamhill Street and 2nd Avenue in downtown Portland, Oregon. With 97 stations, ridership on the MAX is approximately 120,000 people per week.

HIGH-SPEED RAIL

In 1964, Japan celebrated the Olympics by inaugurating the world's first high-speed "bullet" train on the Osaka–Tokyo route, a distance of 320 miles. Today, it is the world's busiest high-speed rail line, with more than four hundred thousand passengers per day. According to the International Union of Railways, there are more than 18,500 miles of high-speed rails worldwide. That number will increase by 50 percent when current construction is completed; many more thousands of miles are planned and under consideration. China has by far the most high-speed rail lines—more than 50 percent of the total—followed by Western Europe and Japan. China, Japan, and South Korea have introduced a variation of high-speed rail, the maglev train, which deploys magnets that lift the train off its supporting structure, propelling it at astonishingly smooth and quiet speeds—to the order of 270 miles per hour in the run between Shanghai and its distant airport.

High-speed rail (HSR) is powered almost exclusively by electricity, not diesel. Compared to driving or flying, it is the fastest way to travel between two points a few hundred miles apart and reduces carbon emissions up to 90 percent. HSR's market advantage is on trips of seven hours or less. Train stations are in the middle of cities and major suburbs, and for the time being,

security issues are less burdensome. To boot, the new trains have comfortable cabins, wonderful visibility, and full connectivity. The long-term success of HSR is well established on medium-distance (four-hour) high-density corridors. In certain popular markets in Western Europe and Asia, fast trains have captured more than half of the overall travel business on those routes. HSR virtually owns the London–Paris, Paris–Lyon, and Madrid–Barcelona routes. In 2013, high-speed trains recorded 220 billion passenger miles globally, about 12 percent of the total rail market.

The United States boasts a grand total of twenty-eight miles of high-speed track, in rural Massachusetts and Rhode Island, utilized by Amtrak's Acela service. In California, where enthusiasm for HSR is perhaps the highest in the country, voters approved $10 billion as a down payment on a state-of-the-art system. Projections of a completed California HSR system show it could reduce car travel by 3.6 billion miles annually—equivalent to pulling 300,000 cars off the roads every day, eliminating 2.2 million tons of greenhouse gases. Still, progress is slow and resistance persists. Completion is scheduled for 2028, but no one expects that to happen: Cost estimates have doubled from $33 billion to $68 billion.

1.42 GIGATONS REDUCED CO2	$1.05 TRILLION NET COST	$310.8 BILLION NET SAVINGS

A Central Japan Railway Co. (JR Central) Shinkansen bullet train arrives at Tokyo Station on January 19, 2016. Japanese rolling-stock manufacturers have been collaborating with the Japan Railways group to expand the business worldwide with its technology and standards used for the Shinkansen bullet train system. Texas Central Partners LLC plans to start construction of the Texas Central Railway High-Speed Rail Project between Houston and Dallas next year using the Shinkansen bullet train technology.

Therein lies one of the major hurdles: cost. The trains themselves are expensive, as are any new stations. The tracks typically range from $15 million to $80 million per mile; and then there are bridges, tunnels, and viaducts. In the Northeast Corridor, Amtrak estimates that creating a high-speed rail system rated at 220 miles per hour would cost roughly $150 billion. A lower, 160-mph system would save only a little. Given the numbers, government subsidies and excise taxes are necessary, but opponents of high-speed rail cite subsidies as proof that it is not economical. However, any assessment should include the costs if a high-speed rail line is not built, as all of our transportation systems enjoy significant government subsidies, hidden or otherwise. The public, not private enterprise, pays for new highways, new lanes for old highways, bigger airports, traffic jams, wasted time, and ever more greenhouse gases. The public costs that any HSR project would avoid need to be subtracted from the capital cost of the system.

Proponents of HSR have claimed that high-speed trains will end oil dependency and drive huge cuts in emissions. These are unrealistic expectations. High-speed rail requires many passenger trips to break even. Only certain places in the world have sufficient population density to support HSR. The carbon footprint of an up-and-running HSR is lower than that of planes and cars, but only when it replaces significant air and vehicle trips. Another factor to take into account: There are significant greenhouse gas emissions associated with HSR construction, in particular large amounts of cement required to build railroad tracks strong enough to support trains travelling at high speeds (also true of runways and roads).

One of the advantages that HSR has over air, automobile, and conventional rail is that its energy source is more likely to get cleaner as time passes. As governments push for carbon-free power generation throughout the globe, HSR can become increasingly clean. This advantage must be tempered by the fact that automobile travel is becoming less carbon intensive as electric vehicles become more prevalent. Air travel is less likely to make big gains in efficiency, however, maintaining HSR's per-passenger emissions benefit as long as ridership meets or exceeds expectations.

Moreover, HSR can be an important component of smart growth and help revitalize city centers. Hub-and-spoke designs for HSR, with city-center stations sharing space with mass transit and properly planned mixed-use zones nearby, can contribute to wider climate, health, and social benefits. As part of a sustainable transportation system, HSR can compound its emissions benefits.

There are other economic and environmental benefits that argue for expanded HSR travel. For example, as travelers shift from conventional rail to HSR, more lines will be made available for freight shipping by rail. This may reduce costs and greenhouse gas emissions from transporting goods with diesel-burning trucks, and therefore aid economic growth. Other advantages include the comparative ease and comfort of traveling by high-speed train as opposed to automobiles and airplanes, as well as the accessibility to travel that this may open up for more people. These additional benefits can be difficult to quantify and include in traditional benefit-cost analyses, but further research may reveal that they tip the scales in favor of HSR, making it an optimal choice for infrastructure development. ●

IMPACT: *If HSR construction and ridership continue at their projected pace, this solution can deliver 1.4 gigatons of carbon dioxide emissions reductions by 2050. A global network of 64,000 miles of track, with an average trip length of 186 miles, could support 6 to 7 billion riders annually. Regionally, most impact will come from Asia, especially China. If HSR is concentrated between cities with heavy, short-haul flight routes, impact can be greater. Implementation comes at a steep cost of $1 trillion. Operational savings over thirty years, however, are $310 billion and $980 billion over the lifetime of HSR infrastructure.*

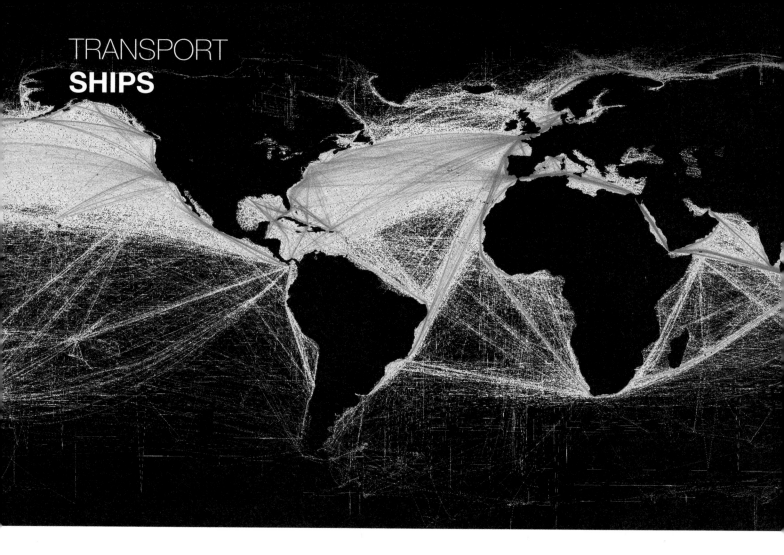

More than 80 percent of global trade, by weight, floats its way from place to place. Ninety thousand commercial vessels—tankers, bulk dry carriers, and container ships—make the movement of goods possible to the tune of more than 10 billion tons of cargo in 2015. Ships are the most carbon-effective way we have to move materials from one geography to another, where an efficient rail system does not exist or cannot be used due to geography. A plane emits forty-seven times more carbon dioxide to transport the same quantity of goods the same distance. Even though shipping is an industry essential to the world's economy, it is largely invisible.

Shipping oil, iron ore, rice, and running shoes across oceans produces 3 percent of global greenhouse gas emissions, and those emissions grow as world trade continues to increase. Forecasts predict they could be 50 percent to 250 percent higher in 2050, depending on economic and energy variables. Though considerable attention has been paid to vehicular emissions, the impact of oceanic freight has not been a climate priority. That is beginning to change. The industry, government, and NGOs are working out how to take to the high seas without such high emissions.

Because of huge shipping volumes, increasing shipping efficiency can have a sizable impact. It begins with design of the ships. The most efficient vessels are larger and longer than others. They trim out unnecessary parts of their structure and use lightweight materials. Some new vessels have ducktails at the

rear—flat extensions that project from the ship's aft to lower resistance—and compressed air pumped through the bottom of the hull to create a layer of bubbles that "lubricate" passage through the water. These two innovations alone can reduce fuel use by 7 to 22 percent depending on the type of boat. Efficient ships may also have additional machinery on board, such as solar panels to provide electricity and automation systems that take the guesswork out of optimizing a ship's performance. Some design and technology approaches are applicable only to new ship builds; others are viable for retrofitting—particularly important because vessels currently in use will remain so for decades.

Key efforts aim to improve ship design and the technology onboard. In 2011, the International Maritime Organization (the United Nations agency tasked with making shipping safer and cleaner) established the Energy Efficiency Design Index (EEDI) for new builds. Like fuel-economy standards for cars, EEDI requires that new ships meet a minimum level of energy efficiency and raises that bar over time. The Sustainable Shipping Initiative is a partnership between fifteen of the leading shipping companies, the World Wildlife Fund, and Forum for the Future, working together to create a completely sustainable shipping industry by 2040. In 2011, a joint effort by RightShip and the Carbon War Room produced an A-to-G Greenhouse Gas Emissions Rating system for commercial vessels new and old, benchmarking each ship against its peers based on carbon dioxide pollution.

RANKING AND RESULTS BY 2050

7.87 GIGATONS
REDUCED CO2

$915.9 BILLION
NET COST

$424.4 BILLION
NET SAVINGS

It takes 5 million barrels of fuel per day to move commercial ships across the routes shown on this map. Added up over the course of a year, international shipping emits more than 800 million tons of carbon dioxide or its equivalent in other greenhouse gases—11 percent of the total emissions from the transportation sector.

© NCEAS/T. HENGL

The rating scheme, like other specialized indices, creates transparency and addresses a key challenge for upping ship efficiency: split incentives. Because companies that send cargo pay the bulk of fuel costs, shipowners have little reason to upgrade their vessels, especially if performance is opaque. The Greenhouse Gas Emissions Rating creates a new point of leverage: Charterers looking to reduce costs and green supply chains can target ships accordingly. Already 20 percent of global trade uses the system, as do banks, insurers, and local port authorities, like two in British Columbia that discount harbor fees for cleaner, better-rated ships.

Maintenance and operations are also vital for marine fuel efficiency. Techniques can be as simple as removing debris from propellers or smoothing the surface of a hull with a sharkskin-like coating. Marine organisms easily plant themselves on the hulls of ships, where they add weight, create drag, and lower fuel efficiency. This biofouling can increase fuel consumption by 40 percent. The rough, toothlike scales of sharks prevent algae and barnacles from attaching to their skin. Harnessing these attributes of sharkskin, University of Florida professor Anthony Brennan developed a biomimetic coating to keep hulls clean for smoother sailing. It is one of many technologies and practices that can make cargo ships more hydrodynamic and energy efficient.

Reducing a ship's operating speed—what the business calls "slow steaming"—lowers fuel consumption more than any other practice, up to 30 percent. An upside of the 2009 global recession is that slow steaming has become standard across much of the industry. Route and weather planning are also critical. When the small gains from design, technology, maintenance, and operations are collectively applied, industry-leading ships can be twice as efficient as laggards. In sum, available efficiency approaches can reduce shipping emissions by 20 to 40 percent by 2020 and 30 to 55 percent by 2030.

In addition to improving climatic health, making oceanic freighting more efficient is important for air quality and human health. Ships are powered by low-grade bunker fuel, the dregs of the oil refining industry, which contains thirty-five hundred times more sulfur than the diesel used in cars and trucks. The port cities where ships congregate suffer most from the nitrous and sulfur oxides and particulate matter they spew into the air. Researchers attribute sixty thousand deaths each year from cardiovascular and lung diseases to particulate matter emitted by ships. Some ports require ships to switch to cleaner-burning diesel fuel when they approach the shore—a practice that can dramatically reduce populations' exposure to hazardous pollutants from ships. Similarly, more ports are requiring that docked ships plug in to onshore power rather than running their oil-fueled generators for power.

Thanks to design innovations and greenhouse gas ratings initiatives, some of the industry is changing. Yet, emissions from marine vessels are not included in global climate change agreements, and no global emissions targets have been established or agreed upon. In October 2016, the International Maritime Organization met and deferred any discussion on capping carbon emissions until 2023—a delay considered too late given that the maritime industry is predicted to generate 17 percent of global carbon emissions in 2050. Considering that trillions of dollars of goods are shipped annually, it may fall to the companies whose goods are being transported to pressure maritime shipping into being a responsible industry. RightShip and Carbon War Room initiatives may be the means to reduce global carbon emissions in a workable amount of time. Cutting shipping's greenhouse gases remains a voluntary act; this alone is not driving change quickly enough. As with fish, buildings, food, and timber, it may be time for a clean shipping certification. Economics work in favor of improvement. Fuel costs are the main expense of ship operation, which means carriers, the companies that use them, and ultimately businesses and consumers that purchase shipped goods all have an interest in fuel use being as low as possible, lowering carbon emissions. ●

IMPACT: *With an efficiency gain of 50 percent across the international shipping industry, 7.9 gigatons of carbon dioxide emissions can be avoided by 2050. That could save $424 billion in fuel costs over thirty years and $1 trillion over the life of the ships.*

ELECTRIC VEHICLES

Electric cars have been romanced for nearly two hundred years, since the first prototype was built in 1828. In 1891, Henry Ford worked for Thomas Edison at the Edison Illuminating Company in Detroit. Edison and Ford became fast friends for life, and it was Edison who supported and encouraged his friend—early in Ford's career—to build a gasoline-powered automobile. Ironically, Edison was hard at work making better, less expensive batteries, some specifically designed for electric vehicles. At one point, he turned the tables on Ford, writing, "Electricity is the thing. There are no whirring and grinding gears with their numerous levers to confuse. There is not that almost terrifying uncertain throb and whirr of the powerful combustion engine. There is no water-circulating system to get out of order—no dangerous and evil-smelling gasoline and no noise."

The young Ford was not convinced and went on to create the Model A and the Model T. Sales of the $360 car surpassed $250,000 in 1914, but in that year, Edison's prodding seemed to take effect. Ford, satisfied that Edison soon would deliver on an inexpensive, lightweight battery, announced that he would manufacture an electric automobile in collaboration with Edison—the Edison-Ford. Months and then years went by and the Edison-Ford never came to pass, because Edison could not make good on that lightweight, durable battery.

In fact, the electric vehicle was not invented by any one person, but rather evolved over time through a series of transcontinental breakthroughs. Early nineteenth-century inventors in Britain, the Netherlands, Hungary, and the United States all created various types of small-scale electrical vehicles (EVs), but the first practical vehicles were not created until the last half of the century. In 1891, William Morrison, a chemist from Iowa, made a six-passenger vehicle capable of reaching speeds up to fourteen miles per hour. By the end of the century, vehicles were available in the United States with gasoline, electric, and steam power trains. Electric vehicles outsold both gasoline- and steam-powered cars for a variety of reasons: They did not require hand cranking to start, it was unnecessary to change gears, and they had a longer range than steam-powered cars. Like electric vehicles today, they were quieter and did not pollute.

By the 1920s, Americans were traveling farther because of an improved road network, so the shorter range of EVs compared to gasoline vehicles started to become a limitation. Meanwhile, gasoline vehicles gained in appeal: Henry Ford commenced mass production, making them cheaper than EVs. Charles Kettering invented the electric starter, eliminating the need for hand cranking, and crude oil was discovered in Texas, making gasoline affordable for the average consumer. Internal combustion engines have dominated the automotive landscape ever since. The atmosphere has paid a high price for the more

than 1 billion cars on the road today. Fortuitously, there are currently more than 1 million electric vehicles on the road, and the difference in impact between the two is remarkable.

Two-thirds of the world's oil consumption is used to fuel cars and trucks. Transport emissions are second only to electricity generation as a source of carbon dioxide, accounting for a 23 percent share of all emissions. As developing nations industrialize, the number of motor vehicles is projected to surpass 2 billion by 2035.

Electric vehicles are powered by the grid or distributed renewables, and this includes hydrogen-powered vehicles employing fuel cells to generate onboard electricity. They are about 60 percent efficient compared to gasoline-powered vehicles, which are about 15 percent efficient. The "fuel" for electric cars is cheaper too. The Nissan LEAF, an all-electric vehicle, will travel 3.3 miles on 1 kilowatt-hour of electricity. If the car is charged in the middle of the night at 7 cents per kilowatt-hour, that is comparable to $.72 per gallon. If the Nissan LEAF gets 23 miles per equivalent gallon, compared to 34 miles per gallon for the Nissan Versa with gasoline at $2.30 per gallon, it achieves 69 percent cost savings.

Carbon dioxide emission per gallon of gasoline is 25 pounds, whereas the emissions for 10 kilowatt-hours of electricity are 12.2 pounds on average—a 50 percent reduction in carbon dioxide if power comes off the grid. If the electricity is derived from solar, carbon dioxide emissions fall by 95 percent.

Increasingly, the electric car is the preferred option. Sales volume has multiplied tenfold in less than a decade. From 2014 to 2015, sales jumped from 315,000 to 565,000 vehicles, thanks mainly to Chinese enthusiasts. Two-thirds of EV sales worldwide are in the three largest markets for passenger cars: the United States, China, and Japan. EV leader Tesla shocked the car industry in 2016, when it almost instantaneously attracted 325,000 advance orders for its compact Model 3, each accompanied by a $1,000 down payment. To bolster its position and reduce costs, Tesla has built the largest factory in the world for lithium-ion batteries in Nevada. Government programs around the world are encouraging the purchase of electric cars, including the United States, which offers a $7,500 subsidy. The United States and China now mandate that at least 30 percent of government car purchases be nonpolluting. India wants to be all-electric by 2030—and it has the incentives to make it happen.

Electric vehicles will disrupt auto and oil business models—the two biggest economic sectors in the United States—because EVs are simpler to make, have fewer moving parts, and require little maintenance and no fossil fuels. But that disruption will not happen quickly. Electrics remain a tiny fraction of total car sales. This imbalance is reflected in the number of models available: There are hundreds of models for gasoline-powered vehicles, but only thirty-five for electric so far. Change is coming

10.8 GIGATONS REDUCED CO2	$14.15 TRILLION NET COST	$9.73 TRILLION NET SAVINGS

more quickly in the heavy-duty market, building on a long tradition of electric trains, subways, and industrial equipment (forklifts and the like). Commercial operators are more able and willing to make the extra capital investment because costs can be amortized. Fleet operators, with depots easily retrofitted for charging purposes, are natural candidates for converting to all-electric trucks, vans, and cars. Thousands of electric buses and delivery trucks, including portions of the UPS and FedEx fleets, ply the streets of North American, Asian, and European cities. China has more than 170,000 electric buses; London's iconic double-deckers will soon join the grid.

What is the catch? With electric cars, it is "range anxiety." In order to keep the first EVs affordable, the batteries on those models were engineered to go less than 100 miles per charge. Typical today is a range of 80 to 90 miles. A hybrid plug-in can make it about 50 miles without a charge. That is long enough, Chevy says of its Volt, to make 90 percent of trips, including the daily commute. The numbers will get better. Carmakers promise a range of 200 miles for 2017.

The ultimate solution to the range issue is the network of charging stations. The global stock more than doubled between 2012 and 2014 to more than a hundred thousand charging points, and their numbers will increase dramatically with demand. The stations themselves are not that expensive, at $3,000 to $7,500 per port. They can charge the car off-peak, when electricity is cheapest, or "fuel" up when the grid has an abundance of solar or wind power. Malls and chains are installing ports at their outlets. Apps will pinpoint the closest charging stations, whether public or private. The charging network will expand, innovate, and improve, alleviating range anxiety while providing the electricity storage that the twenty-first-century power grid needs.

Projections for the electric-car market vary. Will there be 100 million on the road within several decades? A hundred and fifty million? Bloomberg takes the 2015 figure of 60 percent sales increase, projects it for the next twenty-five years, and arrives at 400 million cumulative sales by 2040, including 35 percent of all new sales. What also remains to be seen is how the natural synergy between electric cars and self-driving cars will play out, as both become software platforms on four wheels. Apple and Google are working on car design; you can be sure they will not be your standard EVs, if there is such a thing. The rate of innovation in EVs guarantees they are the cars of the future. The question for those concerned about global warming and carbon dioxide emissions is how soon the future will arrive. ●

IMPACTS: *In 2014, 305,000 EVs were sold. If EV usage rises to 16 percent of total passenger miles by 2050, 10.8 gigatons of carbon dioxide from fuel combustion could be avoided. Our analysis accounts for emissions from electricity generation and higher emissions of producing EVs compared to internal-combustion cars. We include slightly declining EV prices, expected due to declining battery costs.*

Since the Ford Model T hit the streets in 1908, people have deployed their cars' passenger capacity for more than just family and friends. In 2015, the *Oxford English Dictionary* added the verb *ride-share* to its official inventory. A new word for an old practice, ridesharing is the simple act of filling empty seats by pairing drivers and riders who share common origins, destinations, or stops en route. (It excludes taxi-like services in vehicles driven by the average Joe, which often receive the same moniker.) The first example of carpooling for the common good emerged during World War II with the advent of car-sharing clubs. "When you ride *alone*, you ride with Hitler!" Americans were told. To carpool was to conserve resources for the war effort, and employers were responsible for helping riders and drivers connect, typically via a workplace bulletin board. When the 1970s oil crisis hit, concurrent with growing public concern about air pollution, another round of employer-sponsored and government-funded initiatives proliferated. To conserve fuel, high-occupancy vehicle (HOV) lanes incentivized people to ride together, and ad hoc, informal carpools known as "slugging" took hold among commuters in Washington,

.32 GIGATONS REDUCED CO2	ZERO COST	$185.6 BILLION NET SAVINGS

At first glance, there could be no more irresponsible image than this demonstration of ridesharing. Know that the jeep was stopped and the people got on to pose for the humor of it. We show it for another reason: Vehicles and mobility are precious commodities, like timber and fisheries. People in affluent countries tend to take their cars for granted and casually use them for small details and errands. We put this image here to show how valuable mobility is and how we need to share resources if we are to have resources.

D.C., and beyond. In the 1970s, the heyday of ridesharing, one in five people carpooled to work.

By the time the U.S. Census Bureau asked about carpooling again in 2008, the trend of sharing rides to work had slacked off considerably. Just 10 percent of Americans commuted jointly, despite efforts to encourage ridesharing as a way to address traffic congestion and air quality during the 1990s and early 2000s. But thanks to global economic woes, the ubiquity of smartphones and social networks, and declining interest in car ownership among urban millennials, ridesharing is again riding high. This resurgence is timely, given the climate crisis. When trips are pooled, people split costs, ease traffic, lighten the load on infrastructure, and may reduce commuting stress, while curtailing emissions per person. For every one hundred cars being driven to work in the United States today, only five carry another commuter. Imagine the impact of shifting that number just slightly—of drivers becoming passengers a mere one day each week. Ridesharing can also make other forms of transit more viable by addressing the "first and last mile" challenge, closing the gap that often exists between point A, mass transit, and point B.

While it is not a novel idea, a new wave of technologies is accelerating ridesharing today. Smartphones allow people to share real-time information about where they are and where they are going, and the algorithms that match them with others and map the best routes are improving daily. Comfort with social networks buoys trust, so individuals are more likely to hop in with someone they have not met or open the door to strangers. By reaching the critical mass needed to ensure reliability, flexibility, and convenience, popular ridesharing platforms make it possible to find rides when and where required—a persistent limitation for ridesharing in the past. Indeed, matching kindred spirits, whether for a one-off pool trip or on a long-term basis, is the focus of numerous peer-to-peer business models. BlaBlaCar enables its 25 million members in twenty countries to share long-distance trips. UberPool and Lyft Line both group passengers along chains of pickups and drop-offs, using algorithms that link people heading the same way or to neighboring destinations. In China alone, Uber is running 20 million pooled trips each month. With a tech-based take on slugging, Google's Waze has matched commuters for carpooling in Israel since 2015, and is now piloting the concept in San Francisco. (Lyft tested a comparable commuting feature in the Bay Area, with poor results.) With a dense base of users, these companies can try interesting things, betting that if drivers can make money or save time, they will share their seats, and if riders can ride cost effectively and with ease, they will happily go dutch.

Getting people to double or triple up in their cars is not always easy. As evidenced over the past century, when fuel is cheap carpooling declines. An abundance of free or cheap parking also steers people to journey solo. So does the desire for autonomy, privacy, and expedience, although the benefits of carpooling are clear. In that sense, driving alone seems to be one form of the phenomenon sociologist Robert D. Putnam terms "bowling alone," the decline of social capital and community in modern life. Perceived safety risks may also be a deterrent, where strangers are involved. The good news is that when passengers and drivers do link up, community, connection, and engagement are catalyzed along the way. Beyond getting around, ridesharing is an invitation to imagine. For many, cars have seemed indispensable to day-to-day life. But some are beginning to conceptualize mobility as a service to access. When cars are used more collaboratively, as something shared rather than something each person must own, you can catch a glimpse of the future—one with fewer cars overall.

So what can be done to fill a car's empty seats anytime it is on the road? Macro changes in areas such as oil pricing and city design will certainly play a role in ridesharing's future, but its key to success is to become ever more dynamic, flexible, and cost-effective. That means technology will have a significant impact on ridesharing's future, just as it does on its present, not least because it can help achieve a critical mass of users. The best algorithms in the world will not work without multitudes, and though business interests may run counter, sharing data across platforms could enable the most effective matching yet. In addition to entrepreneurs and coders, employers and governments also have roles to play, just as they did in ridesharing's halcyon days gone by. Policies to promote and encourage ridesharing range from pretax programs for ridesharing expenses to reduced tolls and parking fees for carpools. Ultimately, if hopping into a car with someone can be as easy and sensible as taking your own, perhaps more so, ridesharing can become self-reinforcing—and emissions-reducing as well. ●

IMPACT: *Our projection for ridesharing focuses solely on people commuting to work in the United States and Canada, where rates of car ownership and driving alone are high. We assume that carpooling rises from 10 percent of car commuters in 2015 to 15 percent by 2050, and from an average of 2.3 to 2.5 people per carpool. Ridesharing has no implementation costs and can reduce emissions by 0.3 gigatons of carbon dioxide.*

TRANSPORT
ELECTRIC BIKES

Electric bikes are all the rage in China. The trend dates to the mid-1990s, when China's booming cities put strict antipollution rules in place in an attempt to redeem some of the world's dirtiest urban air. Tens of millions of people now commute by e-bike, and Chinese e-bike owners outnumber car owners by a factor of two. According to one expert, this is "the single largest adoption of alternative fuel vehicles in history." It is little surprise, then, that China accounts for some 95 percent of global e-bike sales, but these pedal-motor hybrids are on the rise in many parts of the world, as urban dwellers seek convenient, healthy, and affordable ways to move around their congested cities, curbing carbon emissions in the process.

Half of all urban trips are less than 6 miles, an easy distance for e-bikes. But few people live in the perfectly flat, perfectly temperate locales that make moving around by bicycle a breeze. Some are older or less able. Others face lengthy commutes or time constraints, or need to reach a destination without perspiring heavily. By lending the equivalent oomph of a strong wind at a rider's back, e-bikes make hills manageable, journeys swifter, and longer trips more viable. Those less inclined to take up a conventional bicycle might think again with a motorized boost. Indeed, as they grow more effective and affordable, e-bikes are increasingly drawing people out of more polluting modes of transportation, such as driving solo.

| .96 GIGATONS REDUCED CO2 | $106.8 BILLION NET COST | $226.1 BILLION NET SAVINGS |

The 31 million e-bikes sold in 2012 came in many shapes and forms. Some were beach cruisers with big baskets. Others were sleek and sporty, the two-wheeled analogue to a Tesla. Many looked more like scooters. Regardless of the style, they shared the same underlying technology. On e-bikes, pedals still turn a crank that moves a chain that rotates a wheel. But these quintessential bike parts do not ride alone. They are accompanied by a small battery-powered motor that can add speed—typically capped at twenty miles per hour—or assist legs when they tire. (Without speed caps, e-bikes can be too fast for safe riding in bike lanes.)

That battery, of course, gets its charge from the nearest outlet, which taps into whatever electricity is on hand, from coal based to solar powered. That means e-bikes inevitably have higher emissions than a regular bicycle or simply walking, but they still outperform cars, including electric ones, and most forms of mass transit. (Jam-packed trains or buses can, at times, do better than e-bikes on energy efficiency per passenger mile traveled.) When it comes to carbon, the mode of mobility from which a rider switches makes all the difference. As other forms of motorized transportation shift away from internal combustion engines and the grid shifts more toward renewables, the huge emissions advantage e-bikes currently enjoy will shrink, yet remain.

An e-bike's battery is at the heart of its effectiveness—and its challenges. Electric bicycles are expensive, easily five times the price of a classic bike and often more. The battery is a major driver of cost, though that can range widely depending on the type used. In China, sealed lead acid batteries dominate, making e-bikes cheaper but also creating issues of environmental contamination, especially because battery recycling is typically inconsistent at best. Lithium-ion batteries address those pollution concerns and raise performance, but are significantly more costly. As battery technology improves and achieves scale, and thus prices drop, e-bikes will become more and more attractive. To keep pace, effective battery recycling will be imperative.

Little is known about the man who first filed a patent for an electric bicycle in 1895. He was an Ohio-based inventor named Ogden Bolton, and though it was developed more than 125 years ago, his design was strikingly modern. Others were also working to motorize the popular velocipede. Today, the e-bike is following the nonmotorized bicycle's contemporary climb. In the coming years, e-bikes will benefit from the same infrastructure being built for ordinary two-wheelers and cycling's growing cultural cachet. But they pose regulatory complications their kin do not—specifically, when and where they are allowed. Because e-bikes have such variety of form and function, policy makers have struggled to define rules of the road (or cycle track). Clear, consistent regulations that make them both safe and usable will aid their growth. Electric bicycles are already the most common and fastest-selling alternative-fuel vehicles on the planet. Given that e-bikes are the most environmentally sound means of motorized transport in the world today, that popularity bodes well for their continued growth. ●

Left: A German bicycle mechanic trying out the latest electric bike from his shop in Berlin.

Below: An illustration included in the 1895 patent for an electric bike designed by Ogden Bolton, Jr., of Canton, Ohio.

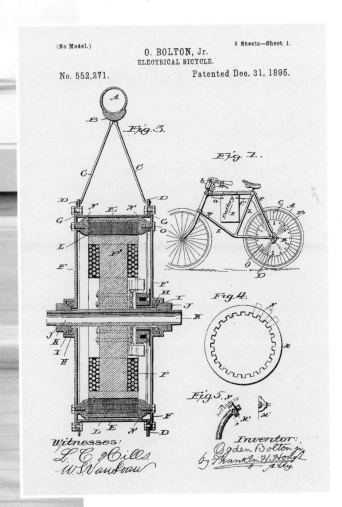

IMPACT: *In 2014, e-bike riders traveled around 249 billion miles, largely in China. Based on market research, we project travel can increase to 1.2 trillion miles per year by 2050. Shifting from cars will drive that growth, which promises to be greatest across Asia and in higher-income countries. This solution could reduce 1 gigaton of carbon dioxide emissions and save e-bike owners $226 billion by 2050.*

TRANSPORT
CARS

Worldwide, some 83 million cars rolled off the assembly line in 2013. Almost all of them contained conventional internal combustion engines—that quintessential creation of the Industrial Revolution that transforms fossil fuels into motion and emits greenhouse gases. In the United States, so-called "light duty" vehicles are responsible for more than 15 percent of annual emissions. Around the world, they play a major role in the transportation sector's overall responsibility for a quarter of emissions stemming from energy use.

Of those cars new in 2013, 1.3 million contained an electric motor and battery as well as an internal combustion engine—hybrid cars hardwired for better fuel economy and lower emissions. This marriage merges strengths and amends shortcomings. Gasoline- or diesel-powered engines excel at sustaining high speeds (highway driving) but have a harder time overcoming inertia to get moving. Electric motors are uniquely efficient at low speeds and going from stop to start. They also can keep a car's air-conditioning and accessories running while idling at a traffic light, sans engine; capture the kinetic energy typically released as heat during braking and convert it back into electricity; and boost the engine's performance, allowing it to be smaller and more efficient. Where the engine is weak, the motor is strong, and vice versa.

The pairing that gives a hybrid car its name means the internal combustion engine need only do part of the work; thus, gasoline need only provide part of the energy required. Battery-stored electricity augments it, enabling a vehicle to travel more miles for each gallon—or kilometers for each liter—and produce fewer emissions along the way. According to the International Energy Agency, hybrid cars realize fuel economy improvements of 25 to 30 percent over engine-only vehicles. (Used primarily in a city, that number moves higher.) Already on the rise, electric cars are the future. But hybrids are a key car *now*, largely because they are unhampered by the issues their full-electric kin face, from limited driving ranges to additional infrastructure needs. Hybridization is the most effective technology we have for driving up vehicle fuel efficiency until society transitions to a fleet that is not powered by fossil fuels.

Nearly synonymous with the word *hybrid*, the Toyota Prius hit car dealerships in Japan in 1997. It was the first commercially available hybrid car, but its earliest predecessor was unveiled nearly a century before. In 1900, Ferdinand Porsche built on the design of his electric vehicle, combining battery-powered wheel-hub motors and two petrol engines. Dubbed the Lohner-Porsche Semper Vivus—"always alive"—it was "able to cover longer distances purely on battery power until the combustion engine had to be engaged to recharge the batteries." The same basic technology can be found today in the Chevrolet Volt and newly minted Hyundai Ioniq. Porsche debuted his hybrid prototype at the Paris Motor Show in 1901, before refining it as the Lohner-Porsche Mixte and selling five of them by the year's end.

The technical complexity of the Mixte kept its price and maintenance costs high, and batteries of that era were expensive and heavy. Ultimately, Porsche's hybrid could not compete with conventional petrol cars.

Hybrid technology languished on the shelf for much of the twentieth century because of issues with technical complexity, batteries, and cost, as well as the cheap price of oil. Its reemergence and growth over the past two decades owe a debt to fuel efficiency standards adopted by the world's established economies and now China as well. Such standards were first established by the United States in 1975—the Corporate Average Fuel Economy (CAFE) standards. As of 2014, 83 percent of the global car market had fuel economy regulations. These obligatory benchmarks have compelled car manufacturers to wrestle with energy inefficiency. Between engine heat loss, wind and rolling resistance, braking, idling, and other drags on performance, only 21 percent of a petrol car's energy consumption propels it forward on average. Of the resulting force, 95 percent powers the car, not the driver. In essence, 99 percent of the energy used in a car is waste: It moves three thousand pounds of steel, glass, copper, and plastic in order to move a 150-pound human being.

Hybrid cars eliminate some of this inefficiency. Beyond hybridization, engines can be downsized, car bodies streamlined and crafted of lighter materials, and moving parts tweaked to reduce friction. Because these additional techniques for reducing fuel consumption result in just a few percentage points here and there, they are better complements to hybridization or full electrification than stand-alone technologies on conventional cars.

Fuel economy standards, the price of oil, new car labeling, and financial incentives such as differentiated tax rates for efficient cars influence adoption of hybrid vehicles. As fuel efficiency regulations raise the bar, hybrids—and full electrics—will command a greater share of the market. Their growth will also hinge on price; specifically, the price of batteries. Hybrids are more expensive than conventional cars, though they are becoming increasingly competitive as battery costs decline. The International Energy Agency estimates that hybridization adds $3,000 in price premium, but owners see overall savings through reduced fuel costs over a car's lifetime. Nonetheless, higher up-front cost can be prohibitive. There also is some concern that hybrids may hasten an increase in vehicle miles traveled, thus overall fuel consumption. Studies have shown, however, that this so-called "rebound effect" is typically small, just a few percentage points where personal transportation is concerned.

More than 1 billion motor vehicles exist worldwide. By 2035, there will be more than 2 billion. Despite growth in carpooling, car sharing, telecommuting, and transit, cars are not going away. People continue to be drawn to the freedom, flexibility, convenience, and comfort they offer. Can we grow the number of cars, especially in emerging economies such as China and India, while drawing down emissions? Hybrids have been called the vanguard of a revolution, catalyzing fuel efficiency and challenging the auto industry to innovate. But that is true only if they pave the way for full-electric vehicles. While 97 percent of the world's cars still contain just internal combustion engines, that number is shifting. It could shift with greater speed, heading toward all-electric motors and no engines at all. ●

IMPACT: *Under some business-as-usual projections, 23 million hybrid vehicles will be in operation in 2050, less than 1 percent of the car market. We estimate growth in 2050 to reach 6 percent of the market, or 315 million hybrid vehicles. Those additional 315 million cars can reduce carbon dioxide emissions by 4 gigatons by 2050, saving owners $1.76 trillion in fuel and operating costs over three decades.*

In 2007, General Motors introduced the Chevrolet Volt Concept Car, a plug-in electric hybrid, at the North American International Auto Show. According to GM estimates upon debut, the car's battery-powered electric motor sustains it solo for up to 40 miles, after which the combustion engine kicks in to create electricity, replenish the battery, and extend range to 640 miles. If charged overnight and driven 60 miles daily, fuel efficiency is an astonishing 150 miles per gallon.

Mobility is an undeniable social good and integral to the global economy. The pollutants that trail movement by flight—carbon dioxide, nitrogen oxides, water vapor in contrails, black carbon—are not. A century after the first commercial flight, a twenty-three-minute trip across Florida's Tampa Bay, the aviation industry has become a fixture of global transport as well as global emissions. More than 3 billion plane tickets were sold in 2013, and air travel is growing faster than any other transport mode. Both passenger and freight traffic are on the rise. (About half of air freight volume travels in the "belly" of passenger planes; the other half, in designated cargo planes.) Some twenty thousand airplanes are in service around the world, producing at minimum 2.5 percent of annual emissions. With upwards of fifty thousand planes expected to take to the skies by 2040—and take to them more often—fuel efficiency will have to rise dramatically if emissions are to be reduced.

Efficiency trends are headed in the right direction, chiefly because fuel represents 30 to 40 percent of airlines' operating costs and aircraft purchase decisions are often driven by efficiency. From 2000 to 2013, the fuel efficiency of domestic flights in the United States increased by more than 40 percent. Over the same period, fuel efficiency of international flights, which use heavier jets, improved by 17 percent. Those gains were largely thanks to fleet upgrades, while airlines also sought to maximize the number of passengers on each plane. Propulsion technologies, aerodynamic aircraft shapes, lightweight materials, and improved operational practices can push efficiency advances further.

As with all modes of transport, engines are a key area of opportunity. Jet engines work by sucking in air, which gets compressed, combined with fuel, and combusted. The energy from combustion both turns the engine's turbines and creates thrust. Industrial-strength turbofans at the front of the engine direct some air into the engine's core to feed that process. They also divert air around the engine core, improving thrust and efficiency and reducing noise. Engines with high rates of air bypass improve fuel efficiency by roughly 15 percent. For the engine maker Pratt & Whitney, adding a gear to its turbofan engine design cut fuel use by an additional 16 percent. That gear allows the engine fan to operate independently of the engine's turbine, so it can spin at an optimal velocity for better air bypass. Other companies are employing composite ceramics to reduce fuel use. Highly heat resistant, they allow fuel to combust more efficiently at hotter temperatures, while also reducing engine weight. Rolls-Royce is using strong, light carbon fiber for its newest generation of lightweight engines. More sweeping change may ultimately come from hybrid and battery-powered engines, assuming weight challenges can be addressed.

When it comes to aircraft design, changes range from minor to wholesale. What Boeing calls "winglets" and Airbus calls "sharklets"—upturned birdlike tips that improve a wing's aerodynamics—trim fuel use by up to 5 percent on both new models and retrofitted older vessels. With one fin curving up and a second curving down, split scimitar winglets (named after the curved scimitar sword) can add an additional 2 percent to that total. Winglets are currently a fundamental of efficient design.

5.05 GIGATONS	$662.4 BILLION	$3.19 TRILLION
REDUCED CO2	NET COST	NET SAVINGS

The U.S. National Aeronautics and Space Administration (NASA) is working with research universities and corporate engineering teams on a host of more sweeping advances: placement of engines, fuselage width, length, width, and placement of wings, and even comprehensive redesign of the airplane body. Boeing and NASA, for example, are collaborating on an aircraft that resembles a manta ray and seamlessly blends wings into the aircraft body. Today, a 6 percent scale model is flying in NASA's subsonic wind tunnel, but the real thing could be ready for use in a decade. The two organizations are also working on a longer, thinner, and lighter wing design with a brace or truss for added support. By moving engines to the rear of a vessel, finer wings become viable. Estimates suggest more dramatic redesigns such as these would result in efficiency gains of 50 to 60 percent. They herald the planes of the not-too-distant future.

Existing aircraft can achieve significant fuel savings with simple operational shifts, treating taxi, takeoff, and landing as uniquely fuel-consuming legs. Research out of the Massachusetts Institute of Technology identifies taxiing on a single engine, rather than both, as the most effective measure for reducing fuel use on the ground, where aircraft spend 10 to 30 percent of their transit time. Fuel burn from gate to runway or vice versa can drop 40 percent and save a single large airline $10 million to $12 million a year. Towing planes with their engines off is another tactic for efficient taxiing, though it is more time consuming. The landing methods of continuous and late descent are gaining traction. They save fuel by reducing the time planes fly at low altitudes, where efficiency is lowest. Increasingly, planes can also communicate with one another via onboard computers, effectively doing some of their own air traffic control and taking inefficient zigzags out of flight paths. Another group of researchers recently investigated the use of behavioral economics approaches with pilots, across all phases of taxi and flight. When airline captains were given monthly data on their fuel efficiency along with targets and personalized feedback, fuel-efficient practices improved by 9 to 20 percent. For each ton of carbon dioxide abated, the airline saved $250.

NASA has long been the leading experimenter in future aircraft design. They believe new designs could save airlines $250 billion in coming decades. Along with reducing fuel and pollution by 70 percent, these prototypes make 50 percent less noise than conventional passenger planes. The aircraft shown here is one of several N + 3 designs — aircraft that can be used three generations into the future. Dubbed the Double Bubble, this MIT model places three engines at the rear of a double-wide fuselage, enabling the wings to be smaller and lighter. Rear engine placement allows for smaller engines and reduced weight. Each optimization on large aircraft has cascading benefits to other components, resulting in groundbreaking efficiency.

Because airplanes will continue to be dependent on liquid fuels for the foreseeable future, investment in jet biofuels, such as those made from algae, is on the rise. The Carbon War Room (CWR) calls sustainable aviation fuels "the most challenging emissions reduction opportunity," as well as "the greatest potential for achieving carbon-neutral growth in aviation." Jet biofuel options exist today, but cost is high, supply is limited, and infrastructure is poor. CWR pinpoints airports as being pivotal to aggregate demand at scale and orchestrate supply, and the organization is working to bring a viable business model to life. For now, though, the impact biofuels could have on aviation emissions remains uncertain.

Despite the clear economic advantages of fuel efficiency for airlines, regulation also has a role to play. When the International Council on Clean Transportation (ICCT) investigated the relationship between fuel efficiency and airline profitability, it found that the relationship was not corollary, much less causal. In fact, the most profitable American airline in 2010 was its least fuel efficient. As the ICCT put it: "Fuel prices alone may not be a sufficient driver [for] efficiency. . . . Fixed equipment costs, maintenance costs, labor agreements, and network structure can all sometimes exert countervailing pressures." Requiring airlines to report their fuel efficiency data would be a first step to inform innovation and policy making. Fuel efficiency ratings by airline and route would help consumers—and investors—make more informed choices. Because operational practices vary widely from airline to airline, policies can facilitate and encourage more consistent adoption of efficient operations.

For many years, the contribution of airplanes (and ships) to climate change escaped international regulation. That changed in October 2016, when 191 nations agreed to curb aviation emissions through the Carbon Offset and Reduction Scheme for International Aviation (CORSIA). Instead of defining a cap or charge for emissions, the accord enlists airlines in a scheme—initially voluntary—to offset aviation's emissions with projects that sequester carbon. (Emissions in 2020 will be the benchmark above which most emissions must be offset.) It is meant to give airlines a greater stake in reducing emissions from their industry: By improving their fuel efficiency, airlines can avoid the cost of offsets, projected to be about 2 percent of aviation's annual revenue. For the industry to make sufficient headway, other levers for change will be needed. ●

IMPACT: *This analysis focuses on adoption of the latest and most fuel-efficient aircraft; retrofitting existing aircraft with winglets, newer engines, and lighter interiors; and retiring older aircraft early. Over thirty years, 5.1 gigatons of carbon dioxide emissions can be avoided, saving $3.2 trillion on jet-fuel and operating costs. Other efficiency measures could provide additional emissions reductions and savings.*

6.18 GIGATONS REDUCED CO2	$543.5 BILLION NET COST	$2.78 TRILLION NET SAVINGS

"The greenest gallon of gas, diesel, heating oil, or ton of coal is the one you don't burn." So said Ray Anderson, the late founder and CEO of Interface and corporate sustainability luminary. Swap the word *greenest* with *cheapest* and the same holds true. The cheapest gallon or ton is the one you do not burn—and do not have to buy. It is this combination of saving money and preempting pollution that lies at the heart of energy-efficiency measures. For the global freight trucking industry, this integration of financial and environmental benefits is particularly pertinent in the era of climate change.

Evolving from its horse-and-wagon and rail predecessors, trucking pitter-pattered along until World War I, when trucks became key to operations of the military. A combination of improved truck technology and better roads made them more viable for transport. Diesel trucks were first introduced in the 1930s, hit their stride in the 1950s, and now move roughly half of land freight. Trucks convey nearly 70 percent of all domestic freight tonnage in the United States—more than 8 billion tons annually. Even when goods move by rail or on water, they typically start and end their journeys on trucks.

Transporting all that freight, in the United States and around the world, requires diesel fuel in mass quantities. In the United States alone, trucks guzzle 50 billion gallons of diesel each year, and the role they play in greenhouse gas emissions is as oversize as they are. Making up just over 4 percent of vehicles in the United States and 9 percent of total mileage, they consume more than 25 percent of the fuel. Worldwide, road freight is responsible for about 6 percent of all emissions. Carbon emitted by transport has ballooned in recent decades, with emissions from trucking substantially outpacing that of personal transportation. Because freight activity appears to be increasing as incomes rise, and road freight emissions are projected to continue climbing, dramatic efficiency improvements are imperative.

There are two main tracks for reducing the ratio of fuel used per ton of freight moved: building it into the design of new trucks and driving it up in rigs already on the road. In 2011, the Obama administration issued the first fuel-efficiency standards for new heavy-duty trucks manufactured between 2014 and 2018. A second round aims to continue innovation and adoption of fuel-efficient technologies. These call for better engines and aerodynamics, lighter weights, less rolling resistance for tires, hybridization, and automatic engine shutdown. Top-notch automatic transmissions can overcome poor driving habits when operating manually. Based on 2010 U.S. prices, investing in a typical package of modernizations for a new truck can cost around $30,000, but save almost that much in fuel costs per year. Payback periods for some technologies are short—as little as one to two years.

Tractor-trailers remain on the road for many years, an average of nineteen in the United States, often more in lower-income countries. In light of trucks' long life, addressing the efficiency of existing fleets is essential. That is especially true in parts of the world where trucks are significantly older—and significantly less efficient. An array of measures can trim energy waste and increase fuel performance: making improvements to a truck's aerodynamics, installing anti-idling devices, making upgrades that reduce rolling resistance, altering transmissions, and integrating automatic cruise-control devices. The effect of each measure in and of itself may be relatively small, but when they are advanced together, they can make a substantial difference.

Improving existing truck efficiency is relatively low cost but delivers a big financial return on investment. According to the Carbon War Room, for a typical heavy-duty truck in the United States, reducing fuel use by 5 percent results in a yearly savings of over $4,000. Compounded cost savings matter in an industry in which the fuel tank and bottom line are tightly tied. Still, capital to make that up-front investment can be a challenge—especially for small players, who often struggle to obtain financing. Split incentives can also pose an issue: When owners who would pay for efficiency upgrades do not cover fuel costs, they have little reason to adopt them. A dearth of available, credible data on the performance of various efficiency technologies poses another barrier to adoption—one that the Carbon War Room and others are working to change.

Along with making new and existing trucks more efficient, optimizing the best routes from point A to point B, avoiding legs with empty trailers, and training and rewarding drivers for fuel frugality can decrease total miles traveled and accelerate miles per gallon. In the long-term, transitioning the industry to trucks that use low-emissions fuels or electric engines will be imperative. Making bigger trucks that can carry heavier loads could also move the needle. Along the way, society will benefit from reductions in air pollution—sulfur dioxide, nitrous oxide, and particulate matter, which plague many urban areas and impact public health. From voluntary truck retrofits to national policies that set fuel-efficiency standards, ongoing efforts to make road freight more efficient will be good for the industry and the climate. ●

IMPACT: *If adoption of fuel-saving technologies grows from 2 percent to 85 percent of trucks by 2050, this solution can deliver 6.2 gigatons of carbon dioxide emissions reductions. An investment of $544 billion to implement could save $2.8 trillion on fuel costs over thirty years.*

The Concept S truck by MAN reduces fuel consumption by 25 percent compared to conventional 40-ton trucks. The integrated truck/trailer combination is aerodynamically designed to reduce drag. It also prevents cyclists from being dragged under the wheels. The front windshield greatly increases driver visibility and safety.

A 1942 short story called "Waldo," by science fiction author Robert Heinlein, helped birth the idea of telepresence, the use of technology to interact from afar. The late Massachusetts Institute of Technology professor Marvin Minsky, a leader in the field of artificial intelligence (AI), took inspiration from the primitive system concocted by Heinlein. Such a muse seemed perfectly appropriate to Minsky, who appreciated that his own work in AI was "carried on in a world as much fiction as science" and embraced the gray area between pragmatism and imagination. He coined the term *telepresence* in a 1980 article and articulated his vision for giving an individual the feeling of being in a distant location and the ability to take action there. "Your remote presence possesses the strength of a giant or the delicacy of a surgeon," he wrote of the technology to come.

Minsky also identified the central issue that the field of telepresence continues to grapple with: "The biggest challenge to developing telepresence is achieving that sense of 'being there.' Can telepresence be a true substitute for the real thing?" Nothing beats face-to-face contact, many would argue, but telepresence aims to come exceptionally close. By integrating a set of high-performance visual, audio, and network technologies and services, people who are geographically separated can interact in a way that captures many of the best aspects of an in-person experience. Imagine Skype or FaceTime on steroids. When it is possible to exist and function remotely, the need to travel becomes less necessary; herein lies telepresence's potential impact on climate. In a world of global business footprints and international collaboration, if people can work together without being in the same place, they can dodge a host of travel-related carbon emissions. According to CDP (formerly the Carbon Disclosure Project), by activating ten thousand telepresence units, businesses in the United States and the United Kingdom could cut 6 million tons of carbon dioxide emissions by 2020—the "equivalent to the annual greenhouse gas emissions from over one million passenger vehicles" —and save almost $19 billion in the process.

The world has not come as far as Minsky imagined in 1980, but telepresence now comes to life in a variety of ways and a diversity of settings. From companies and schools to hospitals and museums, virtual interaction is opening new possibilities. Using a mobile telepresence robot, a surgeon can advise on a rare procedure in real time, without traveling from Austin to Amman. Gathering in telepresence conference rooms in Sydney and Singapore, executives can debate a possible acquisition without taking a single flight. Companies that have embraced telepresence with gusto are finding not all trips can be trimmed, but many can be. Beyond staving off carbon emissions, telepresence affords many other benefits: cost savings from avoided travel, of course, as well as less grueling schedules for employees, more productive remote meetings, the ability to make decisions more quickly, and enhanced interpersonal connection across geographies.

To achieve the fullness of these benefits, a significant initial investment is required, higher than that of standard videoconferencing. But while initial cost and ongoing expenses are higher for telepresence systems, they tend to be used much more heavily, making the cost per use commensurate. Payback happens quickly—in as little as one to two years. Telepresence also depends on strong network infrastructure, skilled technical support, and dedicated space if specific meeting rooms are used. Once telepresence technology is installed, companies can encourage employees to use it by educating them, establishing policies around avoiding travel, and tracking and rewarding its use. Costs are going down, while simplicity, reliability, and efficacy are on the rise, but the adoption of technology and accompanying behavioral change—using it and using it well—still takes time. That upward adoption curve should steepen as these trends continue, improved technologies come online, the pressure to reduce costs and emissions builds, and more people have positive telepresence experiences. More and more, we will be able to go to work without going anywhere at all, and potential carbon emissions will stay put too.

IMPACT: *By avoiding emissions from business air travel, telepresence can reduce emissions by 2 gigatons of carbon dioxide over thirty years. That result assumes that over 140 million business-related trips are replaced by telepresence in 2050. For organizations, the investment in telepresence systems pays off with $1.3 trillion worth of savings and 82 billion fewer unproductive travel hours.*

A staff member from PricewaterhouseCoopers in Toronto, waving to a team member from Prague. The mobile, two-wheeled scooter can travel around the office so that the staff member from Prague can converse and meet with other people in the Toronto office at will.

.52 GIGATONS	$808.6 BILLION	$313.9 BILLION
REDUCED CO2	NET COST	NET SAVINGS

Trains may ride the rails, but they run on fuel. Most rely on diesel-burning engines; some tap into the electric grid. Trains have steadily improved their fuel-use efficiency in recent decades. Between 1975 and 2013, energy consumption declined by 63 percent for passenger rail and 48 percent for freight services. Emissions dropped by 60 percent and 38 percent respectively. Still, in 2013, rail was responsible for 3.5 percent of emissions within the transport sector—more than 260 million tons of carbon dioxide. Continuing to improve the efficiency with which railways move 8 percent of the world's passengers and goods is essential.

Railway companies already employ a range of technical and operational measures. As locomotives are retired, more efficient models replace them, many with more aerodynamic designs. In some cases, those models include hybrid diesel-electric engines and batteries, which gain efficiencies similar to those of hybrid cars, saving 10 to 20 percent on fuel. Some trains are being equipped with regenerative braking systems that capture and use energy otherwise lost as heat, as well as "stop-start" technology that curbs fuel use during idling—much like efficient cars do. The U.S. passenger service Amtrak reduced energy consumption by 8 percent with regenerative braking. Distributing the power of locomotives throughout a train also improves fuel use.

Better locomotives, more strategically placed, are enhanced by better cars—lighter, more aerodynamic, able to hold more cargo, and equipped with low-torque bearings. Eliminating gaps between cars reduces drag, while longer, heavier trains often prove more efficient. The rails themselves can be better lubricated to reduce friction. Even with hyperefficient design, how a train is driven remains critical. Software can control train speed, spacing, and timing, as well as provide efficiency information and "coaching" to locomotive engineers, improving performance.

The number of electric trains is increasing, but to what extent that reduces emissions depends on the efficiency of the grid supplying the power. According to the International Energy Agency, "rail electrification can lead to an efficiency gain of around 15 percent on a life-cycle basis." As electricity production shifts to renewables, rail has the potential to provide nearly emissions-free transport.

In the meantime, improving the fuel efficiency of trains, whether diesel powered or electric, reduces cost and makes them more competitive, especially for moving freight. As the Rocky Mountain Institute notes, "[Trains], one of the [world's] oldest transportation platforms . . . can move four times more ton-miles per gallon than trucks, typically at a lower cost." Cost advantages may encourage companies to move freight by train rather than by truck, thereby reducing emissions from the mass movement of goods. (Of course, until electricity generation shifts to renewables, a central paradox will persist: Many freight trains carry

A General Electric Evolution Series Tier 4 locomotive before being painted at its factory in Fort Worth, Texas. This family of diesel-electric locomotives is among the most efficient in the world with respect to emissions, achieving a 70 percent reduction in particulate matter and nitrous oxide compared to Tier 3 predecessors. (Tier 4 is the U.S. Environmental Protection Agency standard for new locomotives, effective since January 1, 2015.) This 440,000-pound behemoth can move one ton of freight 500 miles on one gallon of fuel. Sensors throughout the engine gather real-time data to diagnose and improve performance and efficiency. Among other places, many Tier 4 locomotives can be found hauling freight along the rail corridor that runs between Los Angeles and Seattle.

coal and oil, so greater efficiency could benefit fossil fuel companies' bottom lines.)

When the steam locomotive came into public use in England in the early nineteenth century, a single locomotive could transport six coal cars and 450 passengers nine miles in under an hour. Compared to a horse-drawn vehicle, the speed was staggering. Today, a diesel locomotive can carry one ton of freight more than 450 miles on a single gallon of fuel. In 1980, a gallon of diesel would have taken that freight just 235 miles. Together, China, the European Union, India, Japan, Russia, and the United States are responsible for some 80 percent of rail-sector emissions, which means a small set of policy interventions could have an outsize impact. As trains continue to transport 28 billion passengers and more than 12 billion tons of freight annually, it is time for the whole industry to follow its most efficient leaders. ●

IMPACT: *Globally, electrification of rail comprises 166,000 miles of track length. If that increases to 621,000 miles by 2050, emissions from fuel use for freight operations alone can be reduced by 0.5 gigatons of carbon dioxide. This additional electrification could cost $809 billion, saving $314 billion over thirty years and $775 billion over the lifetime of the infrastructure. Prioritizing high-usage corridors can lower net costs.*

MATERIALS

The most important insight about materials in the twentieth century was biologist John Todd's, when he coined the phrase "Waste equals food." That happens to be the exact practice of all living systems, but at the time Todd made his observation, it was in stark contrast to the realities of the manufacturing world. Industry has come a long way since then, with responsible companies now paying close attention to where they source their materials and what happens to them after the useful life of their products. That being said, society is at the very beginning of redesigning and reimagining the materials used in products and structures, as well as the means by which they can be reduced, reused, and recycled. The newest discoveries are not covered here, of course, but this section details the common techniques and technologies critical to the effort to reverse global warming. To underline that fact, the number-one solution is contained within this sector.

HOUSEHOLD RECYCLING

Recycling did not need a name before the twentieth century. In an effort to stretch limited resources, people avoided waste, fixed things that broke, and found ways to give other items a second life. Using the term in the context of waste management did not start until the 1960s, but it quickly became a hallmark of the modern environmental movement thereafter. Pollution Probe, an early and influential Canadian environmental group, coined the phrase "Reduce, reuse, recycle." The "3Rs" became the mantra for addressing the challenge of consumer waste and limiting the stream of materials headed for landfills and incinerators—first reduce, then reuse, *then* recycle. Household recycling is now a meaningful way to direct materials back into value chains and, in the process, mitigate climate change.

As quickly as the world is urbanizing, urban waste is growing faster. Waste production multiplied tenfold over the past century, and experts expect it to double again by 2025—a by-product of rising incomes coupled with rising consumption. Half or less of that waste is generated at the household level, and managing it tends to be the responsibility of local government. Lower-income cities are the exception to that rule, with largely informal systems of waste pickers rather than high-touch and high-tech collection and processing. The stream of discarded

---

The Dassanach people of Sudan are one of the more intact cultural groups in the world. Once pastoralists, they now are primarily farmers due to the loss of their native grazing lands. Traditional or not, the Dassanach women are astonishingly creative in recycling throwaways into headdresses and necklaces made from bottle caps, watch bands, and SIM cards. With small towns and bars springing up near the Omo River settlement, bottle caps have become abundant—so abundant that the women are beginning to sell their headdresses to visiting tourists.

| 2.77 GIGATONS REDUCED CO2 | $366.9 BILLION NET COST | $71.1 BILLION NET SAVINGS |

items includes food, yard waste, paper, cardboard, plastics, metals, clothes, diapers, wood, glass, ashes, batteries, household electronics, paint cans, motor oil, bulk items, and then some. Though the mix varies widely from place to place, in high-income countries paper, plastic, glass, and metal comprise more than half the waste stream—and they are all prime candidates for recycling. (Many less prevalent items should be recycled because of their toxic nature or high-value components.)

Whether recyclable household waste *is* recycled has implications for greenhouse gas emissions, as producing new products from recovered materials often saves energy, in addition to reducing resource extraction, minimizing other pollutants, and creating jobs. Forging recycled aluminum products, for example, uses 95 percent less energy than creating them from virgin materials. Of course, even the most efficient recycling, such as aluminum, is not without emissions of its own. Collection, transport, and processing are, for the time being, largely powered by fossil fuels. Still, when that pollution is taken into account, recycling remains an effective approach to managing waste while addressing emissions.

The process of diverting and recovering waste material is sometimes called *valorization*. The term refers to extracting the value that an item retains when it is thrown away ("away" being a misnomer). Recycled materials actually have two sources of value, as commodities but also as sinks. The first is what typically comes to mind: the fiber remaining in paper, for example, that can be reprocessed into recycled pulp. This commodity value keeps waste pickers picking, sparks recycling start-ups, and sends pallets of compressed plastic bottles from Boston or Buenos Aires to China. It spurs global markets for recyclable materials. The second, often-ignored value of recycling is as a sink. It absorbs the economic, social, and ecological costs otherwise incurred by sending waste to landfills or incinerators. In both of these ways diversion creates value, saving on a range of costs and—especially in the case of metals and paper—creating income.

Recycling rates—which measure the fraction of waste successfully redirected and often include compost—vary widely across the world's cities. Bringing the laggards into line with the front-runners is the opportunity at hand. Interestingly, the recycling rates of many low-income cities and their informal systems are already competitive with what formal systems accomplish in high-income countries. Both Delhi, India, and Rotterdam, Netherlands, hover around a third. San Francisco and Adelaide, Australia, are often cited as leaders that achieve 65 percent or more, but so do Quezon City, Philippines, and Bamako, Mali. It is important to note that informal recycling often supports livelihoods of the urban poor (though not without health implications) and saves resource-strapped cities money on waste management. Microenterprises such as Nigeria's Wecyclers, which provides household recycling services by cargo bike, are increasingly important actors.

Pioneering high-income cities have learned a lot about making formal residential recycling a success. Raising public awareness is necessary but never sufficient. While there is no surefire formula, the most effective systems make collection easy and use incentives to nudge behavior. Pay-as-you-throw programs, such as the one used in San Francisco, bill households for rubbish sent to landfill but carry away recycling and compost for free. (San Francisco also includes clothing, a rapidly growing but often-overlooked waste stream, in its recycling mix.) Mechanisms that require consumers to pay a redeemable deposit at purchase can be applied broadly, from bottles to electrical goods, and also raise recovery rates. One common approach has produced mixed results. Many municipalities now provide large curbside bins, in an effort to accommodate greater volumes of single-stream recycling with all materials intermingled. The additional space has fostered more "creative" and "wishful" recycling—a garden hose here, Styrofoam containers there—producing contamination that makes recycling more expensive to process.

Household recycling faces another emerging challenge: the makeup of garbage itself. Packaging from soda bottles to baby food containers has undergone "light-weighting." New designs require fewer raw inputs and reduce shipping costs (and often greenhouse gas emissions). At the same time, they can be hard to recycle and it takes many more of them to reach the same volume of salable commodities. Once a reliable staple for recyclers' income, newspaper volumes have plunged. These changes couple with the inevitable volatility of global commodity markets to keep the industry on its toes. Nonetheless, the movement for "zero waste" continues. Adoption of the Green Dot or Der Grüne Punkt labeling system, introduced in Germany, continues to grow, gathering funds from manufacturers to cover recovery and recycling costs. Also on the rise are stronger targets for municipal recycling rates, such as the European Union's proposed 65 percent by 2030. Recycling and the other two Rs—efforts to reduce and reuse, first and foremost—will be key elements of managing waste without further warming the world. ●

IMPACT: *The household and industrial recycling solutions were modeled together and include metals, plastic, glass, and other materials, such as rubber, textiles, and e-waste. Paper products and organic wastes are treated in separate waste management solutions. Emissions reductions stem from avoiding emissions associated with landfilling and from substituting recycled materials for virgin feedstock. With about 50 percent of recycled materials coming from households, if the average worldwide recycling rate increases to 65 percent of total recyclable waste, household recycling could avoid 2.8 gigatons of carbon dioxide emissions by 2050.*

MATERIALS
INDUSTRIAL RECYCLING

In 2012, the Interface Corporation, a global company that makes carpet tiles, formed a partnership with the Zoological Society of London to explore an unusual question: How could making carpets address inequality in the world? The answer can be found in these two images. Working with people in coastal communities in the developing world, Interface purchased discarded fishing nets strewn on reefs and atolls—some of the 640,000 tons of abandoned fishing gear in the sea that continue to catch and kill fish (ghost fishing). Until now, local communities had no sustainable way of recycling or disposing of used fishing nets.

The initiative is called Net-Works and at the heart of the program are community banks that help manage funding, loans, coastal cleanups, deposits from sales, and finance of local conservation projects. The discarded fishing nets are processed by Aquafil, which converts the nylon from waste into 100 percent recycled carpet yarn. Interface then incorporates the yarn in a series of designs, one of which you see here, simulating the water where the nets were retrieved. As of 2016, Net-Works has been established in 35 communities, collected 137 tons of waste nets, and given 900 families access to micro-loans and banking.

ake, make, waste—the modus operandi of the industrial era. *Take* the resources needed, *make* them into things, discard the by-product, and, eventually, consign the used goods to *waste*. Today, a new circular way of thinking is beginning to replace that logic. In nature, cycles abound. Water and nutrients move in closed loops, and there is no waste. Instead, discards become resources. Drawing on nature's wisdom, circular business models look at old goods and scrap materials as valuable resources for new products. They begin to redirect the linear flows that start with raw materials and end at landfills and incinerators, making the industrial system function more like an ecosystem instead. Companies can send their waste for recycling but also can be recyclers themselves. By reducing material use to begin with and recycling and reusing waste, they can reduce greenhouse emissions from extracting, transporting, and processing raw materials. And because the global economy currently uses far more of these materials far more quickly than the earth can regenerate, such practices address parallel challenges of resource scarcity.

At least half of waste is generated outside households, and sometimes much more. The sources of industrial and commercial waste are myriad: manufacturing of all stripes, construction sites, mines, energy and chemical plants, stores, restaurants, hotels, office buildings, sports and music venues, schools, hospitals, prisons, airports, and more. They are all sites of use and discharge. The stream of waste they produce includes the usual from food and landscaping, as well as textiles, paper, cardboard and other packaging, plastic, glass, and metal. It also contains huge volumes of industrial solid waste, such as concrete, steel, wood, ashes, and tires, as well as electronic waste—the computers, screens, printers, phones, and more that are the detritus of the Information Age and that contain toxics including mercury, lead, and arsenic. (The majority of the world's e-waste lands in low-income countries, where both regulation and enforcement are lax and black markets are rampant.) Not all of this waste can find a second life, at least not yet, but much of it can.

A suite of efforts is helping to close the loop on commercial and industrial waste (some impact household waste as well). Extended producer responsibility (EPR) is an increasingly popular policy approach that makes companies responsible not just for creating goods but for managing them post-use. Otherwise, the public bears the brunt of disposal. EPR can be purely financial, charging producers for the cost of recovery and recycling; it can be physical as well, getting them directly involved in that process. Since 2006, the Dutch have used EPR for packaging. Where they exist, producer "take-back" laws help address e-waste. Companies such as carpet tile manufacturer Interface voluntarily seek to retrieve their product, so discarded tiles can provide feedstock for new ones. The outdoor clothing company

Patagonia collects "worn wear" for repair or, if too far gone, for recycling. But voluntarily taking such responsibility is unusual. Formalizing it encourages companies to think *now* about what will happen *then* and make their products longer lasting, easier to fix, and as recyclable as possible. In other words, while recycling happens at end of life, it is best considered from the beginning.

Enhancing the exchange of recyclable and reusable goods is essential. As a step in this direction, the U.S. Materials Marketplace was launched in 2015 as a matchmaker for secondary materials. The initiative actively identifies opportunities and links the relevant parties, brokering transactions between companies if need be. In parallel, the science and processes of recycling have to evolve. Writing in the journal *Nature*, Swiss architect Walter Stahel urges, "To close the recovery loop we will need new technologies to de-polymerize, de-alloy, de-laminate, de-vulcanize, and de-coat materials." Innovative conversion technologies can increase recycling rates significantly. Of course, recycling is just one piece of the integrated strategy needed: swapping virgin materials with recycled ones, making more efficient use of materials,

and extending product life through good design and solid construction. Trash cannot always become treasure, but a growing body of evidence suggests significant environmental and economic gains can be realized when that transformation is managed and circularity is embedded into industry. ●

IMPACT: *As mentioned above, household and industrial recycling were modeled together. The total additional implementation cost of both is estimated at $734 billion, with a net operational savings of $142 billion over thirty years. On average, 50 percent of recyclable materials come from industrial and commercial sectors. At a 65 percent recycling rate, the commercial and industrial sectors can avoid 2.8 gigatons of carbon dioxide by 2050.*

Women from the collection hub in the Bantayan Islands are examining the fruits of their labor, carpet tiles made from 100 percent recycled fishing nets. The women clean, weigh and sort the nets, after which they are baled and stored, ready for export to Cebu City.

MATERIALS
ALTERNATIVE CEMENT

Centuries before the Hoover and Grand Coulee dams were constructed in the American West, feats of concrete engineering gave rise to Roman bridges, arches, coliseums, and aqueducts. Roman concrete was used in creating the magnificent Pantheon temple in Rome. Completed in 128 AD, it is famed for its five-thousand-ton, 142-foot dome made of unreinforced concrete—still the world's largest almost two thousand years later. If it had been built with today's concrete, the Pantheon would have crumbled before the fall of Rome, three hundred years after its dedication. Roman concrete contained an aggregate of sand and rock just like its modern kin, but it was bound together with lime, salt water, and ash called pozzolana, from a particular volcano. Blending volcanic dust into the mixture of *opus caementicium* even enabled underwater construction.

The art and science of concrete largely fell away with the Roman Empire itself, until it was revived and evolved in the nineteenth century. Today, concrete dominates the world's construction materials and can be found in almost all infrastructure. Its basic recipe is simple: sand, crushed rock, water, and cement, all combined and hardened. Cement—a gray powder of lime, silica, aluminum, and iron—acts as the binder, coating and gluing the sand and rock together and enabling the remarkable stonelike material that results after curing. Cement is also employed in mortar and in building products such as pavers and roof tiles. Its use continues to grow—significantly faster than population—making cement one of the most used substances in the world by mass, second only to water.

While cement is a source of strength in infrastructure, it is also a source of greenhouse gas emissions. To produce Portland cement, the most common form globally, a mixture of crushed limestone and aluminosilicate clay is roasted in a giant kiln at about 2,640 degrees Fahrenheit. Doing so triggers a reaction to break apart the limestone's calcium carbonate, splitting it into calcium oxide, the desired lime content, and carbon dioxide, the waste. What comes out the kiln's other side are small lumps called "clinker," which are then cooled, combined with gypsum, and milled into the flourlike powder we know as cement. Decarbonizing limestone causes roughly 60 percent of the cement industry's emissions. The rest are the result of energy use: Manufacturing a single ton of cement requires the equivalent energy of burning four hundred pounds of coal. Add those emissions up and for every ton of cement produced, nearly one ton of carbon dioxide puffs skyward. In total, the industry produces roughly 4.6 billion tons of cement each year, more than half of it in China, and generates 5 to 6 percent of society's annual anthropogenic carbon emissions in the process.

More efficient cement kilns and alternative kiln fuels, such as perennial biomass, can help address the emissions from energy consumption. To reduce emissions from the decarbonization process, the crucial strategy is to change the composition of cement. Conventional clinker can be partially substituted for alternative materials that include volcanic ash, certain clays, finely ground limestone, and industrial waste products, namely: blast furnace slag, a by-product of making iron that was used in constructing the Empire State Building and Paris Metro, and fly ash, a powdery residue from coal-burning power plants that found its way into the Hoover Dam. Because these materials do not require

6.69 GIGATONS REDUCED CO2	-$273.9 BILLION FIRST COST	DATA TOO INDEFINITE TO BE MODELED

ash cement is a good use of the by-products—far better than sending them to a landfill or holding pond. Availability is a key factor. Regionally, it varies, and where coal-fired power plants are going off-line, fly ash can be hard to come by. Mining landfills for fly ash from the past may be a potential future source, albeit a more costly one. The cost of transportation and inconsistency of quality are also determinants in giving fly ash new life as a clinker substitute. Questions persist around the implications of fly ash for human health. As a coal by-product, it contains toxins and heavy metals. Scientists continue to research whether those components are held safely within concrete or might leach out, as well as what risks might arise at the end of a structure's life.

According to the United Nations Environment Programme, the average global rate of clinker substitution could realistically reach 40 percent (accounting for all alternative materials) and avoid up to 440 million tons of carbon dioxide emissions annually. Depending on their particular composition, alternatives to Portland cement offer benefits beyond the atmosphere: They can be more workable, less water intensive, denser, more resistant to corrosion and fire, and longer lasting. Though they can be slower to set and not as strong early on, their ultimate strength can actually be higher.

Governments and corporations have begun to concretize the possibilities of clinker substitutes. With regional standards, the European Union reuses most of its available fly ash. Prior to those policy changes, utilization rates varied widely and were as little as 10 percent in some places. New York City has embraced ground bottle glass as an emerging substitute that can be sourced regionally and saves landfill space—an innovation that may be poised for growth. From municipal to international levels, standards and product scales are key for shifting practices within the construction industry and advancing the use of alternative cements in sidewalks and skyscrapers, roads and runways. ●

IMPACT: *Because fly ash is a by-product of burning coal, each ton created is accompanied by 15 tons of carbon dioxide emissions. Using fly ash in cement can offset only 5 percent of those emissions. Even so, if 9 percent of cement produced between 2020 and 2050 is a blended mix of conventional Portland cement and 45 percent fly ash, 6.7 gigatons of carbon dioxide emissions could be avoided by 2050. The production savings of $274 billion are largely a result of longer cement life span.*

kiln processing, they leapfrog the most carbon-emitting, energy-intensive step in the cement production process. Already, more than 90 percent of blast furnace slag is used as clinker substitute. One-third of fly ash is, and that portion could grow. Fly ash and Portland clinker can be mixed together at various ratios depending on the cement's final use and type of fly ash used, with fly ash regularly comprising 45 percent of the blend.

Ultimately, the world will move away from coal power and its attendant emissions, but as long as coal is being burned, fly

The Pantheon was a Roman temple commissioned during the consulship of Marcus Agrippa 2,000 years ago and completed by the emperor Hadrian about 128 AD. After nearly two millennia, the dome remains the largest unreinforced concrete dome in the world. What is more remarkable is that the concrete remains intact, strong and almost ageless. Standing in what is now a church, the oculus at the center of the dome rises 142 feet. Six million people visit it every year.

MATERIALS
REFRIGERATION

Every refrigerator, supermarket case, and air conditioner contains chemical refrigerants that absorb and release heat, making it possible to chill food and keep buildings and vehicles cool. Refrigerants, specifically chlorofluorocarbons (CFCs) and hydrochlorofluorocarbons (HCFCs), were once key culprits in depleting the stratospheric ozone layer, which is essential for absorbing the sun's ultraviolet radiation. Thanks to the 1987 Montreal Protocol on Substances That Deplete the Ozone Layer, CFCs and HCFCs have been phased out of use (along with the ozone-depleting chemicals that used to be standard fare in aerosol cans and dry cleaning). It took two short years from discovery of the gaping hole over the Antarctic for the global community to adopt a legally mandated course of action. Now, three decades later, the ozone layer is beginning to heal.

Refrigerants continue to cause planetary trouble, however. Huge volumes of CFCs and HCFCs remain in circulation, retaining their potential for ozone damage. Their replacement chemicals, primarily hydrofluorocarbons (HFCs), have minimal deleterious effect on the ozone layer, but their capacity to warm the atmosphere is one thousand to nine thousand times greater than that of carbon dioxide, depending on their exact chemical composition.

In October 2016, officials from more than 170 countries gathered in Kigali, Rwanda, to negotiate a deal to address the problem of HFCs. Despite challenging global politics, they reached a remarkable agreement. Through an amendment to the Montreal Protocol, the world will begin phasing HFCs out of use, starting with high-income countries in 2019 and then expanding to low-income countries—some in 2024, others in 2028. HFC substitutes are already on the market, including natural refrigerants such as propane and ammonia.

Unlike the Paris climate agreement, the Kigali deal is mandatory, with specific targets and timetables for action, trade sanctions to punish failure to comply, and commitments by rich countries to help finance the cost of transition. It was a monumental achievement on the path to drawdown, called by then secretary of state John Kerry "the biggest thing we can do [on climate] in one giant swoop." Scientists estimate the accord will reduce global warming by nearly one degree Fahrenheit.

Still, the process of phasing out HFCs will unfold over many years, and they will persist in kitchens and condensing units in the meantime. With adoption of air-conditioning soaring, especially in rapidly developing economies, the bank of HFCs will grow substantially before all countries halt their use. According to the Lawrence Berkeley National Laboratory, 700 million air-conditioning units will have come online worldwide by 2030. All of this means parallel action is requisite: addressing the refrigerants coming out of use, as well as transitioning those going in.

89.74 GIGATONS REDUCED CO2	DATA TOO VARIABLE TO BE DETERMINED	-$902.8 BILLION NET SAVINGS

Refrigerants currently cause emissions throughout their life cycles—in production, filling, service, and when they leak—but their damage is greatest at the point of disposal. Ninety percent of refrigerant emissions happen at end of life. If the chemicals (or appliances that use them) are not disposed of effectively, they escape into the atmosphere and cause global warming. On the other hand, refrigerant recovery has immense mitigation potential. After being carefully removed and stored, refrigerants can be purified for reuse or transformed into other chemicals that do not cause warming. The latter process, formally called destruction, is the one way to reduce emissions definitively. It is costly and technical, but it needs to become standard practice.

In less than a century, air-conditioning in the United States went from being a luxury good to a widespread commodity. Today, 86 percent of U.S. homes have systems that provide cool air. They became common, if not universal, in urban Chinese households in just fifteen years. And why would they not? In seasons of heat and humidity, air-conditioning increases comfort and productivity and can save lives during heat waves. And yet, a great irony of global warming is that the means of keeping cool make warming worse. As temperatures rise, so does reliance on air conditioners. The use of refrigerators, in kitchens of all sizes and throughout "cold chains" of food production and supply, is seeing similar expansion. As technologies for cooling proliferate, evolution in refrigerants and their management is imperative. The Kigali accord ensures a step change is coming, and other practices focused on existing stocks could reduce emissions further. ◉

IMPACT: *Our analysis includes emissions reductions that can be achieved through the management and destruction of refrigerants already in circulation. Over thirty years, containing 87 percent of refrigerants likely to be released could avoid emissions equivalent to 89.7 gigatons of carbon dioxide. Phasing out HFCs per the Kigali accord could avoid additional emissions equivalent to 25 to 78 gigatons of carbon dioxide (not included in the total shown here). The operational costs of refrigerant leak avoidance and destruction are high, resulting in a projected net cost of $903 billion by 2050.*

Left: Mario José Molina-Pasquel Henriquez is a Mexican chemist who was awarded the Nobel Prize for Chemistry in 1995 for his role in unraveling and explaining the threat chlorofluorocarbon gases (CFCs) posed for the ozone layer. His work with Nobel corecipient Sherwood Rowland led to the discovery of how CFCs persist in the atmosphere and how the off-gassing chlorine atoms destroy atmospheric ozone. From their work came the Montreal Protocol on Substances That Deplete the Ozone Layer, banning CFCs. Ultimately, 197 nations have adopted the 2016 Kigali Amendment to the Montreal Protocol, an agreement to phase out hydrofluorocarbons (HFCs) by 2028. HFCs are largely innocuous to the ozone layer, but they are one of the most potent greenhouse gases known to humankind.

Below: Downtown Singapore, showing the ubiquity of air-conditioning units on Asian streets.

Photographer Chris Jordan created a mandala in 2011 from 9,600 mail order catalogs. It represents the number of catalogs printed, shipped, and delivered every three seconds, 97 percent of which are disposed of the day they arrive. This is part of a larger series titled "Running the Numbers: An American Self-Portrait." This piece is called *Three Second Meditation*.

| .9 GIGATONS REDUCED CO2 | $573.5 BILLION FIRST COST | DATA TOO INDEFINITE TO BE MODELED |

Keeping accounts. Capturing stories. Sharing information. Recording history. Exploring ideas. To be human is to communicate, and for two millennia paper has been a prime vehicle for doing so, originating in China and gradually spreading westward. Since the industrialization of papermaking in the nineteenth century, paper has been a widespread, inexpensive commodity. Even with electronic media diverting some need for print, paper use globally is on the rise, particularly for packaging materials. Today, roughly half of paper is used once and then sent to the proverbial scrap heap. But the other half is recovered and repurposed. In Northern Europe, that recovery rate reaches 75 percent. South Korea achieved a recovery rate of 90 percent in 2009. Bringing the rest of the world up to that level of paper recycling, or beyond, presents a significant opportunity to draw down the emissions of the paper industry, which are estimated to be as high as 7 percent of the world's annual total—higher than that of aviation.

Paper recycling rewrites the typical life cycle of paper. It makes paper's journey circular, rather than a straight line from logging to landfill. For the standard piece of paper, created from a pine tree's biomass, there are emissions at every stage of its journey: sourcing, manufacturing, transportation, use, and disposal. But recycled paper can intervene and change the emissions equation, especially at the beginning and at the end, by linking those stages. Instead of relying on fresh timber to feed the pulping process—and releasing carbon with each tree cut—recycled paper draws on existing material, either discarded before reaching a consumer's hands or, ideally, after serving its intended purpose as a magazine or memo. Instead of releasing methane as it decomposes in a dump, wastepaper finds new life. It is viewed not as trash but as a valuable resource—too valuable to send to the landfill or incinerator.

Once recovered, used paper can be reprocessed. Shredded, pulped, cleaned, and rid of contaminants such as staples and coatings, paper that might have been buried in a landfill can become any number of products, from office paper to newsprint to toilet paper rolls. Unlike some recyclable materials, such as aluminum, paper cannot be recycled indefinitely into the same quality of product. Its fibers break down over time, so wastepaper intrinsically becomes a lower-quality product, for which shorter, weaker fibers are suited. A particular piece of paper can be reprocessed roughly five to seven times. Even so, recycling is an effective and efficient alternative to making paper solely from virgin materials.

The benefits of recycled paper are many. Forests are spared, keeping habitats intact and perhaps protecting ancient ecological treasures. Water use is reduced, relieving pressure on a resource that is increasingly threatened. And fewer bleaches and chemicals find their way into waterways. Studies show recycling creates more jobs and produces more economic value than landfilling or incineration. Most important, recycled paper produces far fewer greenhouse gas emissions than its virgin counterpart. Exactly what those climate savings are depends on materials used, the feedstocks they supersede, and what end-of-life treatment is avoided. Of course, making any paper requires energy of some kind, as does transportation of raw material and final product. It matters equally for virgin and recycled pulp, whether mills run on renewables and sustainable transportation options are elected.

A study of studies, conducted by the European Environmental Paper Network, calculates that virgin-fiber paper emits an average of 10.67 tons of carbon dioxide (or its equivalent in other greenhouse gases) per ton of paper product, while recycled paper comes in at just 2.92 tons. That is more than a 70 percent difference. A recent life cycle assessment compares postconsumer recycled paper to its virgin alternatives. The analysis finds that production of recycled paper generates just 1 percent of the climate impacts virgin paper creates. Moreover, it consumes a quarter of the amount of water required for the same quantity of product, and requires 20 to 50 percent less energy for pulping and papermaking.

As a complement to reducing paper use overall, the case for recycled paper is clear. The process is more efficient, requiring fewer resources upstream and producing less waste and emissions downstream. As more wastepaper gets recovered and recycled, the need to log and landfill or incinerate drops. But to take recycled paper to the scale that is possible, cost has to come down. That can happen as production grows. Policies that make conventional waste disposal less attractive and more expensive can boost recycling. And those that disadvantage recycling, such as subsidies for less sustainable alternatives, should be addressed. Customer demand, from retail to wholesale, is also vital to shift the industry's investments in that direction. If the chorus of concern grows, there is no reason recycled paper cannot claim a dominant share of the market. ●

IMPACT: *Over thirty years, recycled paper can deliver .9 gigatons of carbon dioxide emissions reductions. Two key assumptions inform that conclusion: (1) recycled paper produces about 25 percent fewer total emissions than conventional paper, and (2) the percentage of recycled paper being used to produce paper would rise from 55 percent to 75 percent by 2050. Although increasing recycled paper content uses more electricity, the emissions related to harvesting and processing—and the total emissions from pulping and manufacturing—are higher for paper using virgin wood feedstock. The emissions reductions for this solution do not include carbon sequestration from standing trees that would not be harvested if the use of recycled paper grows.*

MATERIALS
BIOPLASTIC

From the Stone Age to the Iron Age to the Steel Age, we delineate society's epochs by their primary material for fabrication. Ours could be called the Age of Plastic. Globally, we produce roughly 310 million tons of plastic each year. That is 83 pounds per person, and plastic production is expected to quadruple by 2050. The material is everywhere, from clothing to computers, furniture to football fields, and almost all of it is petro-plastic, made from fossil fuels. In fact, 5 to 6 percent of the world's annual oil production becomes feedstock for plastic manufacturing. But the polymers that make up plastic exist everywhere in nature, not just as fossilized forms, and experts estimate that 90 percent of current plastics could be derived from plants or other renewable feedstock instead. Such bio-based plastics come from the earth and many can return to it, often with lower carbon emissions than their fossil fuel–based kin.

The Greek verb *plassein*, the root of plastic, means "to mold or shape." What affords plastics their malleability are polymers—substances with chainlike structures, made of many atoms or molecules bound to one another. Most have a backbone of carbon, linked with other elements such as hydrogen, nitrogen, and oxygen. We can synthesize polymers, but they also occur naturally all around and inside us; they are part of every living organism. Cellulose, the most abundant organic material on earth, is a polymer in the cell walls of plants. Chitin is another abundant polymer, found in the shells and exoskeletons of crustaceans and insects. Potatoes, sugarcane, tree bark, algae, and shrimp all contain natural polymers that can be converted to plastic.

Although petro-plastics now dominate the market, the material for the earliest plastics was plant cellulose. In the nineteenth century, playing billiards was de rigueur for the well-to-do in the United States and Europe, and the balls that adorned billiard tables were 100 percent solid ivory. The market was voracious; elephants were slaughtered by the thousands for their tusks, each the source for merely a handful of billiard balls. The trend prompted public outcry, while driving up costs for the billiards industry. Billiards player and tycoon Michael Phelan issued a challenge: $10,000 in gold to anyone who could develop an alternative to ivory. The offer prompted printer and tinkerer John Wesley Hyatt to begin testing possibilities. He developed a substance from the cellulose in cotton, dubbed "celluloid." Celluloid turned out to be less than ideal for billiard balls—Hyatt never got the money—but it was just right for products such as combs, hand mirrors, toothbrush handles, and movie film.

Henry Ford also played with the possibilities of bioplastics, establishing a significant research and development program focused on constructing car parts from soybeans. In 1941, Ford unveiled his soybean car, but he could not overcome rock-bottom fossil fuel prices or the all-consuming focus of World War II. In addition to being the maiden bioplastic, celluloid sparked the invention of Bakelite, Leo Baekeland's petroleum-based plastic—the first of its kind. Along with the emergence of the petrochemical industry, Bakelite ushered in a petro-polymer explosion in the early twentieth century. Suddenly, it was possible to create products of various sizes and shapes—high in durability, low in weight—and to do it on the cheap.

Like so many fossil fuel alternatives, bioplastics were sidelined until the oil crisis of the 1970s rekindled some interest. With the advent of green chemistry in the 1990s, alongside rising oil prices, commercial bioplastic production began in earnest. Today, a wide variety of bioplastics, with various recipes, properties, and applications, are in production or under development. Most are used in packaging of one kind or another, but they also are finding their way into everything from textiles to pharmaceuticals to electronics. Those that are "bio based" are derived, at least partially, from biomass. However, bio-based plastics may or may not be biodegradable. Polyethylene (PE) shopping bags made from sugarcane or corn are not. But bioplastics such as

The first and only bioplastic car was unveiled by Henry Ford in 1941 in Dearborn, Michigan. The car was inspired by the growing shortage of metal due to the war, as well as by the idea of combining industry with agriculture. He already had established the Soybean Laboratory in Greenfield Village at the time, and had made the fuel for the car from hemp oil. The frame was tubular steel, the body was plastic, the windows were acrylic, and it was powered by a conventional 60-horsepower engine. The finished car weighed 1,000 pounds less than its conventional, all-steel counterpart. Though it was created in part to aid the war effort, most car manufacturing ceased for the duration of the war and the bioplastic car was never revived.

polylactic acid (PLA), like you might find in a disposable cup, and polyhydroxyalkanoates (PHA), which can be used for sutures, are both bio based *and* biodegradable under the right conditions. (PLA degrades only at high temperatures, not in the ocean or home compost bins.) Research on bioplastics continues to push the bounds of their feedstocks, formulations, and applications. Finding the right sustainable feedstock and avoiding petrochemical-intensive agriculture is essential.

In contrast to petro-plastics, bioplastics can reduce emissions and sequester carbon. This is especially true when feedstocks draw on waste biomass, like what is left over from pulp and paper or biofuel production. To maximize climate benefits, bioplastics' entire lifecycle should be considered—from growing feedstock to end-of-life disposal. Beyond decreasing greenhouse gases, bioplastics offer other benefits petro-plastics do not. Some have technical advantages, such as thermal properties ideal for 3-D printing. Those that are biodegradable at low temperatures may help address the world's plastic trash crisis, particularly in rivers and seas. Currently, a third of all plastics end up in ecosystems, while just 5 percent are successfully recycled. The rest are landfilled or burned. If current trends continue, plastic will outweigh fish in the world's oceans by 2050.

Perhaps the biggest problem facing bioplastics is that they are *not* conventional plastic. Bioplastics cannot be composted unless separated from other plastics, and few will compost in the garden bin. They require high heat to be broken down or special chemical recycling. If bioplastics are intermixed with conventional plastics, conventional recycled plastic is contaminated, rendering it unstable, brittle, and unusable. Without source separation and appropriate processing, bioplastic is all dressed up with nowhere to go in most municipal waste streams except into the dump.

And yet, a swift transition is possible: DuPont, Cargill, Dow, Mitsui, and BASF are investing in bio-based polymers because they believe they have a strong platform for expansion. Because bioplastics are a replacement technology—something that can be swapped in for existing materials—they benefit from the demand for plastic worldwide. At the same time, the biggest challenge for bioplastics to overcome is the fossil fuel–based plastics industry. When oil prices are low and because economies of scale are often lacking, bioplastics struggle to compete beyond niche markets. Petro-plastics also have the benefit of pipelines and tankers for more centralized production. To realize advantages, the distance between feedstock production and bioplastic manufacturing has to be proximate. Bio-preferred programs and targeted plastic bans can also support the growth of biopolymers and the evolution of the plastic industry. ●

IMPACT: *We estimate the total production of plastics to grow from 311 million tons in 2014 to at least 792 million tons by 2050. This is conservative, with other sources estimating over 1 billion tons if trends continue. We model the aggressive growth of bioplastics to capture 49 percent of the market by 2050, avoiding 4.3 gigatons of emissions. While technical potential is even higher, this solution is constrained by limited biomass feedstock available without additional land conversion. The cost to produce bioplastics in this scenario is $19 billion over thirty years. While the financial costs are currently higher for producers, they are dropping quickly.*

4.61 GIGATONS REDUCED CO2	$72.44 BILLION NET COST	$1.8 TRILLION NET SAVINGS

MATERIALS
WATER SAVING—HOME

Using water at home—to shower, do laundry, soak plants—consumes energy. It takes energy to clean and transport water, to heat it if need be, and to handle wastewater after use. Hot water is responsible for a quarter of residential energy use worldwide. In addition to conservation measures that can be taken at the municipal level, efficiency can be tackled household-by-household and tap-by-tap.

At home, the average American withdraws ninety-eight gallons of water each day—much more than is typical worldwide. Roughly 60 percent is used indoors, primarily for toilets, clothes washers, showers, and faucets. Thirty percent is used outdoors, almost entirely for watering lawns, gardens, and plants—more than any other residential use, even though irrigation is nonessential. Another 10 percent is lost to leaks.

For cutting back indoors, two technologies are key: low-flush toilets and water-efficient washing machines, which can reduce use by 19 and 17 percent respectively. Switching to low-flow faucets and showerheads and installing a more efficient dishwasher also have contributions to make. In total, water-efficient appliances and low-flow fixtures can reduce water use within homes by 45 percent. Measures that affect hot water have an outsize impact on associated energy use. The U.S. Environmental Protection Agency (EPA) estimates that if one American home out of every one hundred switched an older toilet out for a new, efficient one, the country would save more than 38 million kilowatt-hours of electricity—sufficient to power 43,000 households for a month.

These technologies have the advantage of being one-time upgrades. If homeowners or landlords are willing to make the investment and wait out the payback period, no further action is needed. But individual behaviors can also curtail indoor water use. Reducing average shower time to five minutes, washing only full loads of clothes, and flushing three times less per household per day can each reduce water use by 7 to 8 percent. The downside, of course, is that those shifts must become habit to have an impact over the long term, and developing good habits is notoriously challenging.

Outdoors, water use for irrigation can be reduced or eliminated by using captured rainwater, shifting to plants that do not require it, installing drip irrigation, which is more efficient, or turning off the spigot entirely.

Water conservation success stories attest to what works. Local restrictions on water use and policies requiring efficient plumbing are highly effective. Product labeling, such as the EPA's WaterSense program, can inform consumers, while incentives, namely rebates on purchases of efficient appliances and fixtures, can encourage voluntary action. All of these measures have a twofold benefit: reducing energy use and water consumption simultaneously. Communities have a stake in doubling up, as more and more are struggling with water availability. The impacts of climate change are compounding population pressures. During droughts, for example, demand for irrigation goes up, while quality and quantity of supply declines.

This solution focuses on direct reductions of water consumption inside the home, but other domestic choices and technologies have indirect impacts. Energy use is a prime example: Nuclear and fossil fuel plants use enormous quantities of water for cooling—nearly half of all withdrawals in the United States. A single kilowatt-hour of electricity can have twenty-five invisible gallons associated with it. The tight link between water and energy means enhancing efficiency in one often affects the other.

IMPACT: *Ninety-five percent adoption of low-flow taps and showerheads by 2050 could reduce carbon dioxide emissions by 4.6 gigatons, by reducing energy consumption for heating wasted water. Scaling other water-saving technologies would drive additional reductions. We model hot water only in order to calculate energy savings.*

The Nebia showerhead was five years in design and development and employed aerospace engineering for its microatomizing technology. The showerhead produces hundreds or more droplets dispersed over five times the area of a regular shower. It is thirteen times more thermally efficient (the heat you feel on your body) and reduces water use by 70 percent compared to conventional showerheads and by 60 percent compared to the United States Environmental Protection Agency's WaterSense showerheads.

COMING ATTRACTIONS

Containing previews of the world to come, Coming Attractions is among our favorite parts of the book and could have been considerably longer. When it came to the other eighty extant solutions, we drew a bright line: They had to be well entrenched with abundant scientific and financial information about their performance and cost. However, in focusing on solutions that are already scaling, we did not want to imply that our capacity to solve global warming relies solely on what we already know and do. This section provides a window into what is forthcoming and close at hand. The rate of invention and innovation in all the featured sectors is staggering, and we doubt anyone knows the full extent. Many promising ideas are science projects and will never go further. Yet, as you will see, there are technologies and solutions described here that could be veritable game changers.

REPOPULATING THE MAMMOTH STEPPE

The Yakut is a hairy, short, stocky Siberian horse that looks as if it could be cast in a *Star Wars* movie. With their thick layers of fat, an extraordinary sense of smell, and big, rock-hard hooves, Yakutian horses survive temperatures of minus-100 degrees Fahrenheit above the Arctic Circle by scraping away snow and nibbling tiny bits of shriveled grass in the winter darkness. In this, they offer a clue as to how to prevent the melting of permafrost.

To keep the planet cool, you want grasses in subpolar regions, not trees, and you get grasses when you reintroduce herbivores. That is what Sergey and Nikita Zimov have witnessed in their experimental Pleistocene Park: the return of grasses and suppression of shrubs and trees. Grazing animals create pastures just as pastures create grazers. What if animals protected the permafrost and helped the Arctic region reverse its warming trend and start to cool?

Buried in the circumpolar region of the Arctic are 1.4 trillion tons of carbon, two times more than in all the forests on the planet. Permafrost is a thick subsurface layer of perennially frozen soil that covers 24 percent of the Northern Hemisphere. Its name presumes permanence—*perma*—a condition that is no longer true. It is thawing. At warming of 1.5 degrees Celsius (2.7 degrees Fahrenheit), permafrost will release significant amounts of carbon and methane into the atmosphere. If melting continues beyond 2 degrees Celsius (3.6 degrees Fahrenheit), the emissions released from the permafrost will become a positive-feedback loop that accelerates global warming.

Migratory woodland reindeer being driven by an Evenks herder through a valley in the Oymyakon region of the Sahka Republic in the Indigirka River Basin, Russia. The Evenks are famed reindeer riders and pastoralists. Their unique saddles are situated on the reindeer's shoulder and employ no stirrups; their balance is guided by a long stick that you see in the photograph.

When horses, reindeer, musk oxen, and other denizens of the frozen north push away the layer of snow and expose the turf underneath, the soil is no longer insulated by its snow cover and is 3 to 4 degrees Fahrenheit colder, a margin of safety the world needs while it transitions away from fossil fuels. The Zimovs, father-and-son scientists who direct the Northeast Science Station near Cherskii, Russia, have studied and analyzed the permafrost extensively. They created the Pleistocene Park in the Kolyma River basin of Siberia to demonstrate the conclusion of decades of research: If the diverse species of herbivores that once populated the subpolar region of the Arctic are brought back, permafrost melting can be prevented. Some perspective on the scope and implication of this proposal: If it came to pass, it would be the single largest solution or potential solution of the one hundred described in this book.

The road to the Kolyma River basin, the Kolyma Highway, is known as the Road of Bones. Prisoners exiled to Kolyma were expected to die during or after one brutal winter. Besides human bones, the basin shelters the tens of thousands of bones of its prior inhabitants. Bone counts reveal the average population on a square kilometer of pasture: 1 woolly mammoth, 5 bison, 8 horses, and 15 reindeer 20,000 to 100,000 years ago. More widespread were musk oxen, elk, woolly rhinoceros, snow sheep, antelope (saiga), and moose. Roaming among them were predator populations of wolves, cave lions, and wolverines. Twenty thousand pounds of animal life thrived on each square kilometer of pasture, an astonishingly high number that attests to the productivity of an area considered marginal and largely uninhabitable.

Today, as frozen carcasses melt under warming temperatures, swarms of bugs and bacteria devour the rotten remains. The foul odor from the melting permafrost is premonitory, an omen of greater dangers to come if melting is not prevented. Thaw ponds bubble like freshly poured soda water. If you turn a bowl or jar upside down and capture the gas, the methane can be lit up like a gas lamp. Ten-meter-deep ice-rich soils—an immense reservoir of organic matter—are heating up in much the same way. Defrosted microbes are coming back to life and releasing carbon dioxide and methane as they decompose the organic waste.

The Kolyma basin is part of a larger biome called the mammoth steppe, at one time the largest community of flora and fauna existing in any major habitat on earth. It extended from Spain to Scandinavia, across all of Europe to Eurasia and then on to the Pacific land bridge and Canada. For a hundred thousand cool, dry years, the steppe comprised mostly grasses, willow, sedges, and herbs, and was home to millions of herbivores and the carnivores that stalked them. In fairly quick succession, it changed 11,700 years ago. Temperatures rose, rainfall increased, and the woolly mammoth became extinct except for two remnant

The Yakutian horse is a rare breed that is Siberian hardy. At fourteen hands, it is short, compact, and sturdy. This picture shows a subtype of Yakut called the Middle Kolyma. It was brought to the Kolyma Valley by the Yakut people in the 1200s, and quickly adapted to the extreme cold. It survives the winters by pushing snow away with its hooves to get at the browse beneath. The Yakut have a legend that says that when the Creator was distributing the riches of the world, his hands froze when he got to Siberia and dropped everything he had. This explains the abundance, riches, and the extraordinary creatures in a land full of diamonds.

populations on islands created by the rising seas. The steppe contracted to the subpolar area, and dwarf birch, larch, moss, and berries largely replaced the grasses that had nourished animal life. Until recently, scientists assumed that the depopulation of the mammoth steppe was caused by the change of climate and the loss of pasture. Sergey Zimov has walked and explored the basin in minutiae and sees a wholly different past.

Zimov believes the theory of extinction is upside down and backward. Before the end of the Ice Age, approximately thirteen thousand years ago, hunters spread across Eurasia and into the Americas. Animals were tracked down for food and extirpated. Within a relatively short time, fifty species of large mammals were hunted to extinction in Russia, North America, and South America—in particular, the slow-moving and meat-abundant woolly mammoth. Once the grazers and ruminants were gone, the flora of the steppe changed. Away went the grasses, and in their place came the dwarf trees and thorny shrubs that are inhospitable to grazing herbivores.

To Zimov, it was obvious that the mammoth and herbivores were extirpated first, thus altering the landscape. Because the depopulation of the mammoth steppe took place so long ago, his conclusion is a theory. However, it is one based on decades

spent walking and exploring the icy regions of Siberia. Alexander von Humboldt's description of climate change in 1831 was concluded after a long journey through Russia and Eurasia, not a theory based on a hypothesis. In observational science, what something means is less important than what has happened or is occurring. You figure out what something means after you have thoroughly examined, surveyed, and become more intimate with a phenomenon, species, or ecosystem. Sergey Zimov is precisely such a scientist. As fellow scientist Adam Wolf observed, Zimov's peregrinations and excursions in the mammoth steppe were not tainted by groupthink or published papers about what happened there. He could see that the theory that climate change precipitated the extinction of the woolly mammoth was incorrect. A mammoth's weight and inertia could crush larches, bramble, and dwarf birch, and along with herbivore pressure, would have prevented changes in the composition of flora.

The northward spread of the taiga, the coniferous boreal forests, is changing climate dynamics. Instead of heat being reflected back into space by snow, trees and leaves soak it up and reradiate it to the soil. Although the atmosphere is warming evenly at sixty thousand feet, at ground level the Arctic regions are warming much faster than temperate and equatorial regions, and changes in flora are a cause.

To populate the Pleistocene Park, Sergey has had to beg, borrow, and buy. The woolly mammoth was wiped out long ago. The Beringian bison and native musk oxen are likewise missing. He brought in the Yakutian horses from the south. The Canadian government donated bison. He hopes to secure reindeer from Sweden and more musk oxen from Alaska. He purchased an aging Russian tank. Driven in the preserve, it crushes the shrubs and larch as a mammoth would and produces a grassy trail of brome for the years that follow. Zimov would like a shipload of five thousand Canadian bison and a worldwide carbon tax that would finance the repopulation of the mammoth steppe. At the low price of $5 per ton of carbon dioxide, the frozen mammoth steppe is worth $8.5 trillion.

As with advanced multipaddock grazing and regenerative agriculture, the Zimov proposal to repopulate the mammoth steppe is a land-use practice that reverses a long-term trend of degradation. It is difficult to imagine that the wildness of the subpolar regions is actually a degraded landscape, but that is what Zimov has shown. Today, the biomass of all the animals being raised, most of which are entrapped and caged in industrial factories, totals close to one billion tons. The cost: vanishing resources, loss of biodiversity, degraded soils, unhealthy meat, and a changing climate. Repopulating the mammoth steppe may appear to be an esoteric pursuit at first glance. Actually, it is no different from other restoration practices—just bigger. Regeneration of the land can be brought about by rewilding the abandoned lands of the north, returning the animals that created the great, once-dominant, carbon-sequestering grasslands. When herbivores were free to roam, the earth supported twice the number and weight of animals that humans raise today in ranches, feedlots, and animal factories. In the mammoth steppe, considered unlivable to all but a hardy few, the benefit of returning it to its wild origins would be immense. ●

COMING ATTRACTIONS
PASTURE CROPPING

Revelations can happen when your two-thousand-acre farm burns to the ground—outbuildings, trees, twenty miles of fencing, three thousand sheep, and all. Colin Seis inherited Winona, his grandfather's farm located in New South Wales, Australia, from his dad in the 1970s. As a kid, he watched his father apply new agricultural techniques to improve yield and productivity, but the fertilizers, herbicides, and plowing slowly wore the farm out. The soil became compacted and acidic, topsoil bottomed out at four inches, and carbon measured less than 1.5 percent. Costs soared, more chemicals were used, trees turned brown, and the farm lost money. Then, in 1979, a bush fire reduced three generations of work to ash.

When Colin recovered from the burns he suffered during the fire, he found himself at a pub with fellow farmer Daryl Cluff. They each grew grain (annuals) and grazed sheep on pastures, with both activities taking place on separate parts of their farms. Grasses over there, grains here. But why? The pastures tended to be overgrazed, and the grain acreage was plowed and disked every year, drying and decarbonizing the soil. Ten beers later they both wanted to know: Why couldn't annuals and perennials be grown on the same land at the same time? Why couldn't the land be fertilized through grazing between crops?

A vision emerged that night that would become the basis for what is now known as pasture cropping. On pasture-cropped land, the soil is never broken. Planting annual crops in a living perennial pasture creates an ecosystem that gets healthier every year. A complex relationship between the forbs, fungi, grasses, herbs, and bacteria reknits the web of life, increasing the health, resilience, and vitality of the soil, crops, grasses, and animals. And the farmer reaps two crops from the same land: grain and wool or meat.

The next morning Seis and Cluff were sober and it still seemed like a good idea. Seis stopped using fertilizers, herbicides, and pesticides immediately—an easy decision because he was broke. Then came a few years of transition. The land was like a recovering alcoholic; it was addicted to ammonium phosphate. At first, yields were not great as Seis allowed native grasses to repopulate the fields. Because the perennials were lower in protein, the animals did not fare as well with them at the outset. The neighbors were not impressed. Seis kept going. He began to employ rotational mob grazing in his paddocks. And things started to turn around—profits, productivity, and animal and soil health. Soon the regeneration of the farm was evident to all. Costs went down. Seis was saving $60,000 a year on fuel and chemical inputs he no longer needed. Water retention and soil carbon increased threefold. Insect infestation virtually disappeared. Profits from his sheep ranching went up along with yields and the quality of wool. Birds and native animals returned.

Pasture cropping is now practiced on more than two thousand farms in Australia and is spreading throughout the temperate farming world. As dependent as the world has become on annual crops, and as unthinkable as it may be to agriculture schools and Big Ag, at some point, farming must change to sustainable and regenerative methods if it is to recover lost fertility and soil carbon. Pasture cropping is singular in its methodology in that it increases the use of the land by double-cropping (grains and animals) while reducing impact and increasing carbon sequestration.

Colin Seis

COMING ATTRACTIONS
ENHANCED WEATHERING OF MINERALS

Billions of years ago, there were no oxygen molecules in Earth's atmosphere. It consisted of nitrogen, water vapor, and carbon dioxide (and possibly some methane). Cyanobacteria that photosynthesize carbon dioxide arrived and began exhaling oxygen. Myriad life forms, from phytoplankton to pine trees, have been inhaling carbon dioxide and converting it to solid matter and depositing some of that back into soils or ocean sediments. Cycles of biologically sequestered carbon are partially responsible for ice ages: As carbon dioxide levels dropped, less heat was trapped into the atmosphere and temperatures plunged. The consequent ice ages greatly reduced microbial activity, eventually stopping the drawdown of carbon dioxide. Over eons, active volcanoes released carbon dioxide back into the atmosphere—warming the planet—and the cycle repeated. In other words, biology plays a hand in the relationship between global warming and cooling.

Today, thanks to the work of NASA, the public can watch simulations of the fluctuations in the annual carbon cycle. These animations vividly show carbon dioxide being emitted during the late fall, winter, and early spring as Northern Hemisphere vegetation goes dormant and people turn on their fossil-fueled heating systems. In the late spring until the early fall, it is just the opposite. Despite ongoing emissions from deforestation, cars, and electrical use, large amounts of carbon dioxide—equivalent to five to six parts per million—are sequestered by grasses, shrubs, trees, and in the warming waters by the same cyanobacteria whose ancestors started the carbon cycle. The total is in the range of 40 billion tons of carbon dioxide annually.

There is also a slow carbon cycle. The quiet, less-told story is that during this 3.7-billion-year journey to extraordinary biodiversity, rocks have sequestered many trillions of tons of carbon dioxide from the air. Natural rock weathering removes approximately 1 billion tons of atmospheric carbon dioxide annually. Various types of silicate rock on the surface of the earth are weathered by mildly acidic carbon dioxide and dissolved in rainwater, which transforms the carbon dioxide into dissolved inorganic carbonates. These carbonates find their way into streams, rivers, and oceans, eventually becoming calcium carbonate.

Enhanced weathering of minerals refers to a suite of technologies that aim to hasten this process sustainably. One type of silicate that would work well for enhanced weathering is olivine, a greenish mineral, rich in magnesium and iron. A conventional pathway to enhanced weathering involves the mining and milling of silicate rocks containing olivine and applying the rock powder to land and water so that the soil, oceans, and biota can act as "reactors" for accelerated weathering. The rock powder can be strategically distributed over various landscapes, particularly agricultural land, beaches, and shallow energetic seas. The key technologies required for enhanced weathering are already being used on a regional scale for the fertilization and acidity management of farm and forest soils.

To fully halt carbon dioxide accumulation through enhanced weathering would take a staggering effort, involving billions of tons of mineral spread over a significant fraction of the earth's surface. Careful site selection, and the use of existing surface resources, such as tailings piles from previous mining operations, can offer opportunities to durably sequester a meaningful fraction of emissions while minimizing cost and risk. The environmental impact of enhanced weathering could have unpredicted and unwanted side effects on the environment and biological activity, and so careful monitoring and risk management would be required.

One potentially high-impact area to apply olivine minerals is believed to be on agricultural land in the tropics, where the soils are warmer and wetter and have fewer minerals that would inhibit dissolution. Broadly, if olivine was applied to one-third of tropical land, it could lower atmospheric carbon dioxide by thirty to three hundred parts per million by 2100. A key advantage of agricultural soils is that they are already intensively managed, could be monitored with relative ease, and are already served by infrastructure. Implementing enhanced weathering of minerals on tropical croplands as a soil amendment has potential co-benefits for agro-ecosystems because rock powder can act as a fertilizer of crops.

One to two tons of powdered olivine will continue to sequester carbon for approximately thirty years in a temperate climate. Other studies suggest that the optimal places to apply olivine are acidic soils or where there is acid rain, because the lower pH accelerates the rate of mineral dissolution. Those areas include large parts of Europe and some parts of the United States and Canada. Similarly, weathering could be used to regenerate damaged forests in Eastern Europe, where decades of lignite coal combustion has produced some of the most acidic rain over many years on the planet. In areas where mines have closed or have been abandoned, using minerals from the residual tailings could be a useful economic development tactic to help communities.

Some scientists believe that olivine weathering rates are regularly underestimated because weathering in nature tends to proceed much faster than in a laboratory. One study demonstrated that previous assumptions about enhanced weathering dissolution rates are overly pessimistic. It showed sequestration of carbon dioxide to be ten to twenty times greater in nature than what was being found in the laboratory. Biotic factors that accelerate weathering include the effects of lichens, soil bacteria, and mycorrhizal fungi, which provide sugar-based exudates to bacteria that accelerate mineral dissolution.

Significant limiting factors are the carbon cost of implementing enhanced weathering and the capital cost of the infrastructure required to scale production. To produce and then reduce olivine to a size that would optimally dissolve carbon dioxide may require so much energy as to negate up to 80 percent of its positive effect. The required infrastructure would include new mines, railroads, and shipping facilities. To give a sense of scale, one ton of olivine can displace two-thirds of a ton of carbon dioxide. Sequestering eleven gigatons of carbon dioxide, which is about 30 percent of fossil fuel emissions, would require 16 billion tons of rock being mined, powdered, and shipped per year, a bit more than twice the output of the coal industry.

There is an alternative to "conventional" enhanced weathering in which silicate dust is spread across the land (and oceans) to capture carbon dioxide. The technology currently has no name, but it does have proof of concept. In tests conducted in Iceland by Reykjavik Energy and in the United States by the Pacific Northwest National Laboratory, a branch of the U.S. Department of Energy, liquid carbon dioxide was placed underground in caverns of volcanic rocks called basalt. As with olivine weathering, the carbon dioxide combined with the basalt and formed solid carbonates called ankerite. The scientists dub this process high-speed weathering. Professor Klaus Lackner, who directs the Center for Negative Carbon Emissions at Arizona State University, called the results "immense progress." He went on to say that "basalts on land and below the ocean floor are so abundant that if they can be pulled in, we have indeed unlimited storage capacity [for carbon dioxide]."

As yet, no field testing of enhanced mineral weathering has taken place. All numbers and predictions are based on laboratory data, natural analogs, data analysis, and simulations. The basic assumption is that approximately a ton of carbon dioxide could be sequestered for every ton of olivine mined and applied. The cost per sequestered ton is high, between $88 and $2,120, given current analyses. As with several solutions contained herein, it would seem that globalizing this solution creates uncertainty, impacts, and potential negatives that override its benefits. However, it is hardly different from applying lime or silicon ores to the soil, a practice employed around the world. Starting with trials on tropical farmland and acidified temperate land, the application of olivine may prove to be productive and beneficial. ●

Layered ultramafic olivine rock, Duke Island, Alaska

"The number of living creatures of all Orders, whose existence intimately depends on kelp, is wonderful. A great volume might be written, describing the inhabitants of one of these beds of seaweed . . . I can only compare these great aquatic forests . . . with terrestrial ones in the intertropical regions. Yet if in any country a forest was destroyed, I do not believe nearly so many species of animals would perish as would here, from the destruction of the kelp." — Charles Darwin, *from* Voyages of the Adventure and Beagle

In his 1989 book, *The End of Nature*, Bill McKibben describes how nature is no longer a force independent of human activity but a process subordinate to human alteration, most of which is destructive to life. Recently, scientists have announced that civilization has entered a new epoch, the Anthropocene, a period defined by human domination of earth's physical environment. It marks the end of the Holocene, an 11,700-year "Goldilocks" era of benign and stable climate—not too cold and not too hot—just right for the birth of human civilization.

The usual assumption about human activity is that it makes nature worse, however well intentioned. But that has not always been the case. The productivity of the tallgrass prairies of the Great Plains region can be attributed to the fire ecology practiced by Native Americans. In Norman Myers's book *The Primary Source*, he describes going into a forty-thousand-year-old "untouched" primary forest in Borneo with an ethnobotanist. Both stayed in one spot for the day while the ethnobotanist identified the towering dipterocarps and other flora for Myers. It turns out the entire forest had been placed and planted by human beings before the last ice age. The Swiss agroecologist Ernst Gotsch works with deforested and desertified lands in Brazil and restores them in a matter of years to lush forest farms bountiful with food. In a video segment in which he describes his work, Gotsch picks up dark, moist soil and proclaims, "We are growing water."

In other words, human intervention can increase wildlife, fertility, carbon storage, diversity, fresh water, and rainfall. This entire book asks whether, as a species, we can reverse global warming. To do that, the demise of living ecosystems needs to be reversed. Marine permaculture may be one of the most extraordinary ways to answer that question affirmatively.

We usually do not speak of oceans and forests in the same sentence, but what if you could reforest the ocean? Dr. Brian Von Herzen devotes his life to this proposition. With a physics degree from Princeton University and a Ph.D. from California Institute of Technology, he had a fruitful career as a consultant specializing in electronic design and systems engineering. He created solutions for Intel, Disney, Pixar, Microsoft, HP, and Dolby. For adventure,

he would pilot his twin-engine Cessna 337 Skymaster across the Atlantic.

The 337s are used extensively by firefighters as spotter aircraft. At the request of friends who were glaciologists, Von Herzen looked for melt ponds as he flew over the Greenland ice sheet in 2001. He spotted a few small ones. Two years later when he flew over again, there were hundreds. In 2005, there were thousands. By the next year, there were lakes exceeding six miles long and a hundred feet deep. By 2012, 97 percent of the ice sheet surface had melted. This led Von Herzen to focus on reversing global warming using the only means possible: increasing the primary production of living systems, specifically the oceans. Primary production is the creation of organic compounds from aqueous or airborne carbon dioxide through photosynthesis. This is accomplished by kelp and phytoplankton, the microscopic wandering plants that thrive in the oceans—a quarter billion of which fit nicely into a cup of seawater.

We are talking kelp forests, hundreds of thousands of acres of underwater plantations situated offshore, floating forests in the middle of the ocean. Today, kelp forests cover nineteen million acres. Ultimately, floating kelp forests could provide food, feed, fertilizer, fiber, and biofuels to most of the world. They grow many times faster than trees or bamboo. Von Herzen wants to restore the subtropical ocean desert and its fish productivity with thousands of new kelp forests. He calls this *marine permaculture*.

The situation in the oceans is dire. Half of the carbon dioxide that is recaptured from the atmosphere goes into oceans, causing surface acidification. And over 90 percent of the heat caused by global warming is absorbed into the surface waters, a trend that is steadily erasing the marine food chain. What makes oceans productive are upwellings of cold, nutrient-rich water from deep in the sea. Natural upwellings occur around the world, such as in the Grand Banks of Newfoundland—the richest fishing ground in the world—where the icy Labrador Current meets the warm Gulf Stream. This phenomenon is known as *overturning circulation*.

As waters have heated up, ocean deserts have expanded. Ninety-nine percent of the subtropical and tropical oceans are largely devoid of marine life. The oceans' wind- and current-driven pumps are being turned off one by one. In the Atlantic, satellite imagery is detecting a 4 to 8 percent per annum decline in biological activity, a number that exceeds predictions in global warming models.

Warm water reduces overturning circulation across thermoclines, the temperature gradients in the ocean. As heating of surface water increases, currents slow or are thwarted, and upwelling of nutrients decreases or stops altogether. Phytoplankton and seaweed production drops; subsequently, the aquatic food chain declines. Phytoplankton are minute, but the 1 percent

---

The number of creatures in a kelp ecosystem is extraordinary. Corallines, a branching coral like seaweed, may incrust every frond and leaf; cuttlefish dart in and out; multicolored ascidia, tiny invertebrate filter feeders, dot and cling to the waving leaves. On flat surfaces you find sea snails, limpets, mollusks, and bivalves. Permeating this undulating landscape, attached or unattached, you may find krill, shrimp, barnacles, woodlice, cuttlefish, and crabs. Sea urchins will be gnawing away at the stems, and wolf eels, starfish, and triggerfish will feed on them. Among them all will be tiny forage fish, the smelt, halfbeaks, and silversides. And circling the waters around the dense kelp growth, shimmering game fish will feed on the prey fish.
(Inspired by Darwin)

annual decline in the oceans' plankton and kelp is massively significant: They comprise half of the organic matter on earth and produce at least half of the earth's oxygen.

What Von Herzen proposes would restore overturning circulation in the subtropics. Employing marine permaculture arrays (MPAs) .4 square mile in size—situated offshore and far from land—would re-create entire marine ecosystems. It would be like reforesting a desert—in this case, the ocean desert. Imagine a lightweight latticed structure made of interconnecting tubing, submerged 82 feet below sea level, to which kelp can attach. MPAs can be tethered near land, or self-guiding on the open sea. They are far enough below the surface that the largest cargo ships and oil tankers can pass right over them with no damage save some shredded kelp.

Buoys attached to the MPAs rise and fall with the waves, powering pumps that bring up colder waters from hundreds or thousands of feet below sea level. As the nutrient-laden waters come to the sunlit surface, seaweed and kelp soak up the nutrients and grow. What soon follows is what is called a *trophic pyramid*. With phytoplankton come algae, more kelp, and sea grass. These feed populations of herbivorous forage fish, filter feeders, crustaceans, and sea urchins. Carnivorous fish feast on the smaller herbivores, and seals and sea lions and sea otters feed on them. On top of this are seabirds, sharks . . . and fisher folk. The phytoplankton and kelp that is not consumed dies off and the majority drops into the deep sea, sequestering carbon for centuries in the form of dissolved carbon and carbonates.

Often the ocean is thought of as a single fluid entity, but nothing could be further from the truth. Most of the carbon emitted by human activity is contained within the top five hundred feet of the ocean known as the photic zone. It is accumulating carbon significantly faster than the rest of the ocean. In its entirety, the ocean stores fifty-five times as much carbon as is contained in the entire atmosphere. Looked at another way, if all the carbon in the atmosphere were removed and stored uniformly throughout the ocean, the increase in ocean carbon would be less than 2 percent. Thus, it is mostly an issue of moving carbon from the near-surface photic zone into the middle and deep ocean. Oceans naturally do an exquisite job of sending carbon from surface water into the depths, a process known as the biological pump. Marine permaculture supports the functioning of the biological pump so that oceans can do the job they always have.

Kelp harvests can produce food, fish feed, fertilizer (including nitrate, phosphate, and potash), and biofuels. Each dry ton of kelp sequesters a ton of carbon dioxide. Fish populations will soar; these will be the ultimate fish farms (free-range aquaculture), except the fish will be diverse, wild, untainted, and full of omega-3 fatty acids. MPAs in larger groups may seasonally protect coastlines from the worst effects of hurricanes by lowering the surface water temperature and the energy upon which hurricanes depend. It is possible to seasonally protect reefs from thermally induced bleaching. Given that Hurricane Katrina alone cost $108 billion, and that 2015 saw twenty-two Category 4 or 5 hurricanes, this may be a cost-effective solution. The material costs are estimated at $2.6 million per square mile. With a million MPAs active for thirty years, the carbon dioxide reduction would equal 12.1 parts per million, or 102 billion tons. The economic return would exceed $10 trillion. On paper, the protein from restored fisheries could supply the protein needs of most of the earth's people. Perhaps with the implementation of MPAs, human beings can be agents of restoration and increased productivity of fish and kelp forests. ●

COMING ATTRACTIONS
INTENSIVE SILVOPASTURE

Silvopasture is a common form of agroforestry, practiced today on over 350 million acres worldwide. The theory is simple: Combine trees or woody shrubs and pasture grasses to foster greater yields. Cattle fatten faster and provide better-tasting meat than in any other system. Rarely are livestock and climate mitigation used in the same sentence; silvopasture, however, sequesters up to three times more carbon per acre than grazing alone—ranging from one to four tons per acre in the tropics and averaging 2.4 tons in temperate regions.

What happens if you intensify the silvopasture process? Add more cattle, plant different types of trees, and rotate the herd more quickly? It seems counterintuitive that it could have a beneficial effect on land and climate, as well as human health, but it does. There are reams of data showing how conventional cattle-raising systems, involving feedlots and accelerated fattening procedures, are among the more significant contributors to climate change, if not the most. Implausibly, ranchers have developed an intensive silvopasture system that is one of the most effective means known to sequester carbon. First developed in Australia in the 1970s before spreading to the tropics, it looks like chaos to the untrained eye. To someone accustomed to fields neat as a pin, with laser-guided row crops, intensive silvopasture would appear to be an unkempt jungle. In areas where ranching and farming are stressed by volatile and uncertain patterns of rainfall and heat, intensive silvopasture systems teem with life. Extremes in climatic variation make livestock farming riskier, if not ruinous, because grasslands are completely dependent on available natural resources, including rainfall. In contrast, intensive silvopasture creates resilience by increasing the density of flora and fauna.

Most intensive silvopasture systems revolve around a quickly growing, edible, leguminous woody shrub. *Leucaena leucocephala*, planted four thousand per acre, is intercropped with grasses and native trees. These intensive systems require rapid rotational-grazing regimes. They employ electric fences that allow for one- to two-day paddock visits, with forty-day rest periods between. Trees keep the wind in check and improve water retention, which causes increases in biomass. The combination of flora can reduce ambient temperatures in the tropics by fourteen to twenty-three degrees Fahrenheit, which enhances both humidity and plant growth. Species biodiversity doubles in intensive silvopasture systems. Stocking rates nearly triple. Meat production in pounds per acre per year is four to ten times higher than in conventional systems. The tannin content in *Leucaena leucocephala* seems to protect protein degradation in the rumen of cattle, reducing methane emissions, which partially explains the significant weight gain of animals raised via intensive silvopasture. And during the dry season, *Leucaena leucocephala* seeds can be harvested—netting another $1,800 per acre in income. *Leucaena leucocephala* is an invasive in Florida and many other

places, and is toxic to animals with a single stomach, like people and horses. In the United States and in tropical highlands around the world, other species are being trialed. The key to intensive silvopasture is a fast-growing, high-protein woody plant that can handle heavy browsing and re-sprout quickly. In tropical Australia and Latin America, Leucaena is one that has passed the test so far.

Today, intensive silvopasture is practiced on more than five hundred thousand acres in Australia, Colombia, and Mexico. In Colombia and Mexico, producers are cultivating fruit, palm, and timber trees to further boost income. It may sound too good to be true, but there is one more piece of data: In a five-year study of intensive silvopasture in which trees were incorporated with grasses and *Leucaena leucocephala*, the rate of carbon sequestration was roughly three tons per acre per year, a high rate for any land use. ●

COMING ATTRACTIONS
ARTIFICIAL LEAF

For decades, a dedicated group of scientists have attempted to replicate natural photosynthesis in an artificial plant leaf and create fuels directly from the atmosphere, powered by sunlight. The payoff is obvious; almost all energy comes from the sun, and most of that comes from photosynthesis. (We get energy in the form of food from plants or plant derivatives such as oil, gas, peat, coal, wood, and ethanol.) Photosynthesis seems simple: water, sunlight, and carbon dioxide in; carbohydrates and oxygen out. Trying to satisfy the world's growing appetite for energy using natural photosynthesis alone, however, is not feasible.

When you grow corn, poplars, or switchgrass to create biofuel, you are at a significant disadvantage in terms of energy efficiency. Plants convert sunlight effortlessly and without fail; however, when it comes to converting photons into useful stored energy, they are about 1 percent efficient. In the case of corn, the farmer has to till the fields with diesel-powered tractors, probably use herbicides to control weeds, harvest the crop with combines, and truck the crop many miles to be processed. At the processing plant, the corn is milled, made into mash, blended with enzymes and ammonia, cooked to kill bacteria, liquefied, and then fermented for a couple days with yeast to convert the sugars to ethanol. From there it is distilled and centrifuged. The solids are separated and the liquids go to a molecular sieve. The carbon dioxide is captured and sold to soft-drink manufacturers. Denaturants are added to make it untaxed and undrinkable, and then it goes to storage tanks. From there it is placed in tanker trucks en route to refineries, where it is blended into gasoline.

The industry calls this a renewable fuel, but that stretches the meaning of the concept. The process is heavily dependent on diesel, oil, gasoline, electricity, and subsidies. When fully calculated, corn-based ethanol produces slightly more energy than was required to produce it. If you add emissions from land use, groundwater depletion, loss of biodiversity, and the impacts of nitrogen fertilizers, the benefit to the atmosphere is debatable. Corn's highest and best use is as staple food for people who are hungry, not as ethanol powering an SUV.

Imagine, then, if you could bypass the farms, fertilizers, tractors, trucks, processing plants, and subsidies and make fuel from water and carbon dioxide, wherever you and the water

Daniel Nocera

reside. That is the goal of the artificial leaf project, founded by Daniel Nocera more than two decades ago.

Nocera is a professor of energy science at Harvard University. He has devoted himself to splitting water into hydrogen and oxygen since graduate school at Caltech in the early 1980s. His work began as a means to jump-start and empower the hydrogen economy. Initial versions of his technology used a slender sheet of silicon coated with a cobalt-nickel catalyst on one side that, when the sheet was dropped into a container of water, caused hydrogen to bubble up to the surface on one side and oxygen on the other. The early press heaped praise and exaggerated the implications of the technology. Nocera himself prophesied the benefits that would be bestowed upon the poor. He described how hydrogen gas could be burned for cooking or turned into electricity with a fuel cell. But what can a poor person do with a canister of hydrogen? Nothing . . . so far, unless they have a fuel cell, which is an expensive technology. It was a scientific breakthrough with no economical application.

Hydrogen is the lightest element in the universe and disperses like a will-o'-the-wisp. Although a pound of hydrogen contains three times more energy than a pound of gasoline, getting a pound of hydrogen is a tricky process and requires equipment, high-pressure tanks, and compressors. Generating enough energy for a family would require a slice of silicon the size of a sheet of plywood and a tank equivalent to three bathtubs. Nocera was focused on providing affordable energy for the poor, but little thought was given as to how the poor could actually make electricity. Nevertheless, he was determined to come up with an energy source and technology that could be shared by everybody,

a concept he attributes to being a Deadhead in the 1970s. The Grateful Dead were decades ahead with their concept of music sharing, an idea that eventually undid the industry. The band allowed and encouraged people to record their own tracks from their concerts, and to this day there are sites devoted to sharing and exchanging tracks. Is this concept possible with an energy technology?

Nocera thinks so.

He believes that by focusing on technologies that benefit those who have the least, all of society will benefit the most. For many years, he answered skeptics by pointing out that if as much money was invested in artificial photosynthesis as is invested in batteries, a breakthrough would occur sooner.

A breakthrough did come. On June 3, 2016, Nocera and his colleague Pamela Silver announced that they had successfully created energy-dense fuels by combining solar energy, water, and carbon dioxide. Employing two catalysts, they produced free hydrogen from water, which is fed to bacteria, *Ralstonia eutropha*, that synthesize liquid fuels. When the bacteria are fed pure carbon dioxide, the process is ten times more efficient than photosynthesis. If the carbon dioxide is taken from the air, it is three to four times more efficient.

Until recently, Nocera had focused on inorganic chemistry to create hydrogen gas. By seeing the hydrogen not as an energy source for people, but as a feedstock of energy for bacteria, he and his team at Harvard made a giant step toward his original goal: inexpensive energy made with sunshine and water. Oh yes, and bacteria. Perhaps economically viable artificial photosynthesis will not be so artificial after all. ●

I walked all around the curious vehicle, and I finally decided to get into the car . . . I climbed in and sat down, with a queer feeling at the complete absence of the steering wheel and gear-shift levers. However, on the dashboard were a great many dials; and something was ticking quietly somewhere inside the machine. Then there was a "clickety-click" and a whirr of the motor, and the car moved gently away from the curb. It swerved out into the street, gathered speed, and then turned to the right around a corner. It slowed down for two women crossing the street, and avoided a truck coming toward us. It gave me an eerie feeling to sit in the thing and have it carry me around automatically. Then it suddenly dawned on me, that here I was alone in the thing, on an unknown street, in an unknown city, racing along at too high a speed to jump out, and rapidly getting farther away from places with which I was familiar. — Miles J. Breuer, M.D., Paradise and Iron (1930)

Autonomous vehicles (AVs) may be the ultimate disruptive technology. The origin of the word *autonomous* is from the Greek *autonomos*, "having one's own laws." Applied to a vehicle, it means the vehicle has its own laws and rules, not yours. Self-driving vehicles are being programmed, designed, tested, and readied as fast as any technology ever has. There are literally trillions of dollars at stake. Though the idea of self-driving vehicles goes back more than ninety years, it is the recent convergence of motion sensors, GPS, electric vehicles, big data, radar, laser scanning, computer vision, and artificial intelligence that will radically change cities, highways, homes, work, and lives. The Institute of Electrical and Electronics Engineers predicts AVs will make up 75 percent of road vehicles by 2040, though there are many legal and regulatory hurdles to overcome before that can become a reality. Whether they will have a benign, neutral, or negative impact on society is not clear. Expert opinion is arrayed on both sides.

How cars are owned and utilized today could not be any less efficient. About 96 percent are privately owned; Americans spend $2 trillion per year on car ownership; and cars are used 4 percent of the time. The contemporary car is not a driving machine but a parking machine for which 700 million parking spaces have been built—an area equivalent to the state of Connecticut. If the populace were to undergo a shift and view mobility as a service—rather than private ownership of expensively insured, two-ton assemblages of steel, glass, plastic, and rubber that emit carbon dioxide and health-destroying pollutants—the material, infrastructure, and health-care savings would be immense. But that is not a given. Electrics are at least four times more efficient than gasoline-powered vehicles in overall energy use, which will be the main greenhouse gas benefit of autonomous vehicles.

It would be hard to discuss the basic technological capabilities of AV technology without acknowledging three other parallel and complementary areas of research and practice: shared vehicles, on-demand vehicles, and connected vehicle technology.

- *Shared vehicles* enable higher vehicle occupancy by facilitating shared trips, in which riders are headed in similar directions. Lyft Line and UberPool are two common platforms that provide this service already.
- *On-demand vehicles* can be requested by customers and are expected to show up within a reasonable amount of time with a driver, a service that exists today with apps. Autonomy means your car will arrive without a driver.
- *Connected vehicles* will be equipped with vehicle-to-vehicle and vehicle-to-infrastructure communications capabilities, allowing those vehicles to collect and share data with other vehicles, roads, traffic lights, and so on in real time, to smooth traffic flow and increase safety. So far, companies competing in this market have no agreement to have vehicle-to-vehicle or vehicle-to-infrastructure communication. That would be a loss because this communication combined with onboard artificial intelligence would equip cars to learn constantly and get increasingly smarter about geography, streets, situations, and destinations.

The potential ecological advantages of autonomous vehicles are numerous but not inevitable. Most current AV demonstration models are built on existing production vehicles with aftermarket sensor packages. Concept models of autonomous vehicles being tested and proposed are smaller, more aerodynamic, and can form a platoon, a group of vehicles following closely behind one another, if they have dedicated lanes, benefiting from draft as cyclists do in a peloton. The transition to dedicated lanes may take decades, however. If autonomous vehicles are shared between several people, congestion will shrink. Cars will no longer circle the block looking for parking places—they will pick up another passenger instead. Autonomy will accelerate the adoption of electric vehicles because most trips are local, thus in battery range. Smaller, efficient vehicles may pare road width and release land for other uses.

However, the shift to autonomous vehicles may be messy. There are innumerable obstacles to the transition. The technology is expensive and must perform to exacting tolerances in all conditions; no mistakes are acceptable when the lives of drivers, passengers, and bystanders are on the line. The back-and-forth between AV capability and the regulatory environment may be slow, and bylaws may differ from state to state. For a considerable period of time, AVs will be interacting with drivers in non-autonomous vehicles, with no way to communicate or receive

---

A woman uses a mobile phone as she walks in front of an autonomous self-driving vehicle, as it is tested in a pedestrian zone during a media event in Milton Keynes, north of London, on October 11, 2016. On that date, driverless vehicles carrying passengers took to Britain's streets for the first time in a landmark trial that could pave the way for their introduction across the country.

The Navly self-driving shuttle on the Lyon Confluence, Lyon, France. Driverless, autonomous, and fully electric, the shuttle carries passengers between the Confluence shopping area and the tip of the peninsula. Equipped with lasers, cameras, and highly precise GPS, the Navly shuttle will reach 25km/hour, but be safe for passengers or pedestrians.

communication. The greatest impediment may be how powerfully embedded the desire to possess one's own car is. Privately owned, traditional automobiles are likely the most meaningful competitors for AVs, both culturally and functionally. They are symbols of personal freedom—not just in the United States—and displacing them will be no small task for the four-wheeled robots of tomorrow. It may require a generational shift in attitude. People without a car at home may feel marooned or trapped.

There could be a significant populist revolt against autonomous vehicles, as augured by the angry pushback against Uber by taxi drivers in European cities and in California. Costs plunge if your taxi does not have a driver; nothing will stop that. On the other side, a time could come when people are banned from driving because in a world of self-directed, connected vehicles, individual drivers are a danger to everyone else. Futurist Thomas Frey has made a list of what will disappear in the driverless-car era, and at the top of that list are drivers. Drivers not wanted: taxi, Uber, UPS, FedEx, bus, truck, and town car. Also eliminated:

insurance agents, auto salesmen, credit managers, insurance claims adjusters, bank lending, and traffic reporters on the news. What goes the way of the cassette tape: steering wheels, odometers, gas pedals, gas stations, AAA, and the many outlets for individuals to service their own cars, from body shops to car washes. Good riddance to: road rage, crashes, 90 percent or more of all injuries and auto-related deaths, driving tests, getting lost, car dealers, tickets, traffic cops, and traffic jams.

The auto-and-truck industry has a disproportionate impact on climate. Automobiles and trucks account for one-fifth of all greenhouse gas emissions, and that does not include construction and maintenance of streets, highways, and other infrastructure. Along with a reduction in greenhouse gases could be a reduction in jobs for millions. (Compare now-defunct Blockbuster to Netflix for a sense of what this could mean for overall employment.)

Just as the freeway and auto industry transformed cities, so too will AVs. Actual miles traveled could go up, not down. The reason is simple: When the cost of a service or object goes down, consumption invariably increases. Automated bookable cars at one's door could see individuals moving farther away from the city, especially if they can work within the car rather than drive.

An optimistic vision of the convergence of car sharing and autonomy is common to the companies pioneering the field. There are estimates that the total U.S. auto fleet would decline by 50 to 60 percent. John Zimmer, cofounder of Lyft, calls it the "third transportation revolution." It describes a transformed urban and suburban landscape that is built for people, not cars. On-demand autonomous vehicles will allow a great majority of city dwellers to abandon car ownership at significant savings to themselves and their cities. Given the sheer hassle of owning a car in an urban setting and the average U.S. ownership cost of $9,000 per year, the pay-as-you-go model for on-demand vehicles will appeal to rich and poor alike. The catch in all of this is rush hour. Unless people are willing to use autonomous carpooling, such as the existing service Lyft Line, then the number of idle autonomous vehicles in dense urban environments or large suburban corporate headquarters would overwhelm the advantages.

The other shift is urbanization. By 2050, 100 million more people will live in American cities. What will those cities be like? Clearly, they will be denser. Arguably, there will be fewer vehicles per capita, though there are persuasive arguments that counter that conclusion. The urban landscape could morph into people-oriented areas, with broader sidewalks, narrower streets, more trees and plants, voluminous bike lanes, and parking lots converted to parks. The emphasis will shift from transport to community.

The urban form of cities—the layout, roads, structures, and physical patterns of cities—could change dramatically if autonomous mobility was a well-planned, functional service. Today, all cities are noisy and crowded, and the overwhelming source of that noise and crowding is vehicles. In contrast, electric vehicles make little noise. If autonomous vehicles are single or zero occupancy, they will be of little help to cities, or the planet. If they are brought into service in dedicated lanes absent human drivers, their impacts could be significant and beneficial, what urban planner Peter Calthorpe calls "autonomous mass transit." ●

COMING ATTRACTIONS
SOLID-STATE WAVE ENERGY

The kinetic energy of oceans, which surge with roughly 80,000 terawatt hours of power, is extraordinary. It is a staggering amount of energy—enough to power human needs four times over. A single terawatt is the equivalent of 1 trillion watts and sufficient to provide electricity to 33 million U.S. homes. Because water is nearly one thousand times denser than air, aqua turbines are technically more efficient than wind turbines. The problem with wave energy technologies is economic inefficiency. They require moving parts that can withstand the stress and corrosion of the deep sea. The raw energy found in the ocean can easily become wave power's downfall.

A company in Seattle, Oscilla Power, has created a wave-energy technology that converts the kinetic energy of the ocean without external moving parts. The technology is simple in principle. It consists of a large solid-state float on the water's surface. Inside the float surface are magnets; outside are rods made of an iron-aluminum alloy. The rods, when compressed and decompressed, undergo stress changes, which are converted to electricity by coils wrapped around the rods. What causes the compression is a large, concrete heave plate tethered below the water by cables. This acts like an anchor that prohibits the solid-state float from moving with the rise and fall of the surface waves,

thereby creating compressive pulses within it. The heave, pitch, troughs, crests, and roll of oceanic surface water create a constant flow of compression and thus electricity. The computation of the weight of the heave plate, the configuration of the magnetic field in which the alloy rods are compressed, and the overall mass distribution of the system in response to the kinetics of the ocean surface are complex calculations aimed at achieving optimum output. However, once the parameters are set, the mechanics are fairly straightforward, thanks to the lack of turbines, blades, motors, and other moving parts.

A technology that captures a tiny fraction of the ocean's kinetic energy would be an astonishing achievement—if it were affordable. Affordability entails maintenance, replacement parts, servicing in high seas, and underwater cables to transfer the power. The qualities of ocean energy that make wave power so attractive are the same qualities that may take it out of human reach: It is an intense, random, and powerful force. Solid-state wave energy eliminates some of the key issues that have plagued other start-ups in the field. It may be the breakthrough. Or, perhaps, the wave-energy breakthrough is yet to come. Whether now or later, the ocean remains the largest untapped source of renewable energy on earth. ●

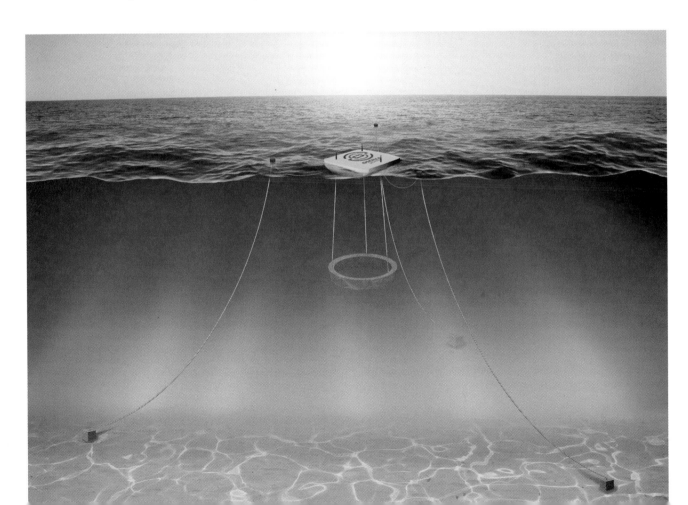

n 2000, the U.S. Green Building Council unveiled its Leadership in Energy and Environmental Design (LEED) certification program as a way to measure and mark more sustainable buildings. LEED, and its various metallic honors (silver, gold, and platinum), cultivated and challenged the building industry to change how it measures a building's value, and developed prescriptive credits in an attempt to quantify and evaluate a building's impact upon the environment and its inhabitants. LEED certification encompasses design, construction, maintenance, and operations. The metrics include lumens, water, energy use, cleaning products, daylighting, indoor air quality, renewable energy, and much more.

Six years after LEED standards were established, a different set of criteria was put forth by architect Jason McLennan and the Cascadia Green Building Council: the Living Building Challenge (LBC). (LBC is now owned and operated by the International Living Future Institute.) It too is a building certification program with core principles and performance categories. These seven categories are called "petals": Place, Water, Energy, Health and Happiness, Materials, Equity, and Beauty. LEED is about sustainability, the reduction of the negative environmental impacts caused by the built environment. LBC is based on regeneration, buildings that can reanimate and renew the environment, for both the natural world and human communities.

Fundamentally, LBC is not about leading, but about living. Buildings can function more like a forest, generating a net surplus of positives in function and form and exhaling value into the world. Buildings, in other words, can do more than simply be less bad. They can contribute to the greater good. LBC lays out criteria for what a living building is and does to benefit both people and planet. Each of the seven petals is populated by imperatives a building ought to fulfill—twenty in total. The imperatives are not a checklist. They are performance expectations that define a holistic approach to buildings based on a simple question: How do you design and make a building so that every action and outcome improves the world?

For example, living buildings should grow food, produce net-positive waste (a waste stream that nourishes living systems or land), create net-positive water, and generate more energy with renewables than they use. They need to incorporate biophilic design, satisfying humankind's innate affinity for natural materials, natural light, views of nature, sounds of water, and more. On the unnatural side of things, living buildings have to avoid all "red-listed" materials, such as PVC and formaldehyde. They are required to cater to the human scale, rather than the car scale, and intentionally educate and inspire others—building as teacher rather than container.

When it comes to greenhouse gas emissions, living buildings make their greatest impact by producing more energy than they consume and offsetting all embodied carbon as well. To provide energy to the world, the buildings are highly efficient, requiring significantly less energy to begin with than conventional "green" buildings, and integrate on-site renewable energy, such as solar or geothermal.

The path to achieving net-positive energy and the nineteen other imperatives is not prescribed, so each living building is shaped and tailored to local conditions and allows for local genius. Checking any given box is a matter of context. Ultimately, LBC accreditation is not based on meeting prescriptive design specifications or projected building performance. What matters instead, based on at least twelve months of occupancy and actual performance, is how a living building comes to life.

As with many innovations, initial uptake was slow to begin for the Living Building Challenge. It delivered on its name and proved a challenge that was nearly insurmountable for designers, architects, engineers, building inspectors, banks, and contractors. A steep learning curve flattened the initial adoption curve. Today, however, there are more than 350 buildings in various stages of certification, encompassing several million square feet in two dozen countries. Just as with LEED, as designers and contractors master the means and methodologies to achieve certification, the costs are reduced and confidence increases. Recent economic studies demonstrate that the initial cost of living buildings is going down and at the same time the provable return on a dollar-to-dollar basis is showing them to be economical, not just visionary.

Building the LBC way is not without its challenges. It requires up-front investment, a long-term view on returns, and significant technical expertise to tackle the unique dynamics of each project. Sometimes it entails overcoming restrictive building codes that make living buildings illegal in some places (e.g., processing sewage on-site is not allowed in all places). Addressing those hurdles, through incentives, policy change, and developing a deeper bench of experts, will be key to realizing the persuasive promise of this approach to the built environment; numerous positive regulatory changes have already been made thanks to the program. If society sees that the structures we build are actually human habitats—ecosystems made for us, by us—buildings that live are the ones that truly make sense.

And there is that final petal: beauty. Buildings that are LBC certified are spectacular to look at and be in. Architect David Sellers summed it up perfectly when he said the pathway to sustainability is beauty, because people preserve and care for that which feeds their spirit and heart. All other buildings are torn down sooner or later.

The Imperatives

1. *Limits to growth.* Only build on a previously developed site, not on or adjacent to virgin land.
2. *Urban agriculture.* A living building must have the capacity to grow and store food, based on its floor area ratio.
3. *Habitat exchange.* For each acre of development, an acre of habitat must be set aside in perpetuity.
4. *Human-powered living.* A living building must contribute to a walkable, bikeable, pedestrian-friendly community.
5. *Net positive water.* Rainwater capture and recycling must exceed usage.
6. *Net positive energy.* At least 105 percent of energy used must come from on-site renewables.
7. *Civilized environment.* A living building must have operable windows for fresh air, daylight, and views.
8. *Healthy interior environment.* A living building must have impeccably clean and refreshed air.
9. *Biophilic environment.* Design must include elements that nurture the human and nature connection.
10. *Red List.* A living building must contain no toxic materials or chemicals, per the LBC Red List.
11. *Embodied carbon footprint.* Carbon embodied in construction must be offset.
12. *Responsible industry.* All timber must be Forest Stewardship Council certified or come from salvage or the building site itself.
13. *Living economy sourcing.* Acquisition of materials and services must support local economies.
14. *Net positive waste.* Construction must divert 90 to 100 percent of waste by weight.
15. *Human scale and humane places.* The project must meet special specifications to orient toward humans rather than cars.
16. *Universal access to nature and place.* Infrastructure must be equally accessible to all, and fresh air, sunlight, and natural waterways must be available.
17. *Equitable investment.* A half percent of investment dollars must be donated to charity.
18. *JUST organization.* At least one entity involved must be a certified JUST organization, indicating transparent and socially just business operations.
19. *Beauty and spirit.* Public art and design features must be incorporated to elevate and delight the spirit.
20. *Inspiration and education.* A project must engage in educating children and citizens. ●

The Brock Environmental Center was built by the Chesapeake Bay Foundation at Pleasure House Point in Virginia Beach, Virginia. Completed in 2014, it produces all of its drinking water from rainfall, uses 90 percent less water than a commercial building of the same size, and generates 83 percent more energy than it consumes. The Brock Center is the first commercial building in the United States allowed to treat and process rainwater to federal potable standards.

On Care for Our Common Home
POPE FRANCIS

Thousands of books and articles have addressed climate change in the past forty years; however, when Pope Francis penned "On Care for Our Common Home," his encyclical letter on the environment, it was as if a veil of obscuring jargon was lifted. The scientific issue of global warming was given a fully human dimension, thoughtful and caring. An encyclical is a papal letter to the 5,100 bishops of the Roman Catholic Church and is intended to guide its leaders on how to teach and steward its adherents. "Laudato Si" is a message from the Church, to be sure, and it is a message from the heart, suffused with compassion and unflinching in its analysis of the causes of global warming and its unjust and inequitable impact on the poor. In this message, global warming is illustrated—perhaps for the first time—as a universal and moral issue, not only an environmental issue. This excerpt is 1,353 words taken from the 37,000-word encyclical. — PH

The climate is a common good, belonging to all and meant for all. At the global level, it is a complex system linked to many of the essential conditions for human life. A solid scientific consensus indicates that we are presently witnessing a disturbing warming of the climatic system. In recent decades this warming has been accompanied by a constant rise in the sea level and, it would appear, by an increase of extreme weather events, even if a scientifically determinable cause cannot be assigned to each particular phenomenon. Humanity is called to recognize the need for changes of lifestyle, production and consumption, in order to combat this warming or at least the human causes which produce or aggravate it. It is true that there are other factors (such as volcanic activity, variations in the earth's orbit and axis, the solar cycle), yet a number of scientific studies indicate that most global warming in recent decades is due to the great concentration of greenhouse gases (carbon dioxide, methane, nitrogen oxides and others) released mainly as a result of human activity. As these gases build up in the atmosphere, they hamper the escape of heat produced by sunlight at the earth's surface. The problem is aggravated by a model of development based on the intensive use of fossil fuels, which is at the heart of the worldwide energy system. Another determining factor has been an increase in changed uses of the soil, principally deforestation for agricultural purposes.

Climate change is a global problem with grave implications: environmental, social, economic, political and for the distribution of goods. It represents one of the principal challenges facing humanity in our day. Its worst impact will probably be felt by developing countries in coming decades. Many of the poor live in areas particularly affected by phenomena related to warming, and their means of subsistence are largely dependent on natural reserves and ecosystemic services such as agriculture, fishing and forestry. They have no other financial activities or resources which can enable them to adapt to climate change or to face natural disasters, and their access to social services and protection is very limited. For example, changes in climate, to which animals and plants cannot adapt, lead them to migrate; this in turn affects the livelihood of the poor, who are then forced to leave their homes, with great uncertainty for their future and that of their children. There has been a tragic rise in the number of migrants seeking to flee from the growing poverty caused by environmental degradation. They are not recognized by international conventions as refugees; they bear the loss of the lives they have left behind, without enjoying any legal protection whatsoever. Sadly, there is widespread indifference to such suffering, which is even now taking place throughout our world. Our lack of response to these tragedies involving our brothers and sisters points to the loss of that sense of responsibility for our fellow men and women upon which all civil society is founded.

Given the complexity of the ecological crisis and its multiple causes, we need to realize that the solutions will not emerge from just one way of interpreting and transforming reality. Respect must also be shown for the various cultural riches of different peoples, their art and poetry, their interior life and spirituality. If we are truly concerned to develop an ecology capable of remedying the damage we have done, no branch of the sciences and no form of wisdom can be left out, and that includes religion and the language particular to it.

The natural environment is a collective good, the patrimony of all humanity and the responsibility of everyone. If we make something our own, it is only to administer it for the good of all. If we do not, we burden our consciences with the weight of having denied the existence of others.

Ecology studies the relationship between living organisms and the environment in which they develop. This necessarily entails reflection and debate about the conditions required for the life and survival of society, and the honesty needed to question certain models of development, production and consumption. It cannot be emphasized enough how everything is interconnected. Time and space are not independent of one another, and not even atoms or subatomic particles can be considered in isolation. Just as the different aspects of the planet—physical, chemical and biological—are interrelated, so too living species are part of a network which we will never fully explore and understand. A good part of our genetic code is shared by many living beings. It follows that the fragmentation of knowledge and the isolation of bits of information can actually become a form of ignorance, unless they are integrated into a broader vision of reality.

When we speak of the "environment," what we really mean is a relationship existing between nature and the society which lives in it. Nature cannot be regarded as something separate from ourselves or as a mere setting in which we live. We are part of nature, included in it and thus in constant interaction with it. Recognizing the reasons why a given area is polluted requires a study of the workings of society, its economy, its behavior patterns, and the ways it grasps reality. Given the scale of change, it is no longer possible to find a specific, discrete answer for each part of the problem. It is essential to seek comprehensive solutions which consider the interactions within natural systems themselves and with social systems. We are faced not with two separate crises, one environmental and the other social, but rather with one complex crisis which is both social and environmental. Strategies for a solution demand an integrated approach to combating poverty, restoring dignity to the excluded, and at the same time protecting nature.

What kind of world do we want to leave to those who come after us, to children who are now growing up? This question not only concerns the environment in isolation; the issue cannot be approached piecemeal. When we ask ourselves what kind of world we want to leave behind, we think in the first place of its general direction, its meaning and its values. Unless we struggle with these deeper issues, I do not believe that our concern for ecology will produce significant results. But if these issues are courageously faced, we are led inexorably to ask other pointed questions: What is the purpose of our life in this world? Why are we here? What is the goal of our work and all our efforts? What need does the earth have of us? It is no longer enough, then, simply to state that we should be concerned for future generations. We need to see that what is at stake is our own dignity. Leaving an inhabitable planet to future generations is, first and foremost, up to us. The issue is one which dramatically affects us, for it has to do with the ultimate meaning of our earthly sojourn.

Many things have to change course, but it is we human beings above all who need to change. We lack an awareness of our common origin, of our mutual belonging, and of a future to be shared with everyone. This basic awareness would enable the development of new convictions, attitudes and forms of life. A great cultural, spiritual and educational challenge stands before us, and it will demand that we set out on the long path of renewal.

We must regain the conviction that we need one another, that we have a shared responsibility for others and the world, and that being good and decent are worth it. No system can completely suppress our openness to what is good, true and beautiful, or our God-given ability to respond to his grace at work deep in our hearts. I appeal to everyone throughout the world not to forget this dignity which is ours. No one has the right to take it from us. May our struggles and our concern for this planet never take away the joy of our hope. ●

COMING ATTRACTIONS
DIRECT AIR CAPTURE

For hundreds of millions of years, plants have been harnessing the power of photosynthesis to capture carbon dioxide from air and transform it into biomass—the building block of the plant world—using renewable solar power. Only recently have humans started engineering similar direct air capture (DAC) systems. Their goal is to "mine the sky" by capturing and concentrating ambient carbon dioxide. Near-term markers for that carbon dioxide are sought in manufacturing and industrial processes. Long-term hopes are to use DAC and carbon dioxide storage to help achieve and maintain drawdown.

Conceptually, DAC machines act like a two-in-one chemical sieve and sponge. Ambient air passes over a solid or liquid substance and its carbon dioxide binds with chemicals in the substance that are selectively "sticky," while other gases in the air are free to go. Once those capture chemicals become fully saturated with carbon dioxide, energy is used to release the molecules in a purified form. Releasing the carbon dioxide restores the chemicals' ability to filter it out. So the cycle repeats over and over again.

The fundamental technical challenge with DAC systems is showing that it can be done efficiently and cost effectively. First, carbon dioxide in the air is very dilute: 0.04 percent. Separating out meaningful quantities of carbon dioxide requires a large volume of air to come into contact with the capture materials. Second, the capture-release cycle consumes energy. So energy sources that are low cost and low carbon, and do not have competing uses (e.g., helping reduce carbon emissions in the first place), need to be found and used wisely.

Nevertheless, innovators around the world are pursuing a range of DAC designs that they believe will one day offer economically viable carbon dioxide capture from the air. For the capture step, many companies are building on the chemistry of amines (ammonia-like compounds) prevalent in traditional industrial carbon dioxide–capture processes. (Engineers have been using amine-based systems to capture carbon dioxide from the concentrated exhaust streams of various fuel and chemical manufacturing operations for decades.) Some DAC innovators are using novel materials for carbon dioxide capture, such as anionic exchange resins. Plus, a range of material science advances in areas such as metal organic frameworks and aluminum silicate materials could open new frontiers in efficient capture of carbon dioxide from the air.

There are significant innovations happening around the processes used to regenerate the captured carbon dioxide—that is, how the DAC system squeezes the capture "sponge." Temperature, pressure, and humidity can be applied to saturated capture materials to release carbon dioxide in purified form. DAC system designers are developing regeneration techniques that use energy as sparingly as possible and/or rely on energy from the wind, sun, or waste industrial heat.

In the near term, the purified carbon dioxide released from DAC units could be used in a wide range of manufacturing applications. For example, some DAC start-up companies are working to make synthetic transportation fuels using air-captured carbon dioxide, while others are looking to use atmospheric carbon dioxide in greenhouses to improve indoor agricultural yields. But that is just the beginning. Carbon dioxide captured from DAC systems has been proposed for use in manufacturing plastics, cement, and carbon fiber—and even for permanently disposing of excess atmospheric carbon dioxide in underground geologic formations.

In the future, DAC systems could play a pivotal role in the fight against climate change. DAC-derived fuels could help meet the growing demand for decarbonized long-haul transportation if sustainable biofuel supplies are limited, and such fuels could displace the use of fossil fuels in a range of manufacturing applications. In addition, DAC systems could provide a robust and scalable offsetting and neutralizing mechanism for difficult-to-decarbonize sectors of the economy, and could eventually help clean up carbon dioxide from the atmosphere as a sequestration technology.

But again, the main business challenge facing DAC entrepreneurs today is economics. Right now, the lack of strong carbon regulation in most geographies creates small markets for companies to use carbon dioxide from DAC. Nobody will pay to build pilot plants for DAC storage.

There are markets for compressed carbon dioxide already. Applications range from enhanced oil production and beverage carbonation to greenhouses and other niche applications. However, there are abundant supplies of inexpensive, concentrated carbon dioxide elsewhere. Natural carbon dioxide deposits in geologic formations, and highly concentrated industrial sources such as ethanol and chemical manufacturing, depress prices customers are willing to pay for carbon dioxide. For example, pipeline-scale quantities of carbon dioxide used for oil production in the United States can cost as little as $10 to $40 per ton of carbon dioxide, well below the $100 per ton (or more) for DAC-captured carbon dioxide in early prototypes.

Academics have calculated that large-scale deployment of DAC systems could reduce costs to competitive ranges. However, entrepreneurs are currently caught in an external cycle of inactivity: Research and development funding is generally lacking, markets are unable to support adoption, and more learning and innovation is needed for systems to technologically mature. In addition, advances in DAC designs may help reduce the cost of competing carbon dioxide capture at more concentrated industrial exhaust systems, which could maintain downward pressure on carbon dioxide prices. While DAC systems can be sited in

This is a carbon capture unit created by Global Thermostat. It uses amine-based chemical sorbents bonded to porous honeycomb ceramics that together act as carbon sponges, efficiently adsorbing carbon dioxide directly from the atmosphere or smokestacks. The captured carbon dioxide is stripped off and collected using low-temperature steam. The output is 98 percent pure carbon dioxide at standard temperature and pressure. Nothing but steam and electricity is consumed, and no other effluents or emissions are created. The entire process is mild, safe, and carbon negative.

flexible locations to reduce costs associated with carbon dioxide transportation, thereby boosting overall cost competitiveness, that benefit will vary from one place to another.

Going forward, DAC developers will have to develop creative engineering and business models—and get more support from policy focused on long-term climate goals—in order to compete against existing low-cost carbon dioxide sources and growing supplies of compressed carbon dioxide captured from power and industry.

Furthermore, DAC will have to make extra effort with regulators to ensure that it works with other carbon-reduction and removal solutions. Few protocols exist today for DAC systems to get climate-related credit for captured, let alone stored, carbon dioxide. The technology will have to fit into policy frameworks that help the world get to net zero emissions and then into drawdown. Navigating among the various stakeholders and perspectives is possible, but it will not be easy.

Despite economic, technical, and political challenges, a number of intrepid entrepreneurs and researchers are hard at work to improve DAC technologies. Many companies aim at commercializing DAC technology in North America and Europe. Professor Klaus Lackner at Arizona State University has launched the Center for Negative Carbon Emissions to research DAC technologies, and the U.S. Department of Energy embarked on its first-ever DAC research projects in 2016.

It will be fascinating to see how these early ventures into DAC research and commercialization evolve. Can these efforts and early markets for DAC stimulate a new, sustainable process-engineering industry for capturing and storing billions of tons of carbon dioxide directly from the air? Time will tell if humankind can make it happen. ●

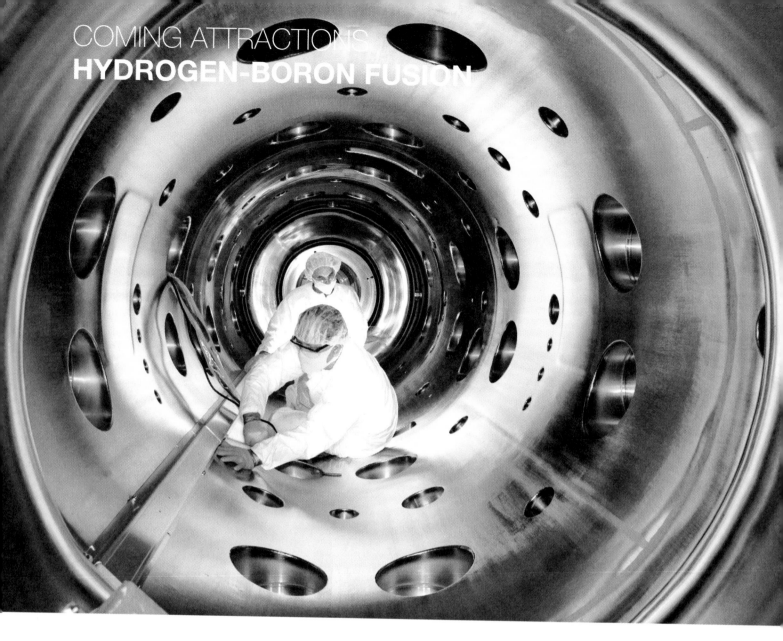

I n 1924, the British physicist Sir Arthur Eddington theorized that nuclear fusion must be at the heart of the sun's radiant energy. Unwittingly, he set off one of the most expensive scientific quests in history: creating the power of a star with a fusion reactor. Unlike nuclear fission, which splits heavy atoms to generate heat, fusion smashes light atoms together to create the energy that powers the stars. One could argue that the world already has a perfectly good fusion reactor, albeit offshore: If captured, one day's energy from the sun would power the earth for many years. Currently, a tiny fraction of that energy is captured by solar photovoltaics, and indirectly with biomass, hydro, wave, and wind. Fossil fuels are themselves stored energy from the giant fusion reactor in the sky; albeit with a production time of many millions of years, and a poor conversion efficiency. (A 2003 study by the ecologist Jeffrey S. Dukes estimated that the average gallon of gasoline requires over 90 tons of prehistoric biomass as raw material.) Renewable energy, however, is variable and utilities want a steady source of energy that does not turn off. To that end, scientists and engineers have been pursuing the Holy Grail of physics

since the 1930s: a clean, virtually unlimited source of energy that would take the world beyond the age of coal, gas, and oil, and power it for millennia into the future. Accomplishing star power would generate "an inflection point in human history," Lev Grossman declared in *Time* magazine in 2015—an "energy singularity" that would spell the end of fossil fuels.

Making starlight on earth is absurdly difficult. For more than fifty years, theorists and engineers imagined and constructed what they hoped would be a working fusion reactor. Millions of experiments were tried, well over $100 billion invested, and no one came close to succeeding. Until recently, that is. In the past two decades, private enterprise entered the field. With less money, these organizations had to be nimble by utilizing innovative approaches employed in high-tech start-ups—fail faster and fail better at greatly reduced cost.

In June 2015, a company that had been considered a maverick because of its unorthodox approach announced it had achieved one-half of the Holy Grail, the more difficult half, nicknamed "long enough." The company, Tri Alpha Energy (TAE), had been secretive for most of its eighteen-year history. And for good reason: The history of fusion energy is littered with hype, fantasy, and claims that fell flat. Better to be quiet and do the work, and that is what TAE did. By the time of its announcement, TAE had already completed more than forty-five thousand experimental runs.

TAE's visionary cofounders, the late Norman Rostoker and chief technology officer Michl Binderbauer, started the company with the end in mind. They asked what might seem like the obvious question: What a utility needs, rather than what a plasma physics journal would want to publish. Utilities want safe, compact, affordable, dependable energy generators that could be built and placed anywhere they are needed. Safety is critical. Although fusion reactors do not produce radiation in the same way as fission reactors, fusion reactors to date have been based on tritium and deuterium fuels, isotopes of hydrogen that produce free neutrons. Neutrons cause a reactor to become radioactive over time, which means its working components decay and have to be replaced every six to nine months.

Rostoker and Binderbauer went out on a limb and chose hydrogen-boron as their fuel because of safety, practicality, and availability. Hydrogen-boron does not produce neutrons of any significance. The reactor will hold up for decades, if not a century. It can be placed safely anywhere. If it shuts down, nothing happens. Or to put it another way, if something happens, it shuts down. If it shuts down, it can be restarted with a household generator. Whereas tritium and deuterium are scarce, there is at least a hundred-thousand-year supply of boron and it is cheap. To make the point, TAE half-jokingly says they will give you the fuel for free if you buy the reactor.

Hydrogen-boron fusion produces three helium atoms, and a fractional portion of remaining mass converts to energy . . . a lot of energy. Atoms can make energy in two ways: divide or unite; fission or fusion. Einstein predicted that given the right conditions, mass can become energy or vice versa, and that the amount of energy contained in a tiny bit of mass, in human terms, is astounding. Hydrogen-boron fusion produces three to four times more energy per mass of fuel than nuclear fission, with virtually no waste: That means no plutonium, no radiation, no meltdowns, no proliferation.

Some plasma physicists scoffed at TAE's choice of fuel because hydrogen-boron fusion requires thirty times more heat than the "mere" 180 million degrees Fahrenheit required in a conventional fusion reactor—5.4 billion degrees Fahrenheit, to be precise. For hydrogen-boron, this is "hot enough," the other half of successful fusion. When you put long enough and hot enough together, you have made starlight on earth.

Long enough refers to the ability of a fusion reactor to sustain plasma indefinitely. Plasma is the fourth state of matter, completely unlike any other (solid, liquid, and gas being the other three). When you see images of cloudlike galaxies, the sun, or the northern lights dancing on the horizon, you are seeing plasma. It is an ionized gas, and when it is heated it becomes virtually impossible to control. If plasma touches anything, it disappears in a nanosecond. It is pretty much like trying to pick up a cat by the tail. Plasma is a cloud of subatomic particles with the electrons stripped out. It constitutes 99 percent of the universe. In order to achieve

fusion, plasma has to be contained and controlled and then heated to supercritical temperatures. Those are two opposing forces: The hotter plasma gets, the more violently unstable it becomes. To corral it has been the challenge of plasma physicists and engineers.

Binderbauer achieved *long enough*—that is, a plasma state that could be indefinitely sustained—by an ingenious method. By placing six particle beam injectors that fired hydrogen atoms at the periphery of the plasma field, he created the equivalent of a spinning plasma top. Every child knows the faster a top rotates, the more stable it becomes. Similarly, plasma becomes more stable as it spins, heats up, and generates its own magnetic field. In the TAE reactor, the plasma is self-confining as long as its rotational speed is maintained. The faster it spins, the hotter it gets; the hotter it gets, the more stable it becomes—the opposite of every fusion technology previously promoted and funded.

By late 2017, TAE will have built the fourth reactor in its history, one large enough to achieve fusion. With their theory of *long enough* plasma stabilization accomplished, they now have to achieve *hot enough*. How do you create 5.4 billion degrees Fahrenheit when the sun tops out at 25.2 million Fahrenheit? According to Binderbauer, you let the plasma do it. The Large Hadron Collider in Switzerland is creating temperatures in the trillions of degrees, a thousand times what TAE requires. These numbers are attained in the Hadron particle accelerator, thanks to the high energy at which particles travel around its sixteen-mile circumference. Hence, for TAE, Binderbauer believes the remaining challenge is one of engineering, not science. You can work out what the temperature will be in the new TAE reactor with a scratch pad (and a degree in plasma physics) because you know the circumference of the plasma field.

What abundant, clean energy produced by a fusion reactor would do is speculative. In terms of energy, a viable fusion reactor could be the power station of the future: Hydrogen-boron fusion is carbon-free, sustainable, and safe. At this time, the company is predicting costs of ten cents per kilowatt-hour, which will drop to five cents. The latest power purchase agreements in wind energy are coming in at two cents per kilowatt-hour, and solar is not far behind. However, renewable energy is enabled by dispatchable power or storage. Until there is a reliable substitute for gas and coal or effective energy storage on a mass scale, the demand for dispatchable power provided by carbon-heavy fuels will persist. There is an energy revolution afoot, however, whether fusion works or not. If fusion joins the pack with other renewable energy technologies, it will be a rout in terms of fossil fuels for electricity. In time, these sources of energy may come to underpin drawdown pathways across all industries.

Inside TAE's company lobby in Irvine, California, there is a basket of pink rubber pigs with wings that exemplify the company's attitude toward a skeptical world. Apparently, pigs may fly soon. ●

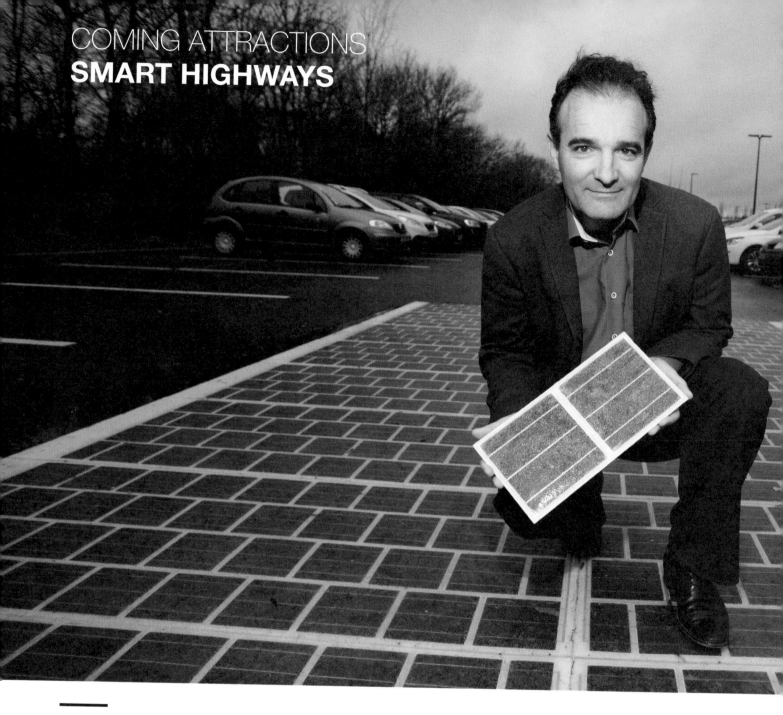

COMING ATTRACTIONS
SMART HIGHWAYS

Philippe Raffin poses with a section of a Wattway solar road, solar tiles that adhere to existing roadways in order to produce electricity. Developed in France, a ten-by-twenty-foot section can supply the electricity requirements for an average French home.

More than 160,000 miles of asphalt comprise the U.S. National Highway System. On eighteen of them, a stretch south of Atlanta in west Georgia, an initiative called The Ray is working to reimagine what a highway could be. The Ray is named for Ray C. Anderson, the late founder and CEO of Interface, a company that makes carpet tiles and has, since the mid-1990s, charted a course for sustainability in business. Anderson and the Interface community fundamentally reworked the way they operate, transforming a petroleum-based manufacturing company into a restorative enterprise. Their first sustainability mission was for Interface to do no harm; the next, to create net good.

True to its namesake, The Ray will similarly upend business as usual. At present, highways are the epitome of unsustainable. Cars and trucks burn petroleum fuels and emit pollutants as they speed across energy-intensive asphalt surfaces or, worse, idle in traffic. Highways themselves fragment ecosystems and enable sprawling, car-centric development. View a highway at rush hour and you cannot help but wonder if this is the best society can do, especially in the era of climate change. Designed to be a living laboratory, The Ray aims to prove better is attainable. Motor vehicles and the infrastructure they require will continue to be important pieces of mobility and connectivity, even as

transform auto-based transport have focused on cars. The team behind The Ray posits that the infrastructure those cars rely on, namely highways, must also evolve to make clean transportation a reality. Tapping local and national expertise, The Ray is beginning to pilot that evolution.

Electric vehicles (EVs) are a focal point for this living lab. Currently, more than one hundred thousand tons of carbon dioxide are emitted along the eighteen-mile corridor each year. To shift that statistic, The Ray is creating infrastructure on which EVs, the cleanest cars, rely. A roadside visitor's center along its stretch of highway now houses a solar photovoltaic (PV) charging station, where EVs can power up for free in less than forty-five minutes. Eventually, The Ray aims to integrate special lanes for EVs that charge them as they pass through, no stop required. The state of Georgia already has the second-highest count of EV registrations in the United States. More EV infrastructure will mean more EV travel; more EV travel will mean lower emissions. The next generation of cars is already arriving. Smart highways face the task of catching up and thinking ahead.

Also central to The Ray's design is the future of energy. Solar technologies are an ideal fit for the unused open space that flanks highways, so The Ray will house a 1-megawatt PV farm along its right-of-way—an approach already in use elsewhere. Exposed 90 percent of the time, road surfaces themselves are also prime for solar generation. The aptly named Wattway photovoltaic pavement, a French technology, will allow The Ray to produce clean electricity for uses from LED lighting to EV charging, while improving tire grip and surface durability. A noise barrier lined with PV panels may be another win-win solution along The Ray, simultaneously creating energy and containing the sound pollution currently endured by local communities.

The Ray has a kindred spirit in innovation across the Atlantic. Designer Daan Roosegaarde and Heijmans, a European construction-services company, have partnered on an award-winning smart highway pilot in the Netherlands. Among its technologies, the "Route 66 of the future" incorporates energy harvesting, weather sensors, and dynamic paint, including bioluminescent "glowing lines" that absorb sunlight during the day and glow at night. No streetlights, and their attendant energy use, required. The Dutch efforts are now expanding within the Netherlands and beyond to China and Japan.

Since modern motorways first emerged, they have seen remarkably little advancement in design. Now, climate change and the emergence of electric and autonomous vehicles are placing new demands on them. Highways need a smarter way forward. Efforts such as Roosegaarde's and The Ray provide early evidence that this dirty infrastructure can become clean, as well as safe, efficient, and even elegant. Because highways have remained stagnant for so many decades, there exists an oversize opportunity for innovation. They are highly regulated, though, so realizing that opportunity means mobilizing bureaucracy and seeing sustainability join safety as a key roadway priority. The term *smart highways* calls attention to technology, but greasing the wheels of institutional change may prove equally essential to their success.

transit alternatives grow. Understanding that, The Ray aims to morph this stretch of road into a positive social and environmental force, the world's first sustainable highway. Prove it can be done there, and this corridor of "smart" highway may spark the same kind of revolutionary change Interface has.

Vehicles and the surfaces upon which they travel tend to evolve concurrently. A network of paved roads, a third the size of the contemporary American highway system, enabled wheeled vehicles to move armies and goods throughout the Roman Empire. When mass production of automobiles emerged in the twentieth century, so did motorways; for example, the Dwight D. Eisenhower National System of Interstate and Defense Highways in the United States. In the face of climate change and an energy revolution, efficient, electric, and autonomous vehicles are beginning to join modern roadways. Indeed, almost all efforts to

COMING ATTRACTIONS
HYPERLOOP

Most people are too young to remember how vacuum tubes were employed to propel messages, deposits, and documents in steel canisters within buildings and cities. In New York City, a pneumatic tube mail system connected the West Side to East Harlem until 1953. The loop, underneath the street, was manned by operators called Rocketeers and could shunt parcels and mail from Grand Central Terminal to the General Post Office in four minutes.

Now imagine autonomous pods—7.5 feet in diameter and replete with ergonomic chairs, mellow world-beat tunes, and shoulder belts—thrusting you up to 760 miles per hour through steel conduits from San Francisco to Los Angeles in thirty-five minutes, for the cost of a bus ticket. That is the vision of the Hyperloop, seven hundred miles of low-pressure tubes running up and down California. It is based on a paper written by Elon Musk in 2013, "Hyperloop Alpha," that evangelized for a fifth mode of transportation. Musk challenged the concept of high-speed rail and called for a worldwide, open-source design collaboration to build a $6 billion solar-powered system in California. It worked: There are now several companies in the world striving to create complete Hyperloop systems.

Robert Goddard, known for his rocketry science, first imagined vactrains in 1910, magnetically levitated rockets flying through vacuum tubes at an estimated 960 miles per hour. It never went further than a paper, but the system Musk imagined a century later is not that different. Hyperloops, as proposed, are extremely efficient; one reason is the absence of air. Every mode of transportation takes place in air or water, and the higher the speed, the greater the resistance. At 600 to 700 miles per hour, sea-level air is thicker than water in terms of resistance. Every child has placed his or her hand outside a speeding car and felt that force. The challenge of a vacuum-enabled system is removing the last 10 percent of the air.

It takes a great amount of energy to create and maintain a complete vacuum, so Musk and others backed off and are engineering systems that operate in partial vacuums. A fan is placed at the front of the pod to eliminate the buildup of air, with some of it exhausted through to the rear and the balance streamed along the sides as a bearing to prevent the pod from touching the inner wall of the tube. The passenger capsule would be pressurized and sealed.

The promise of a Hyperloop is speed; the virtue of a Hyper-

if needed. A more difficult challenge may be the forces passengers are subjected to on turns. At over 700 miles per hour, even tiny changes in direction could subject passengers to G forces similar to a fighter pilot. Commercial aircraft make slow turns over many miles to minimize the forces on their passengers; hyperloops, which need to follow the terrain, may not have that option.

In addition to safety, Hyperloop may face challenges from infrastructure costs, permitting, and more. After all, building high-speed rail has proven expensive and difficult; Hyperloop shares many of the same design requirements—straight flat track, durable foundations, high peak power demand—only to a greater degree. This is not to say it is impossible or even not worthwhile. The United States built a lot of freeways after World War II, but look what it did to the cities and suburbs. What will a Hyperloop network do? Or will it ever be a network? What happens to the city centers it connects to? Where does it go when most rights-of-way are taken? Is going faster and faster helpful anymore? It has yet to have its Kitty Hawk moment. One recalls the skepticism the Wright brothers faced with their fixed-wing craft barely able to attain an altitude of 10 feet and distance of 120 feet. The French ridiculed them as *bluffeurs* (bluffers). That all changed with their first successful flights on the coast of North Carolina.

Hyperloop companies are busy. Hyperloop One has already made a successful test run at its track in North Las Vegas, at 330 miles per hour in open air. It has signed an agreement with Dubai's Port of Jebel Ali to explore how the 18 million containers landing at the port annually could be moved quickly and safely. And Hyperloop is proposing a door-to-door pod system whereby passengers in Dubai are picked up at their home in an autonomous pod and whisked away to a Hyperloop that reaches Abu Dhabi in twelve minutes. The same company is proposing routes from Los Angeles to Las Vegas, Helsinki to Stockholm, and Moscow to St. Petersburg for cargo. The minister of economy in Slovakia is proposing Hyperloop routes from Bratislava to Budapest and Vienna. Perhaps the most innovative of all is Hyperloop Transportation Technologies, a crowdsourced virtual company of more than five hundred unpaid scientists and engineers from around the world who receive shares in their start-up in lieu of compensation.

Proponents believe that information technologies have sped up communications and brought the world closer together, and that it is now time to do the same with transportation. "Transportation is the new broadband" is their mantra. In the proposed Hyperloop system for California, you could live in L.A. and work in Silicon Valley. And therein lies what is known as the Jevons paradox: As a service or product becomes more affordable, human beings do not necessarily save the money; instead they use more, as in the case of cheap electricity, or purchase something else—another car, a vacation home, or maybe flat-screen televisions for every room. The paradox is that saving costly energy provides cash to spend more. The energy savings can be partially or wholly lost through consumer behavior. In other words, the Hyperloop can create the most effective and renewable transport system imagined, or it can catalyze yet another flood of materialism that already engulfs much of the world.

When Elon Musk wrote "Hyperloop Alpha," a paper challenging the world to accelerate the development of a functional hyperloop system, a group of engineering students at Delft University dived in to enter the competition for the best pod design. The Delft team placed second to the MIT team. Ten of the thirty-three-member team took a year off and are building a pod to compete with other winners on Musk's hyperloop test track in Hawthorne, California.

loop is how little energy it uses to move people and cargo. Estimates per passenger mile are 90 to 95 percent less than planes, trains, or cars. At the speeds envisioned, wheels are literally a drag. Hyperloops are levitated by magnets powered by solar and wind power, with the only real friction being the residual amount of air in tubes. Linear-induction motors, the same kind used in airport shuttle systems, would be used to start and accelerate the passenger pod. The pods would be constructed of carbon fiber, weighing less than one-third of the passengers and luggage. A center rail with magnets on both sides would act as a stabilizer at high speed and an emergency braking system if needed. Some designs incorporate virtual windows with LED screens showing an ersatz panorama of the passing scenery.

Not everyone is a fan. The prospect of tubing to Los Angeles without an obvious way to stop and escape in an emergency sends claustrophobic chills up some people's spines. However, that is exactly what an airplane is: a pod moving at high speed from which you cannot exit, subject to uncontrollable forces such as wind shear, lightning, icing, and flocks of birds. Conversely, the Hyperloop pods do have doors that unseal and linear-induction motors that take you to the nearest escape hatch

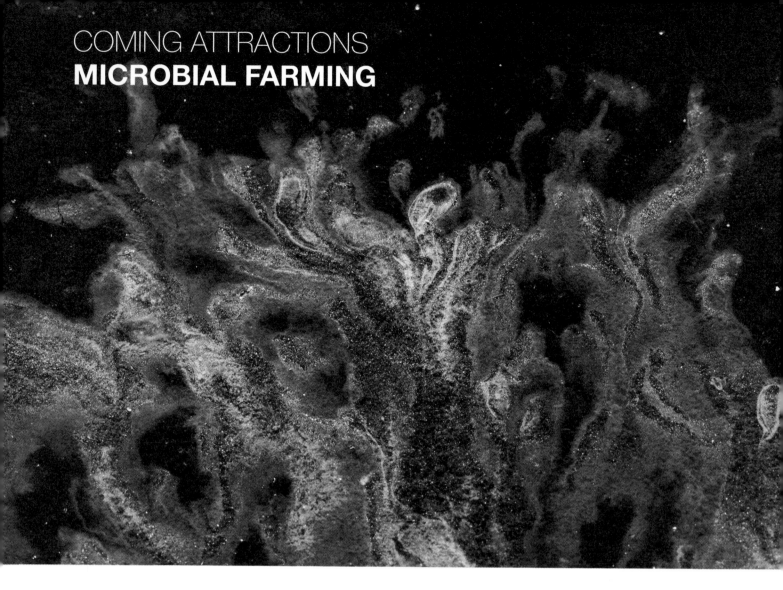

COMING ATTRACTIONS
MICROBIAL FARMING

Imagine: A farmer drives to the local fertilizer store in a four-ton pickup and leaves with a ten-pound bag of nitrogen-fixing bacteria that can fertilize 150 acres of wheat by making bioavailable nitrogen from thin air. A nitrogen-fixing bacteria for wheat has yet to be discovered, but science has just begun to look. Legumes such as soy, alfalfa, and peanuts already have anaerobic bacteria that can break down atmospheric nitrogen into usable nitrates. The roots of legumes cosset these bacteria, protecting them from oxygen and providing them with sugar exudates. In return, the plant receives vital nitrogen. Just as David Montgomery and Anne Biklé make clear in their book *The Hidden Half of Nature*, excerpted in this book, the exploding awareness of and research into the soil microbiome parallels the discoveries in the human microbiome. Both are ecosystems of unimaginable complexity; both are the foundation of health and well-being.

In one gram of soil there can be up to 10 billion denizens, and between 50,000 and 83,000 different species of bacteria and fungi. A gram is 0.035 ounce, and in that thimble of soil is among the most diverse living systems in the world. There can be dramatic variation in these underground ecosystems within just a few feet, depending on whether the soil is under sorghum, an oak tree, or a molehill.

This much is known: The potential of the bacteria, viruses, nematodes, and fungi in the soil is currently immeasurable and sweeping in its possibilities to address agriculture's impact on global warming. Their climate significance is rooted in the prospect of microbes to dramatically reduce the need for synthetic fertilizers, pesticides, and herbicides, while improving crop yields, plant health, and food security.

The soil microbiome has every big agricultural company in the world researching, partnering with, and gobbling up start-ups in the field of soil-microbe identification and testing. They are looking for microbes that will help them do what they have always done—make industrial agriculture more profitable. Ironically, that search trends toward identifying microbes that, well, kill. Researchers within this camp describe microbes as "weapons" against armyworms, rootworms, aphids, mites, cabbage loopers, and weeds. Genetically modified corn and soy stitch in *Bacillus thuringiensis* to create crystal proteins that kill caterpillars, moths, and butterflies. Microbes that are weed killers are already being commercialized.

Even though Big Ag dreams of weaponizing microbes, the nature of the microbial world is the opposite. It is primarily about mutualism—activities between two organisms that are beneficial to each other—rather than competition, the idea of one species prevailing over another.

A healthy soil biome is rich in carbon because soil microbes feed on sugar-rich exudates from the roots of plants; in turn, the bacteria dissolve rock and minerals and make those nutrients bioavailable to plants. A healthy biome is suffused with organic matter, which retains three to ten times more water than degraded soils, providing resilience and drought tolerance. It creates healthier plants and greater aboveground biodiversity as well. The solutions in *Drawdown* outlining regenerative farming and conservation agriculture, as well as those that address agroforestry, tree intercropping, and managed grazing, all feed the soil microbiome, reap the benefits thereof, and significantly reduce or eliminate the need for fossil fuel–derived fertilizers.

At present, converting nitrogen to ammonia for fertilizer requires 1.2 percent of the world's energy use. The process creates emissions from fossil fuel energy generation, and much of that nitrogen ends up in the sky as nitrous oxides—a greenhouse gas 298 times more powerful than carbon dioxide over the course of a century. Or it leaches into groundwater and waterways, causing the overgrowth of algae and dead zones where marine life suffocates from a lack of oxygen.

To be restorative, agriculture aligns with biology and nature, as opposed to warring with it. When a seed is tucked into the soil, a complex array of soil organisms mobilize and support its growth, coevolving as it matures, blooms, fruits, and seeds. The soil microbiome invites agriculture to do a much better job of getting what is wanted from the soil—healthy, tasty, abundant food—by harmonizing farming with the needs of the soil. It comes down to a simple fact: Plants and soil feed upon each other. If that cycle is interrupted by synthetics, whether they be fertilizers or pesticides, the plant is weakened and the soil is diminished in fertility and life.

The microbial farming revolution could not come at a better time. Estimates vary, but agriculture contributes approximately 30 percent of total greenhouse gas emissions. In the past, given what was known and the techniques at hand, reducing agricultural emissions might have meant decreasing world food production. With humanity marching toward more than 9 billion people in 2050, that is not an option.

Soil quality is declining in the world, presenting humankind with a choice: Try to correct this with yet more chemicals or rebuild a healthy soil ecosystem. By inoculating degraded and diminished soils with combinations of organisms that are symbiotic with the crops and foods people want, agriculture can create a virtuous circle, doing what life does. In the words of biologist Janine Benyus, life creates the conditions conducive to life, and there is reason to believe that a new era in agriculture is beginning, one that will fulfill both of its mandates: clean, plentiful, nutritious food along with truly sustainable farming practices that continuously create a more vibrant and nurturing planet for all. ◉

Left: Iron and manganese bacterial oxidizers in the mud of a fishpond.

Below: Scientists sampling bacteria in the Amboseli National Park in Kenya.

COMING ATTRACTIONS
INDUSTRIAL HEMP

Calling industrial hemp a "coming attraction" may seem odd given that it was used ten thousand years ago to spin fiber for human clothing. Its inclusion here is less for what it can do than for what it can replace. The United States effectively banned the cultivation of all types of hemp in 1937, after a campaign led by news stories and documentaries containing lurid descriptions of how hemp, as a narcotic, would precipitate violence and insanity. Because people were comfortable with hemp rope and products made from industrial hemp, the psychoactive varieties (*Cannabis sativa*) were named *marihuana*, Mexican slang that contained implicit racial overtones about its destructive effects. As recreational and medical marijuana continue to be approved by more and more states today, the cultivation of industrial hemp remains obstructed in the United States due to lack of approval from the federal Drug Enforcement Agency. Elsewhere in the world, hemp is a commodity crop that has many uses. Industrial hemp has negligible amounts of the cannabinoids associated with recreational or medical marijuana.

Hemp attracted attention thousands of years ago because of its fibrous stalks. The inner bast, the bark of the stem, contains long, strong fibers that can be spun and woven on their own or combined with flax and cotton to make garments. In the 1840s wood pulp began to be used to make paper; before then, paper was made almost entirely from discarded hemp garments. Ragpickers, looking for scraps of fabric, crisscrossed European cities sorting through street refuse to eke out a living. These rags were sold to what are now called recycling centers, where hemp was sorted, cleaned, and bundled for papermakers.

Hemp produces a strong, sustainable fiber. Uses include paper, textiles, cordage, caulking, carpets, and canvas. The word *canvas* is derived from cannabis (the French *canevas*). The yield of bast, the valuable fibrous part of the plant used in textiles and cordage, is between 800 and 2,400 pounds per acre—more than that of cotton. The difference in impact between the two plants is remarkable. Cotton is the dirtiest crop in the world with respect to chemical use and is largely dependent on fossil fuel inputs. Though cultivated on 2.5 percent of all cropland, cotton accounts for 16 percent of annual insecticide use. When you add to that the estimated twenty thousand people who die each year from pesticide poisoning, water contamination, pesticide-induced disease, the intense use of synthetic fertilizers and herbicides, and the salinization of soils caused by irrigation in arid lands, you begin to get a sense of the social, environmental, and climate impacts of this one crop. Nearly 1 percent of global greenhouse gas emissions stem from cotton production. Total emissions for a white cotton shirt from field to customer are 80 pounds of carbon dioxide.

When bast is removed from hemp, what remains are seeds and what is known as the hurd. A large array of products can be made from hurd, including fiberboard, building blocks, insulation, plaster, and stucco. The versatility of the plant has some believing it is an agricultural panacea. It is not. Hemp is an annual, so it gets rotated to address fertility. However, it does not require the same tillage as a typical annual crop. It is planted close together and grows so quickly that it acts as an herbicide, crowding and shading out weeds such as thistle. And no insecticides are needed or used. At today's prices, it nets two to three times more revenue per acre compared to wheat. It needs quite a bit of water, though, along with deep, nutrient-dense soils, and it is not suitable for restoring degraded land. The environmental benefits of hemp are high, but its affordability is not, at least in the United States. For example, if you harvest hemp with a combine for efficiency, you will damage the bast fibers. As useful as the bast may be, the cost of hemp fiber is close to six times that of wood pulp.

The area where hemp could make a difference is as a substitute for cotton, with the rest of the plant's uses supporting the economics. When Hu Jintao, China's president in 2009, visited his country's hemp processors, he implored them to increase China's cultivation to 2 million acres to avert the harmful impacts of cotton. This kind of growth will depend on producing hemp textiles that are affordable, fashionable, and comfortable. It will not compete with cotton on fiber softness, but if cost competitive it could certainly replace half of the cotton in the world for everyday garments such as jeans, jackets, canvas shoes, caps, and more, and that would have a significant impact on carbon emissions. ●

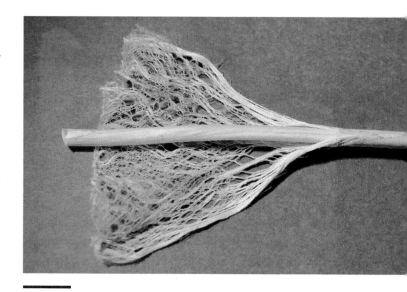

Hemp fiber has been used for thousands of years to make sailcloth, rope, twine, and clothing. It feels like linen but can be combed to have the same texture as cotton. Hemp outproduces cotton or trees by factors of 10 to 100 times in terms of yielding usable fiber.

COMING ATTRACTIONS
PERENNIAL CROPS

Kernza wheat (*Thinopyrum intermedium*) at the Land Institute in Salina, Kansas. The mature wheat will be tied into bundles and fed into a small combine for threshing.

Humans were not always seed eaters. Early diets consisted of meat (including all the organs, marrow, and fat), tubers, mushrooms, seafood (including seaweed, sea mammals, and shellfish), eggs wherever they were found, honey, birds, lizards, insects, berries, and assorted wild vegetables and herbs. Occasionally, wild grains were eaten in some areas. There were no "square" meals, and the seasons and good fortune heavily determined the food of the day. Sometime after the end of the last ice age, eleven thousand to twelve thousand years ago, human beings began to cultivate annuals for food—the first being an early ancestor of wheat called emmer in the Fertile Crescent. Ten thousand years ago rice was being grown in Asia, and nine thousand years ago corn was domesticated in Mesoamerica. All three became the staple crops of the world and remain so to this day. All three are annuals.

What would make a significant difference to soil, carbon, and cost would be perennial grains and cereals. Perennial crops are the most effective way to sequester carbon in any agricultural system because they leave the soil intact. The difference between annuals and perennials is that annuals completely die back every year, roots and all, and regenerate solely through seeds. Herbaceous perennials die back too, but not the roots, which produce new growth beneath the soil. They also can reproduce from seed, and therein lies the possibility researchers around the world are pursuing: grain, cereal, and oilseed plants that are perennial food providers.

Two successful efforts to breed perennial staple crops have emerged at the Land Institute in Salina, Kansas, and the Yunnan Academy of Agricultural Sciences in China. The Yunnan Academy focuses on rice, which has four wild ancestors that spread through roots or aboveground stems (much like strawberries) and yield crops for several years. Rice grows in flooded paddies as well as on upland fields without irrigation. In both cases, a deeper established root system provides drought tolerance and, in the case of upland rice, prevention of erosion. Perennial upland rice would minimize deforestation by farmers who cultivate rice for a few years and then move on, using slash-and-burn techniques because of lack of fertility.

In the case of the Land Institute, there has been an effort to breed perennial wheat for more than forty years, and it seems they may have it in a variety called Kernza. The institute's founder, Wes Jackson, was struck by the difference between the land of local wheat farmers and the rich soils of the native tallgrass prairies. Plant geneticist Lee DeHaan joined the Land Institute in 2001 and developed Kernza from an intermediate variety of wheatgrass native to Europe and Western Asia. The ancestor wheat, known as "tall wheatgrass" by farmers, is extensively planted as a forage grass for animal grazing but was evaluated as a perennial wheat crop for humans by the Rodale Institute in the 1980s. Seeds from the Rodale trials were planted by DeHaan in the early 2000s and have been selected, planted, and reselected for desirable traits ever since. The Land Institute's Kernza is the first to be grown and sold, now served and eaten at select restaurants and bakeries, where it is made into muffins, tortillas, pasta, and ale.

The differences between conventional wheat and Kernza in the field are profound. The carbon sequestered by conventional wheat-farming practices is in the top layer of soil and is released into the air before and after the soil is cultivated. Whereas annual wheat has a spindly three-foot root, Kernza's roots are thick and robust and go down ten feet, sequestering many times more carbon from the air and burying it deep in the earth. Burying carbon might be the wrong way to put it; Kernza's roots exchange it with bacteria that in turn acidify rocks and stones into mineral nutrients for the wheat. It is a good deal for the plants and the soil, with no tilling required.

There is perhaps no greater contribution to soil health and carbon sequestration (or emissions reduction) than the ability to farm without disturbing the soil. Soil nutrient cycles are far more effective in undisturbed soil, regardless of the fertilization method. And perennial croplands are more like watersheds, which means nearby streams are more likely to sustain diverse populations of creatures, leading to greater biodiversity. Furthermore, there is the possibility that perennials can be farmed on abandoned lands.

Kernza is not there yet, nor are the perennial cereal crops that are being developed by Michigan State University, Washington State University, the International Rice Research Institute, and other institutions. The kernels are small, the yield wanting. The good news is that there is a concerted effort around the world to create new perennial staple crops, and Kernza is just fourteen years old. In the plant-breeding world, that is a newborn. ●

A COW WALKS ONTO A BEACH

From ancient Greece to Iceland, seaweeds have been used as livestock feed for thousands of years—especially in winter, when forage is sparse. Graziers and pastoralists have long noticed its fattening effect. On modern-day Prince Edward Island, Canadian dairy farmer Joe Dorgan observed that cows in his seaside paddock were healthier and produced more milk than those pastured farther inland. He began gathering seaweed tossed ashore by storms and feeding it to all his animals. It did not take long for Dorgan to recognize he had a business opportunity on his hands, if he could get his seaweed feed approved for sale. Research scientist Rob Kinley came in to do the necessary testing and found that the seaweed was indeed helping Dorgan's cows digest more efficiently. Methane, the key

waste product generated when cows process their food, dropped by 12 percent on Dorgan's diet. By saving calories on methane production, more efficient digestion resulted in more milk. Having looked at kelp that washed ashore, Kinley wondered if other varieties of seaweed could do an even better job of unhitching a cow's digestion process from methane by-product.

Cows belong to a family of animals known as ruminants, named for the organ they share—a stomach compartment called the *rumen*, where chewed food gets digested by bacteria, then arises as cud to be chewed and swallowed again. This gassy microbial process allows cows, sheep, goats, and buffalo to digest food high in cellulose, such as grass. The result is methane waste expelled from both ends of the animal—90 percent through

seaweeds, delivered in large quantities, had some effect on methane production, but the researchers quickly homed in on *Asparagopsis taxiformis*. This species of red algae grows in warm waters around the world—some where it is native, others where it is invasive—including those off the coast of Queensland. When the test results came in, Kinley and his team wondered if their instruments were broken. In the artificial rumen, *Asparagopsis taxiformis* reduced methane production by 99 percent—and required a dose of just 2 percent of feed to do so. In live sheep, the same dose led to a 70 to 80 percent drop in methane. (The tests have not yet been performed with live cows.)

Asparagopsis taxiformis contains a key compound called bromoform. In a key step of ruminant digestion, bacteria in the rumen typically employ enzymes that create methane as waste. Bromoform reacts with vitamin B12 and disrupts that process. In the absence of *Asparagopsis taxiformis* and its bromoform, ruminants lose 2 to 15 percent of the energy in their feed to waste methane (exact loss varies by diet). Like all waste, methane points to inefficiency in the system: Part of the food ruminants consume is not being converted to body mass. By reducing off-gassing, bromoform may both avert emissions and improve production. Because the efficacy of bromoform will vary with feed type and quality, there is much research still to be done, both inside a test tube and out.

With more than 1.4 billion cows and nearly 1.9 billion sheep and goats inhabiting the planet today, scale is a major challenge for reining in methane emissions with *Asparagopsis taxiformis*. Producing enough of it to treat just 10 percent of Australian livestock would necessitate twenty-three square miles of seaweed farms. Where and how could it be mass-produced? Would drying and storage impact the effectiveness of bromoform? Champions such as Kinley acknowledge the challenge but argue that it is well worth cracking. Widespread seaweed production could be a boon for oceans—absorbing the carbon dioxide causing acidification, emitting oxygen in return, and creating marine habitat. Still, the scale required is extraordinary. Marine permaculture, another coming attraction, promises to scale *Asparagopsis* growth by the square mile, even far offshore. Combining these two solutions can yield synergies with global effects.

It is also worth noting that methane is not the only greenhouse gas caused by ruminants and other livestock. Feed production and processing is the other main culprit, responsible for 45 percent of livestock-related emissions. In addition to helping animals digest more efficiently, that number can be tackled by changing the way livestock are raised—namely through silvopasture and managed grazing—and by reducing overall intake of animal products in human diets. Still, *Asparagopsis taxiformis* shows great promise. In Hawaii, it is called *limu kohu*, which means "pleasing seaweed," and is used as a condiment on raw fish. If fed to ruminants worldwide, it would improve productivity and reduce the amount of soy, corn, and grass required as feed and thus the impact of farming upon the land. Most critically, *Asparagopsis taxiformis* could dramatically reduce livestock methane emissions, which now account for 6 to 7 percent of greenhouse gases released around the world each year. ◗

burps. Around the world, these small discharges add up to 39 percent of all emissions from global livestock production and a quarter of the world's methane pollution. In Australia, methane produced on the country's farms and ranches accounts for nearly 10 percent of all greenhouse gas emissions. The nature of their anatomy means ruminants will necessarily process their food through enteric fermentation, but Kinley's findings on Prince Edward Island suggested enteric fermentation might not inevitably produce so much methane.

At a research outfit in North Queensland, Australia, Kinley joined a team of marine-algae and ruminant-nutrition experts to test a wide range of seaweed species mixed with feed in artificial cow stomachs—essentially small fermentation tanks. Various

For decades environmentalists have campaigned and struggled to save the world's oceans from the perils of overfishing, climate change, and pollution. What if we have it backward? What if the question is not how we can preserve the wildness of our oceans, but how the oceans can be developed to protect them and the planet?

That is what a growing network of scientists, ocean farmers, and environmentalists around the world is committed to figuring out. With nearly 90 percent of large fish stocks threatened by overfishing and 3.5 billion people dependent on the seas as their primary food source, ocean farming advocates have concluded that aquaculture is here to stay.

Rather than monolithic factory fish farms, however, they see small-scale farms, where complementary species are cultivated to provide food and fuel, clean up the environment, and reverse climate change. Governed by an ethic of sustainability, they are reimagining our relationship to the oceans in order to address overlapping climate, energy, and food crises.

Ocean farming is not a modern innovation. For thousands of years, cultures as diverse as the ancient Egyptians, Romans, Aztecs, and Chinese have farmed finfish, shellfish, and aquatic plants. Atlantic salmon have been farmed in Scotland since the early 1600s; seaweed was a staple food for American settlers.

What was once a sustainable fishery practice has been modernized into large-scale industrial-style farming, much in the same way that we now have industrial agriculture. Modeled on land-based factory livestock farms, conventional aquaculture operations are known for their low-quality, tasteless fish treated with antibiotics and fungicides that pollute local waterways. According to a recent *New York Times* editorial, aquaculture "has repeated too many of the mistakes of industrial farming—including the shrinking of genetic diversity, a disregard for conservation, and the global spread of intensive farming methods before their consequences are completely understood."

A small group of ocean farmers and scientists are charting a different course. The new ocean farms are pioneering what is known as multitrophic aquaculture, wherein ocean farmers grow varied aquaculture species that provide for and feed upon the others.

Farmers in Long Island Sound are diversifying small-scale organic shellfish farms with various species of seaweed to filter out the pollutants, mitigate oxygen depletion, and develop a sustainable source of fertilizer and fish meal. In southern Spain, Veta la Palma designed its farm to restore wetlands, and in the process created the largest bird sanctuary in Spain, with more than 220 species of birds.

Seaweed farms have the capacity to grow massive amounts of nutrient-rich food. Professor Ronald Osinga, at Wageningen University in the Netherlands, has calculated that a global network of sea-vegetable farms totaling seventy thousand square miles—roughly the size of Washington state—could provide enough protein for the entire world population. And this is just the beginning, as there are more than ten thousand edible plants in the ocean.

The goal, according to chef Dan Barber, is to create a world where farms "restore instead of deplete" and allow "every community to feed itself." Because ocean farms require no fresh water, no deforestation, and no fertilizer—all significant downsides to land-based farming—they promise to be more sustainable than the most environmentally sensitive traditional farms. And because they can use the entire water column vertically, they have a small footprint, high yield, and low aesthetic impact.

Instead of finfish, the anchor crops of green ocean farms are seaweed and shellfish, two organisms that may well be Mother Nature's secret prescription for addressing global warming. Considered the tree of coastal ecosystems, seaweed uses photosynthesis to pull carbon from the atmosphere and the water, with some varieties capable of absorbing five times more carbon dioxide than land-based plants.

Seaweed is one of the fastest-growing plants in the world; kelp, for example, grows nine to twelve feet long in a mere three months. This turbocharged growth cycle enables farmers to scale up their carbon sinks quickly. Of course, the seaweed grown to mitigate emissions would need to be harvested to produce carbon-neutral biofuels to ensure that the carbon is not simply recycled back into the air, as it would be if the seaweed were eaten or quickly decomposed in the water or on land.

Although oysters absorb carbon, their real contribution is filtering nitrogen out of the water column. Nitrogen is the greenhouse gas you do not pay attention to. It is nearly three hundred times as potent as carbon dioxide, and according to the journal *Nature*, it is the second worst in terms of having already exceeded a maximum "planetary boundary." Like carbon, nitrogen is an essential part of life—plants, animals, and bacteria all need it to survive—but too much has a devastating effect on land and ocean ecosystems.

The main nitrogen polluter is agricultural fertilizer runoff. All told, the production of synthetic fertilizers and pesticides contributes more than 1 trillion pounds of greenhouse gas emissions to the atmosphere each year. Much of this nitrogen from fertilizers ends up in the oceans, where it is now 50 percent above normal levels. According to the journal *Science*, excess nitrogen "depletes essential oxygen levels in the water and has significant effects on climate, food production, and ecosystems all over the world."

Enter oysters to the rescue. A single oyster filters thirty to fifty gallons of water a day. Recent work by Roger Newell, of the University of Maryland, shows that a healthy oyster habitat can reduce total added nitrogen by up to 20 percent. A three-acre oyster farm filters out the equivalent nitrogen load produced by thirty-five coastal inhabitants.

There is an array of projects sprouting up that use a mix of seaweed and shellfish to clean polluted urban waterways and help communities prepare for the effects of climate change. One initiative, spearheaded by Dr. Charles Yarish, of the University of Connecticut, is growing kelp and shellfish on floating lines in New York's Bronx River to filter nitrogen, mercury, and other pollutants out of the city's toxic waterways, with the goal of making them healthier, more productive, and more economically viable.

Then there is the emerging field of "oyster-tecture," dedicated to building artificial oyster reefs and floating gardens to help protect coastal communities from future hurricanes, sea-level rise, and storm surges. Landscape architect Kate Orff, from the design firm Scape, is developing urban aquaculture parks that use floating rafts and suspended shellfish long lines to build more urban green space while improving the environment. She envisions the new urban ocean farmer as part shell fisherman, tending oyster reefs, and part landscaper, tending above-surface floating parks.

In Connecticut, advocates are pushing for an expansion of the state's existing nitrogen credit trading program to include shellfish farms, thereby reimbursing oystermen for the nitrogen they filter from Long Island Sound each year. With new oyster operations sprouting up all around the country, rewarding "green fishermen" for the positive effect their farms have on the environment could be a model for stimulating job growth while creating carbon sinks.

Finding a clean replacement for existing biofuels is becoming increasingly urgent. A report commissioned by the European Union found biofuels from soybeans can create up to four times more climate-warming emissions than equivalent fossil fuels. Increasingly, seaweed and other algae are looking like a viable substitute. About 50 percent of seaweed's weight is oil, which can be used to make biodiesel for cars, trucks, and airplanes. Scientists at the University of Indiana recently figured out how to turn seaweed into biodiesel four times faster than other biofuels are made, and researchers at the Georgia Institute of Technology have discovered a way to use alginate extracted from kelp to ramp up the storage power of lithium-ion batteries by a factor of ten.

Unlike land-based biofuel crops, seaweed farming does not require fertilizers, forest clearing, water, or heavy use of fuel-burning machinery; as a result, it has a negative carbon footprint, according to the World Bank. While the technology is still in development, farmers are eager to begin growing their own fuel and creating closed-loop energy farms.

The DOE estimates that seaweed biofuel can yield up to thirty times more energy per acre than land crops such as soybeans. According to *Biofuels Digest*, "Given the high oil yield from algae, some 10 million acres would be sufficient . . . to replace the total petro-diesel fuel in the United States today. This is about one percent of the total amount of acreage used in the United States today for grazing and farming."

The world's energy needs could be met by setting aside 3 percent of the world's oceans for seaweed farming. "I guess it's the equivalent of striking oil," says University of California, Berkeley, microbial biology professor Tasios Melis.

On their current path, the oceans are in a death spiral. According to the International Programme on the State of the Ocean—a consortium of twenty-seven of the top ocean experts in the world—the effects of climate change, ocean acidification, and oxygen depletion have already triggered a "phase of extinction of marine species unprecedented in human history."

Reversing global warming may be an invitation to develop the world's seas in order to save them. On the other hand, if we do nothing the oceans may die. Marine waters are revered as some of the last wild spaces on earth, ungoverned and untouched by human hands. If we develop them, farms will someday dot coastlines, mirroring our agricultural landscape. But in the face of the escalating climate crisis, we may have to explore new ways of sustaining humanity while protecting the planet.

This means dedicating portions of ocean to farming, while reserving large swaths for marine conservation parks. And rather than building sprawling ocean factories, we need to create decentralized networks of small-scale food and energy farms that grow food, generate power, and create jobs for local communities. While no panacea, ocean farming—carefully conceived—could be a vital part of reversing course and building a greener future. ●

COMING ATTRACTIONS
SMART GRIDS

The twenty-first century is running on a twentieth-century electrical grid. In most high-income cities and regions of the world, three main components make up the complex machine known as the grid: power plants that produce electricity, transmission lines that carry it across distances, and distribution networks that deliver it to residential, commercial, or industrial end users. Designed to ferry electricity from centralized suppliers to broad geographies of consumers, it is fundamentally a one-way system. Reliability, reach, and capacity are real strengths, but last century's grid is struggling with this century's requisite shift to clean, renewable energy. Concentrated fossil fuel generation is predictable and manageable, allowing utilities to align electricity supply with demand. But renewable sources such as solar and wind are variable and much more distributed. They cannot be standardized or readily dispatched when needed. Accommodating their fluctuations and enabling their success necessitates a more nimble, adaptive grid.

Nimble and adaptive are hallmarks of the emerging "smart grid"—a digital refashioning of the traditional grid with the needs of a clean energy economy in mind. A smart grid is smart in the sense that it engages in two-way communication between suppliers and consumers to predict, adjust, and sync power supply and demand. Today, the balancing act between producers and users takes place within utilities' operations centers. Internet connectivity, intelligent software, and responsive technologies can assist and even automate the management of electricity flow, coordinating between the grid's many facets in real time. Smart grids can ensure grid reliability and resilience in an era of photovoltaic panels and wind turbines, while also maximizing energy efficiency throughout the system. This is the root of their climate mitigation potential: Smart grids can reduce overall consumption while facilitating the shift away from centralized fossil fuel plants and their greenhouse gas emissions. They also help manage additional electricity demand from plug-in electric vehicles, enabling that technology to grow. According to the International Energy Agency, smart grids could generate net annual emissions reductions of 0.7 to 2.1 gigatons of carbon dioxide by 2050.

Smart grids are complex systems comprising numerous parts. They have no hard and fast rules, but smart-grid pioneers such as South Korea have helped to define three essential components:

1. High-voltage power lines equipped with sensors to monitor and report on conditions and multidirectional flow
2. Advanced meters that can wirelessly communicate electricity consumption and pricing in real time— to both utilities and end users
3. Web-connected appliances, plugs, and thermostats that can respond to the need to reduce consumption or use available electricity

Together, these and other components of a smart grid make it possible to smooth out peaks in demand and absorb variable, distributed supply from renewables. Demand for electricity varies throughout the day and across seasons, typically peaking in the late afternoon and in the hottest and coldest months. Under the current fossil fuel–based system, those spikes are met by so-called "peakers"—small plants switched on in a pinch to meet surges in demand. They get the job done, but they are expensive and dirty. Instead, smart grids could apply dynamic pricing and signal millions of smart appliances to marginally adjust—freezers allowed to warm by a degree, for example —thereby evening things out. Similarly, they could activate charging of plug-in electric cars at night, when wind turbines whirl but demand is lowest, or tap into energy stored in their batteries when need be. With fewer peaks and troughs in the flow of electricity, carbon emissions drop and both utilities and users save money.

The current grid has been called the largest and most interconnected machine on earth—one of the greatest engineering feats of the twentieth century. Making it smarter is a massive undertaking that will occur in phases over the coming decades, as different technologies within the smart grid roll out. Studies show the investment required will be well worth it, though, thanks to emissions mitigation, financial savings, and improved grid stability. In the United States, for example, an investment of $340 billion to $480 billion in an intelligent grid system would yield a net benefit of $1.3 trillion to $2 trillion over twenty years. It will be crucial to address both security risks from unauthorized access to grid controls and data privacy for individual households. Many people still wonder if renewable energy sources can power the world. But that is a fundamental misunderstanding. The big challenge is not solar and wind generation; it is a grid that can accommodate their unique proclivities. More green requires a wiser grid. ●

BUILDING WITH WOOD

In the words of designer Michael Charters, "the high-rise is tired." His design here is for a site in Chicago at the corner of Harrison and Wells. Charters believes that since Chicago was the birthplace of the skyscraper, it is the suitable birthplace for "Big Wood," mass timber, carbon-neutral structures that change not only the materials of urban buildings but the shape of them. This particular building is a mixed-use complex for the University of Chicago consisting of a library, media hub, three types of housing, retail, a sports complex, parking, a park, and a community garden.

From columns to rafters, flooring to shingles, wood is one of the original building materials. The construction of large timber-framed buildings spans back seven thousand years to China, and it includes the fourteen-hundred-year-old Hōryū-ji Temple complex in Ikaruga, Japan, which has survived seismic threat and a wet environment to endure as the oldest group of wooden buildings. With the Industrial Revolution, steel and concrete became dominant, and wood use declined, mostly relegated to single-family homes and low-rise structures. Nowadays, when one thinks of construction in cities, what comes to mind are cranes swinging steel beams across the skyline. But that is beginning to change: Today, high-rise urban structures are being built almost entirely of wood, sequestering carbon in the process.

Treet in Norwegian means "tree." It is an apt name for the fourteen-story apartment building in Bergen, Norway, that—like ten-story Forté in Melbourne and nine-story Stadthaus in London—is a pioneer in contemporary timber construction. Soon, they will be surpassed by an eighteen-story student housing project at the University of British Columbia and perhaps other aspiring projects with thirty floors and counting. All of these structures are (or will be) made of large wooden beams, modules, and panels, many of which are prefab or precut and quickly pieced together on-site. Glulam, beams of glued laminated timber, can replace steel and was employed 175 years ago in British churches and schools. In the 1990s, a panel technology called cross-laminated timber, CLT for short, emerged in Austria, and has been described as the "new concrete" for its strength and longevity. Both glulam and CLT are now getting more attention as means of reducing the climate impacts of building the places where people work, gather, and reside.

When it comes to climate, building with wood has two key benefits. First, as they grow, trees absorb and sequester carbon, which remains stored in timber construction materials. A unit of dry wood is 50 percent carbon, and that carbon is locked in while the wood is in use. The cycle continues as trees grown to replace sustainably harvested timber sequester additional carbon. Second, the process of producing those materials generates fewer greenhouse gas emissions than producing wood's alternatives. Cement, used in concrete and other building materials, is responsible for producing 5 to 6 percent of global emissions, twice as much as the aviation industry. Steel comes in nearly as high: Manufacturing steel beams requires six to twelve times more fossil fuel than producing laminated timber. Additionally, when a wood building comes to the end of its life, its component parts can find new life in other buildings, be composted, or be used as fuel. Thanks to these layered merits, moderate increases in wood use can bring about sizable benefits for the climate. According to a 2014 study out of Yale University, building with wood could reduce annual global emissions of carbon dioxide by an impressive 14 to 31 percent.

Conventional wisdom suggests that wood and high-rise buildings are incompatible, and that flammability is an issue. Growing knowledge and a renaissance in the processing and manufacturing of wood are challenging those limitations. Whereas steel bends in fire, wood forms a protective char on the outside, keeping its structural integrity within. New high-performance products are more fire resistant, as well as more cost-effective and stronger than ever. Glulam and CLT sandwich together smaller boards to create a composite product with steel-like strength—able to carry greater loads, for use in ever-taller buildings. Another benefit is that they can be prefabricated and then put together like a giant piece of furniture. That means construction can happen quickly, lowering costs and significantly decreasing the waste, noise, and traffic typical of construction sites.

Three key factors impact the benefits of building with wood over the alternatives and need to be addressed. First, if supply is located close to building sites, that proximity limits transportation emissions and cost. Second, harvesting timber with sustainable forestry practices protects ecological integrity and ensures maximum carbon sequestration. If logging is not well managed, using wood as a dominant construction material could spell disaster for forests and the flora and fauna therein. Third, at the end of their life cycle, timber building materials need to be reused, recycled, or disposed of with a method such as composting; doing so prevents stored carbon from being released and wood from undergoing anaerobic decomposition, which produces methane. Ise Jingu, the Grand Shrine of Shinto located in Mie, Japan, is deconstructed and rebuilt every twenty years with hinoki, a wood grown nearby for a ritualistic practice that honors the death, impermanence, and regenerative power of nature. Nothing is thrown away; every scrap of wood becomes a part of other structures and can end up, two hundred years later, as tokens in teahouses on the temple grounds.

Perhaps the biggest challenge to scaling wood construction is perception. Champions such as Vancouver-based architect Michael Green, who has designed a wooden version of the Empire State Building, are working hard to change that. Tall timber buildings themselves may be the most compelling testimony, and competitions like the U.S. Tall Wood Building Prize are helping to propagate demonstration projects from New York City to Portland, Oregon. Though laminated wood technology is well established, it is just beginning to make inroads in many markets. As supply chains develop, these materials will become increasingly cost competitive. Still, many building codes limit the use of wood to four, maybe five stories. Regulation could catch up with engineering and encourage, rather than impede, innovation. Just as the earth grows our food, it can also produce first-rate building material. ●

Reciprocity
JANINE BENYUS

Midway through my forestry degree, I found myself pointing a can of spray paint at the smooth bole of an ironwood tree. I was to mark it as part of a "release cut" in our experimental forest in New Jersey. The orange slash would tell loggers to fell, poison, or girdle anything that might compete with our saw timber crop. We were taught that thinning would help the oaks and walnuts, freeing them to get more water, light, and nutrients. For many in our class, opening up a stand of trees was their favorite part. For me, it was an excruciating, empty choice.

I kept envisioning the historic forest right next to ours that hadn't been cut for two hundred years. I had seen overstory

giants grouped in twos and threes and fours, a middle layer of hardwoods and conifers, and at my feet, trilliums, fiddleheads, and rufous-sided towhees bursting from the duff. Nobody had released these trees from competition, yet all appeared well.

"The old forest is not nearly as open or regimented as this," I told my professor, "but it looks healthier. Do you think the trees might be grouped together for a reason? Do you think they might be benefiting one another in some way?"

He shook his head no, a bit alarmed. "Don't be so Clementsian," he said. "You'll never get into grad school." The reference was to Frederic Edward Clements, an ecologist from the early 1900s who had won and then lost the greatest debate in

forests, and prairies, he saw distinct communities of plants reacting not just to soils and climate but also to each other. He proposed that plants were cooperators as well as competitors, facilitating each other in beneficial ways. Canopy trees "nursed" the saplings beneath their branches, creating more sheltered, nutritious conditions in a plant-helping-plant process called facilitation. They shaded seedlings from the drying sun, blocked the winds, and fertilized the soil with their leaves. As time passed, one community of plants prepared the way for another; annual plants built the soil for perennial shrubs, which nourished saplings that grew into forests. Everywhere Clements looked, he saw communities so tightly interwoven, he called them organismic.

Gleason had a different take. What Clements called communities was simply happenstance, random individuals dispersed by chance and arranged according to how they adapted to water, light, and soil. There was no mutual aid; plants were merely competing for a spot in the struggle. The notion that there might be a connected, interdependent community to be studied as a whole was an illusion; examining the parts would do.

For the first half of the twentieth century, Clements's view prevailed—the ecological literature was full of studies on facilitation. Gleason's work was virtually forgotten until 1947, when a small group of researchers resurrected his individualist views and pitted them against Clements's holism. Gleason's view of plants as individuals allowed them to be studied with neat statistical precision, as if they were atoms.

Within twelve years, the majority of ecologists rejected the idea of positive interactions as a driver of community assembly, and focused instead on negative interactions such as competition and predation. Papers in scientific journals morphed, and if you applied to graduate school there were only certain permissible research questions, beginning with, "How does competition explain . . . ?" Given the times, we shouldn't be surprised. Clements's fall from grace coincided, to the year, with the release of the Truman Doctrine and the onset of the Cold War. For decades, even when talking about plants, you dared not talk about communism.

But here is what I love about the scientific method. Though culture holds its finger on the scale, it cannot stop the restless search for measurable truth. Un-American or not, the math has to work. When fifty years of wall-to-wall research into competition proved inconclusive, researchers went back to the field to find out what else was at play.

The same year I was pardoning an ironwood, ecologist Ray Callaway was in the foothills of the Sierra Nevada, rescuing blue oaks from bad practice. The prevailing wisdom, a gift from Gleason, was that the oaks dotting California's rangelands should be cut to release grasses from competition. Much to Callaway's dismay, thousands of acres of blue oaks were being bucked up for firewood.

The fact that grasses had thrived with blue oaks for eons nagged at him. For two and a half years, he measured the interaction between the oaks and the grasslands—his pans and buckets catching leaves, twigs, branches, and nutrient-laced rainfall dripping from six canopies. His thesis showed that the nutrient totals were twenty to sixty times greater under the oaks than in open

ecological history. Being compared to Clements was a well-known admonition, a sure sign of naïveté.

It was 1977, and ecologists were three decades deep into a paradigm shift that affected our experiments, our narratives about the wild, and most powerfully, our maxims for managing forestlands, ranches, and farms. The precept that trees needed to be released from the struggle of competition was the fruit of a debate between Frederic Clements and his contemporary Henry Gleason. What they both endeavored to describe, in very different ways, was what constitutes a community of vegetation, what determines how plants grow together and why.

When Clements studied bayous, chaparrals, hardwood

grasslands. Those spreading trees, so artfully arranged in the California landscape, are nutrient pumps that lift minerals from the deep and scatter them in an annual leaf drop. Penetrating taproots loosen the dense soil, increasing water storage beneath the boughs and welcoming a profusion of plants. Callaway has gone on to compile more than a thousand studies that describe how plants "chaperone" and enhance their neighbors' survival, growth, and reproduction. To read these examples is to discover a manual for how natural communities heal and overcome adversity, essential reading in a climate-changed world.

Knowing which plants are the helpers, the chaperones in botanical communities, will be important as droughts deepen in the coming years. For example, how and why do Amazonian rainforests create clouds even in the dry season? It turns out that ten percent of the Amazon's annual rainfall is absorbed by the shallow roots of certain scattered shrubs, then pushed downward through taproots deep into the soil bank. When the rainless months come, the taproots lift up the water and pump it out into the shallow roots, distributing it to the whole of the forest. Many species of plants throughout the world perform this hydraulic "lift," watering a multitude of plants under the forest canopy.

The more stressful the environment, the more likely you are to see plants working together to ensure mutual survival. On Chilean peaks, studies of mounded plants huddling together against harmful ultraviolet rays and cold, drying winds reveal complex interactions of support. A single six-foot-wide yareta, or cushion plant, can be thousands of years old and harbor dozens of different flowering species in its mound, tucked like colorful pins in a bright green cushion.

Downslope, if a tree can tough it out and get established on a rockfall it creates a protective refuge where winds calm and snows drift to water sheltered seedlings. Birds roost and mammals hide in what becomes a growing island, importing nutrients and seeds in their excrement. As leaves and needles decay, an organic sponge is built that releases moisture on summer's dry days.

I admit it's counterintuitive to imagine plants growing closer together in the face of scarcity when our competition bias and our economic theories tell us we should do otherwise. For years, careful experimenters tried to explain this as an anomaly, missing the beneficence in their search for the struggle. Now we know that it's not just one plant helping another; mutualisms—complex exchanges of goodness—are playing out above- and belowground in extraordinary ways.

While Callaway was measuring oaks in California, Suzanne Simard was a professional forester wincing through British Columbia's mass clear-cuts. The management protocol of removing paper birch trees that grew in association with Douglas fir seemed off to her—they had been companions for eons. Might they be helping each other in some way?

In a brilliant study, she exposed growing seedlings to two types of radio-tagged carbon dioxide—carbon-14 into Douglas fir and carbon-13 into birch. The seedlings would absorb the carbon dioxide and transform it into sugars. She followed the carbon to see if any would be exchanged. The first results came in an hour's time. She describes a sense of wonder bordering on euphoria when the Geiger counter popped and clicked—

carbon-13 from the birch had traveled to the Douglas fir, while carbon-14 from the fir made its way to the birch.

How? Next time you're in a forest, dig into the duff and you're bound to find white cobwebby threads attached to roots. These are the underground part of fungi that deliver phosphorus to trees in return for carbon. Textbooks describe this as an exchange between one plant and one fungus. Simard's work was among the first to prove that fungi branch out from the roots of a single tree to connect dozens of trees and shrubs and herbs—not only to their relatives but also to entirely different species. The "Wood Wide Web," as Simard calls it, is an underground Internet through which water, carbon, nitrogen, phosphorus, and defense compounds are exchanged. When a pest troubles one tree, its alarm chemicals travel via fungi to the other members of the network, giving them time to beef up their defenses.

Discoveries about the holistic nature of forests have vast implications for forestry, conservation, and climate change. It's time to bring the same penetrating insight to farmlands. Although 80 percent of all land plants have roots that grow in association with mycorrhizae fungi, it's rare to find common mycorrhizal networks in agricultural fields. Plowing and herbicides such as glyphosate disturb the network, and the year-on-year addition of artificial nitrogen and phosphorus fertilizers tell bacterial and fungal helpers that they are not needed—not needed for water transport or pest defense, not needed to absorb the micronutrients our bodies long for.

When communities of vegetation breathe in carbon dioxide, turn it into sugars and feed it to microbial networks, they can sequester carbon deep in soils for centuries. But to do that, the communities need to be healthy, diverse, and amply partnered. If we're to encourage wild and working landscapes to recoup the 50 percent of soil carbon that has been lost to the atmosphere, we'll want to pause before revving a chainsaw, opening a bag of fertilizer, or marking a sapling for removal. We won't want to interrupt a vital conversation.

To help reverse global warming, we will need to step into the flow of the carbon cycle in new ways, stopping our excessive exhale of carbon dioxide and encouraging the winded ecosystems of the planet to take a good long inhale as they heal. It will mean learning to help the helpers, those microbes, plants, and animals that do the daily alchemy of turning carbon into life. This mutualistic role, this practice of reciprocity, will require a more nuanced understanding of how ecosystems actually work. The good news is that we're finally developing a feeling for the organismic, after years of wandering in the every-plant-for-itself world.

One of the fallouts of our fifty-year focus on competition is that we came to view all organisms as consumers and competitors first, including ourselves. Now we're twenty years into a different understanding. By recognizing, at last, the ubiquity of sharing and chaperoning, by acknowledging the fact that communal traits are quite natural, we get to see ourselves anew. We can return to our role as nurturers, one of the many helpers in this planetary story of collaborative healing.

AN OPENING

For both the rich and the poor,
life is dominated by an ever growing current of problems,
most of which seem to have no real and lasting solution.
…the ultimate source of all these problems is in thought itself,
the very thing of which our civilization is most proud,
and therefore the one thing that is "hidden"
because of our failure seriously to engage with its actual working
in our own individual lives and in the life of society.

–David Bohm and Mark Edwards, *Changing Consciousness*

The logical way to read this book is to use it to identify how you can make a difference. How each person thinks and perceives his or her role and responsibility in the world is the first step in any transformation—the base upon which all change depends. As researchers, we were and remain astonished at the impact individual solutions can have, especially as they relate to both the production and consumption of food. What we choose to eat, and the methods employed to grow it, rank with energy as the top causes and cures of global warming. Individual responsibility and opportunity do not stop there: they include how we manage our homes, how we transport ourselves, what we purchase, and more.

However, placing too high an emphasis on the individual can lead to people feeling so personally responsible that they become overwhelmed by the enormity of the task at hand. Norwegian psychologist and economist Per Espen Stoknes has described how individuals respond to being besieged with science that describes climate change in the language of threat and doom. Fear arises and becomes intertwined with guilt, resulting in passivity, apathy, and denial. To be effective, we require and deserve a conversation that includes possibility and opportunity, not repetitive emphasis on our undoing.

That conversation needs to extend beyond the individual, because any idea that we exist as isolated beings is a myth. We are all intricate, interconnected parts of complex social structures and cultures, and more broadly of the entire web of life—the ultimate source of water, food, fiber, medicines, inspiration, beauty, art, and joy.

Arguably, no single person has done more to educate the world about climate change than Bill McKibben. He was the first to write a popular book warning of climate change, *The End of Nature*, a bestseller published in 1989. He is the embodiment of an activist in his nonstop talks, travels, writings, and organizational outreach—the quintessential example of what one person can accomplish. It would be easy for McKibben to exhort us to do more as individuals, to follow his exemplary life, and enact the changes required to reverse global warming. But that is not what he recommends. The problem, he writes, is with the very pronoun "I."

Individuals cannot prevent the torching of Indonesia rainforests by corrupt palm oil corporations or put an end to the bleaching and coral die-off of the Great Barrier Reef in Australia. Individuals cannot stave off the acidification of the world's oceans or foil the onslaught of commercials dedicated to fomenting desire and materialism. Individuals cannot halt the lucrative subsidies granted to fossil fuel companies. Individuals cannot prevent the deliberate suppression and demonization of climate science and scientists by anonymous wealthy donors.

What individuals can do is become a movement. As McKibben writes: "Movements are what take five or ten percent of people and make them decisive—because in a world where apathy rules, five or ten percent is an enormous number." Movements change how we think and how we see the world, creating more evolved social norms. What was once accepted and thought to be normal becomes unthinkable. What was marginalized or derided becomes honored and respected. What was suppressed becomes recognized as a principle. The United States was founded on the premise that there are truths that are self-evident, and one of the unmentioned truths is that we only have one home. If we are to remain here, we must together take great care. To do that means we must become a "we," a movement that is unstoppable and fearless. Movements are dreams with feet and hands, hearts and voices.

This is why, in creating *Drawdown* and its associated website, we sought to do more than merely perform exacting research and inform. We wanted to captivate and surprise, to present solutions to global warming in a new way with an eye towards helping draw the threads and webs of humanity into a coherent and more effective network of people that can accelerate progress towards reversing climate change.

Going forward, the staff, fellows and volunteers at Project Drawdown will be modeling the economics of regeneration—jobs, policy, and economic complexity—mapping climate solutions onto specific national economies and calculating how climate change technologies and processes can generate dignified, socially just, family-wage jobs. The economic data we have collected shows clearly that the expense of the problems in the world now exceeds the cost of the solutions. To put it another way, the profit that can be achieved by instituting regenerative solutions is greater than the monetary gains generated by causing the problem or conducting business-as-usual. For instance, the most profitable and productive method of farming is regenerative agriculture. In the electric power generation industry, more people in the U.S. as of 2016 are employed by the solar industry than by gas, coal, and oil combined. Restoration creates more jobs than despoliation. We can just as easily have an economy that is based on healing the future rather than stealing it.

The word "job" is awkward in that it contains a sense of duty, grind, or drudgery. "Work" may be the better term, as it can imply career, calling, and profession. A friend once addressed a class of third graders and discussed the growing number of unemployed people in the world. A girl raised her hand and asked, "Has all the work been done?" Never has more needed to be done on earth and hundreds of millions of people need those jobs.

It is difficult to watch the accelerating breakdown of our environmental systems or witness the worldwide breakdown of civility into camps, ideologies, and wars. What stands before us, however, is not the choosing of sides but the gift of seeing who we are as stewards of the planet. We will either come together to ad-dress global warming or we will likely disappear as a civilization. To come together we must know our place, not in a hierarchical sense, but in a biological and cultural sense, and reclaim our role as agents of our continued existence. We are surfeited with metaphors of war, such that when we hear the word "defense," we think attack, but the defense of the world can be accomplished only by unifying, listening, and working side by side.

Climate solutions depend on community, collaboration, and cooperation. At the end of the day, every solution in *Drawdown* is initiated and promoted by groups of people forming new and perhaps unlikely alliances: developers, cities, nonprofits, corporations, farmers, churches, provinces, schools, and universities. Food and land-use solutions focus on how to cooperate with nature in order to sequester carbon and improve the quality of all life. Educating girls and family planning are about communities the world over recognizing and supporting the potential of girls and the power of women. Energy and material efficiency arise from architects, engineers, city planners, activists and inventors working as a team. At Project Drawdown, more than 250 people formed a coalition to collaborate—fellows, advisors, funders, expert reviewers, and staff. We are deeply indebted to each and every person who helped create this project.

Science knows that virtually all children exhibit altruistic behavior, even before they can talk. It turns out that concern for the well-being of others is bred in the bone, endemic and hardwired. We became human beings by working together and helping one another. That remains true today. What it takes to reverse global warming is one person after another remembering who we truly are. —Paul Hawken

METHODOLOGY

Project Drawdown collects, analyzes, and presents the best available research and data on social, ecological, and technological solutions that can substantively decrease the concentration of atmospheric greenhouse gases. To work toward drawdown, each of the solutions does one or more of the following:

- Reduce energy use through efficiency, material reduction, or resource productivity.
- Replace existing energy sources with renewable energy systems.
- Sequester carbon in soils, plants, and kelp through regenerative farming, grazing, ocean, and forest practices.

Research for every solution consists of a three-step process:

- Technical Reports: Detailed analyses of solutions that include technical specifications and projection scenarios using financial and climate data.
- Review Process: Careful evaluation of all technical reports and model inputs by experts in the various fields. This ensures that the data is accurate, reliable, and current.
- Integration Models: The solution models are integrated into larger sector models in order to eliminate inaccuracies that can be caused by double counting and interactions between solutions.

We employ multiple datasets, including research from reputable international organizations and agencies, and market reports from global consultancies and industry leaders. Data from the Intergovernmental Panel on Climate Change (IPCC), International Energy Agency (IEA), International Renewable Energy Agency (IRENA), Food and Agriculture Organization of the United Nations (FAO), International Institute for Applied Systems Analysis (IIASA), and other widely cited research organizations and peer-reviewed studies form the core of our global analysis.

Two primary models were developed to evaluate data using statistical methods of analysis. A *reduction and replacement model* was designed to calculate solutions that reduce energy consumption or replace existing fossil fuel–based energy generation. A *land use model* was developed to evaluate the different dynamics of sequestering carbon dioxide from the atmosphere through above- and belowground biomass, while also accounting for avoided emissions that arise from reducing destructive land use practices such as deforestation. The appropriate model was adopted and customized to the solution being analyzed.

Since Project Drawdown's objective was to look at the combined effect of all the solutions, fourteen integration models were developed for groupings of solutions that share common data sets and inputs:

Agriculture
Building Envelope
Building Systems
Electricity Generation
Family Planning
Food Systems
Forest Management
Freight Transport
Heating/Cooling
Lighting
Livestock Management
Passenger Transport
Urban Transport
Waste Diversion

SCENARIOS

The data shown throughout *Drawdown* represents the incremental impact, cost, and/or savings of an ambitious but plausible adoption of the respective solutions when compared to a thirty-year period in which growth is fixed at current levels relative to market size. For example, renewable energy, it currently constitutes 24 percent of world energy use—solar, wind, hydroelectric (large-scale), biomass, waste, wave, tidal, and geothermal. We measure the additional percentage of energy generation created in each category compared to what it is today. Energy generation will increase as a result of population and economic growth. If the percentage of renewable energy remains 24 percent, we measure that as zero. We call this the Plausible Scenario—an optimistic, feasible framework and forecast that models the incremental impacts of increased adoption.

While the scenario is optimistic, it is also realistic. We use conservative estimates when it comes to financial cost and emissions impact, relying on widely cited, peer-reviewed science. We vet sources and incorporate meta-analysis to evaluate a range of potential impacts before settling on one, always with a bias to the conservative. With respect to financial modeling, we purposely chose slower rates of falling costs compared to historical trends.

Projecting global impact for each solution requires an assessment of potential future adoption within markets. Global demand for commodities and services is determined using predictions of both global and regional markets. Examples of market demand include total electricity generation, total passenger kilometers traveled, total square meters of floor space for residential and commercial buildings, and so on. Population and economic conditions, therefore, have a profound impact on the models. *The United Nations' 2015 Revision of World Population Prospects* has three different predictions for 2050: low, medium, and high forecasts. We measure growth, demand, and impact using the medium population forecast, which is 9.72 billion people.

We also completed two other forecasts. The Drawdown Scenario optimizes the conservative carbon and financial assumptions in the Plausible Scenario. The Optimum Scenario represents the maximum potential of major solutions by 2050—notably, the adoption of 100 percent clean, renewable energy. (See next page.)

GENERAL ASSUMPTIONS

Because of the global scope of the project, we made a number of assumptions across solutions. The assumptions, listed below, allowed us to conduct the research within a reasonable time frame. Models for specific solutions have a range of additional assumptions particular to the solution itself; these are described in detail within the individual technical reports available on the Project Drawdown website.

- **Assumption 1:** Future infrastructure required to sufficiently manufacture and scale each solution globally is in place in the year of adoption, and is included in the cost to the agent (the individual or household, company, community, city, utility, etc.). Because we have made this assumption, we have eliminated the need for analysis of capital spending to enable or augment manufacturing.
- **Assumption 2:** Policies required to enable, augment, or regulate solutions at the local, national, and international level are in place in the year of adoption. This assumption eliminates the need for country-level analysis of direct government intervention in promoting solutions.
- **Assumption 3:** There will be no price on carbon. Because of the uncertainty related to carbon pricing and the policies required to ensure its implementation, its potential impact has not been evaluated in our analyses.
- **Assumption 4:** All costs and savings are calculated according to the agency level. For example, the costs associated with household LED lighting are calculated based on the cost to the homeowner, whereas the costs associated with heat pumps are those incurred by building owners, commercial or residential.
- **Assumption 5:** Prices will change as a result of production efficiencies and technological improvements. In the absence of reliable future cost projections, we have adjusted prices according to conservative solution-specific learning rates derived from historical trends.
- **Assumption 6:** Solutions may become outdated, significantly improved, or supplanted by new technologies or practices within the period under analysis. In the absence of reliable forecasts, we have not considered these developments in the analyses.

The general assumptions described here do not necessarily reflect our expectations for the future. For example, while we have assumed for the purposes of this project that there will not be policies in place implementing carbon pricing, cap-and-trade and other carbon pricing mechanisms are already in place and growing. Such policies can greatly accelerate adoption of almost all of the solutions beyond what is modeled here.

SYSTEM DYNAMICS

Solutions operate within complex, interconnected systems. Their effects are not discrete; rather, they are interdependent, interactive, and circular. For that reason, we have tried to map and analyze the extent to which the impacts of one solution affect other solutions. Outputs from one model can be inputs into another solution within this system.

One example is the dynamic between reducing food waste, composting, agriculture, and methane digester solutions. When we reduce food waste, we reduce the amount of organic material available for composting and for processing in methane digesters. Additionally, reducing food waste means existing land can be used to feed a growing population, supplanting the need to cut down intact forests for cropland. Accounting for the impact of one solution requires us to account for impacts to the broader system of solutions.

DOUBLE-COUNTING

Analyzing many diverse solutions requires care in ensuring no two models are counting the same impact. If we calculate the avoided emissions of solar photovoltaics as a solution and solar-powered, net-zero buildings as another, we have counted solar power twice. This is double-counting, a critical issue to address when modeling the combined impacts of solutions, and one we made sure to avoid.

REBOUND EFFECT

The rebound effect is a principle about human nature: If the price goes down for a given product or service, people generally buy and use more of it, negating the efficiency gains. For example, if increased energy efficiency results in a decreased cost to consumers, consumers may use more energy. Evaluating rebound effect is a challenging effort because so much depends on how people behave in response to these changes. While we do not directly model it, we address potential effects in technical reports, which are found online.

MORE ABOUT THE RESEARCH

This is but a brief outline of what went into the research behind *Drawdown*. As you might imagine, there is far more behind the models including millions of data points. If you have further interest and questions, please visit www.drawdown.org. There you will find technical reports for each of the solutions, descriptions of how each solution was modeled, and other helpful information about the methodology. —Chad Frischmann

WHAT DO THE NUMBERS TELL US?

The quantitative results shown in the pages of *Drawdown* represent the total impact of each solution modeled over a thirty-year period using a reasonable yet optimistic forecast for their global rate of growth. We call this the Plausible Scenario. If we apply this method, the total amount of carbon dioxide avoided and sequestered is 1,051 gigatons by 2050.

As shown below, there are two more scenarios. The Drawdown Scenario shows what happens when the conservative bias of the Plausible Scenario is removed: The total amount of carbon dioxide reduced by 2050 increases to 1,442 gigatons. Electrical energy generation is 100 percent renewable; however, it includes biomass, landfill methane, nuclear, and waste-to-energy—solutions on the decline, but still important to achieving carbon drawdown. We call it the Drawdown Scenario because it estimates a net reduction of 0.59 gigatons from the atmosphere in the year 2050.

The Optimum Scenario represents the most aggressive potential of solutions, in particular renewable energy. It projects 100 percent adoption of *clean* renewable energy by 2050—no biomass, landfill methane, nuclear, or waste-to-energy. This scenario reduces carbon dioxide in the atmosphere by a total of 1,612 gigatons. Here, the amount of emissions avoided or sequestered in the year 2050 is considerably greater than emissions released. Drawdown is potentially reached as early as 2045 with 0.99 gigatons reduced from the atmosphere.

Could any of these scenarios actually achieve drawdown? The Plausible Scenario would not. It is possible the Drawdown Scenario would, and it is more likely in the Optimum Scenario. In each case, we do not model the impact of ocean, land, or methane sinks. In order to estimate the point in time when drawdown would actually occur, one would need to know how much carbon the oceans and unmanaged landmasses are absorbing at that time. Because of increased warming, the oceans may not be able to absorb and store as much carbon. Approximately half of the carbon dioxide emitted by fossil fuels has been absorbed in the oceans. This uptake of carbon dioxide, which results in carbonic acid, is now impairing the entire chain of ocean life—the very capacity of the ocean to sequester carbon. The same principle holds with respect to the land: As temperatures rise, soil, grasslands, and forests can dry out and emit more carbon than they sequester. Thus, how the oceans and land will change in the coming decades can only be estimated. Because we do not know how long ocean and land sinks will continue to absorb carbon, achieving drawdown requires that we do everything we can to address global warming as aggressively, completely, and thoroughly as possible now.

On the following pages is a summary of rankings by solution and by sector. One of our questions going into the research was: How much would it cost to reverse global warming? The *first cost* (total cost to implement) of all the modeled solutions is $131 trillion over thirty years, equivalent to $450 per person per year. A more illuminating number, however, is the *net cost*—how much more money would be required to implement climate solutions compared to the cost of repeating business as usual. *Net cost* is a lower number than *first cost*. We calculate the difference in costs between a solar farm and a coal-fired plant, for example, and between an electric transport system and one fueled by oil. Spurred forward by the decreasing cost of renewable energy, net zero buildings, LEDs, heat pumps, batteries, electric vehicles, and so on, the net cost to implement all solutions modeled here is $27 trillion over thirty years. We also look at the *net operating cost* or *savings* from climate solutions compared to continuing business as usual. The *net operating savings* is $74 trillion over thirty years.

Some of the numbers on specific solutions may seem high, low, or confusing. For example, few would predict that solar farms would rank number 8 in climate solutions (if you combine solar farms and rooftop solar, total solar PV moves to number 7). Solar technology has become synonymous with solving global warming, which is overly simplistic. It is a critical solution, but by itself it does not solve the problem. In our models, we exceed the forward optimistic projections of solar uptake used in several prominent models. Nevertheless, there are other solutions that have greater impact. Bear in mind, we need them all.

Solutions ranked number 6 and number 7 are educating girls and family planning. Why are their impacts the same? It is difficult to draw a bright line between the impacts of family planning and educating girls because they are intertwined and both impact birth rates, so we took the total impact of both and divided it in half. Family planning refers to the universal access to contraception and reproductive health care for all women in all countries. Girls educated through secondary school have fewer children; how many less depends on the country. Providing equal access to education for girls levels the playing field, providing women with the freedom and knowledge to engage in family planning throughout their lifetimes. The dynamics between these two solutions are difficult to disentangle, and can rightly be summed up simply as empowering women and girls.

Each of the three scenarios uses different assessments of future growth based on a variety of factors, such as reductions in cost to implement, policy changes, or improved technology efficiency. Because of this, solution rankings in the summary results

---

*The term *carbon dioxide* includes the measure of equivalent greenhouse gases including methane, nitrous oxide, CFC-12, HCFC-22, and other minor gases based on their global warming potential.

below will change from one scenario to the next. For example, electric vehicles jump from number 26 in the Plausible Scenario to number 10 in the Optimum Scenario. The results of family planning and educating girls are the same in each scenario. This is because there should be no more or less aggressive pathway to providing equal rights and freedom to women. There is only one pathway, and it is universal.

Because the data are dynamic and constantly being updated to show changes, what you see on our website is not necessarily the same as the numbers you see here. Sometime before the end of 2017, we expect to have a dashboard up for each solution. At that point, you can modify the major inputs and come to different conclusions about future impacts and costs. In the meantime, below are the top fifteen solutions for each scenario. ●

Solution	RANKING	Plausible Scenario GIGATONS REDUCED	RANKING	Drawdown Scenario GIGATONS REDUCED	RANKING	Optimum Scenario GIGATONS REDUCED
Refrigeration	1	89.74	2	96.49	3	96.49
Wind Turbines (Onshore)	2	84.60	1	146.50	1	139.31
Reduced Food Waste	3	70.53	4	83.03	4	92.89
Plant-Rich Diet	4	66.11	5	78.65	5	87.86
Tropical Forests	5	61.23	3	89.00	2	105.60
Educating Girls	6	59.60	7	59.60	8	59.60
Family Planning	7	59.60	8	59.60	9	59.60
Solar Farms	8	36.90	6	64.60	7	60.48
Silvopasture	9	31.19	9	47.50	6	63.81
Rooftop Solar	10	24.60	10	43.10	13	40.34
Regenerative Agriculture	11	23.15	14	32.23	15	32.08
Temperate Forests	12	22.61	12	34.70	11	42.62
Peatlands	13	21.57	13	33.51	14	36.59
Tropical Staple Trees	14	20.19	15	31.50	10	46.70
Afforestation	15	18.06	11	41.61	12	41.61
Total (all 80 solutions)		1,051.01		1442.27		1612.89

SUMMARY OF SOLUTIONS BY OVERALL RANKING

	Solution	Sector	TOTAL ATMOSPHERIC CO2-EQ REDUCTION (GT)	NET COST (BILLIONS US $)	NET SAVINGS (BILLIONS US $)
1	Refrigeration	Materials	89.74	N/A	-$902.77
2	Wind Turbines (Onshore)	Energy	84.60	$1,225.37	$7,425.00
3	Reduced Food Waste	Food	70.53	N/A	N/A
4	Plant-Rich Diet	Food	66.11	N/A	N/A
5	Tropical Forests	Land Use	61.23	N/A	N/A
6	Educating Girls	Women and Girls	59.60	N/A	N/A
7	Family Planning	Women and Girls	59.60	N/A	N/A
8	Solar Farms	Energy	36.90	-$80.60	$5,023.84
9	Silvopasture	Food	31.19	$41.59	$699.37
10	Rooftop Solar	Energy	24.60	$453.14	$3,457.63
11	Regenerative Agriculture	Food	23.15	$57.22	$1,928.10
12	Temperate Forests	Land Use	22.61	N/A	N/A
13	Peatlands	Land Use	21.57	N/A	N/A
14	Tropical Staple Trees	Food	20.19	$120.07	$626.97
15	Afforestation	Land Use	18.06	$29.44	$392.33
16	Conservation Agriculture	Food	17.35	$37.53	$2,119.07
17	Tree Intercropping	Food	17.20	$146.99	$22.10
18	Geothermal	Energy	16.60	-$155.48	$1,024.34
19	Managed Grazing	Food	16.34	$50.48	$735.27
20	Nuclear	Energy	16.09	$0.88	$1,713.40
21	Clean Cookstoves	Food	15.81	$72.16	$166.28
22	Wind Turbines (Offshore)	Energy	15.15	$545.28	$762.54
23	Farmland Restoration	Food	14.08	$72.24	$1,342.47
24	Improved Rice Cultivation	Food	11.34	N/A	$519.06
25	Concentrated Solar	Energy	10.90	$1,319.70	$413.85
26	Electric Vehicles	Transport	10.80	$14,148.03	$9,726.40
27	District Heating	Buildings and Cities	9.38	$457.07	$3,543.50
28	Multistrata Agroforestry	Food	9.28	$26.76	$709.75
29	Wave and Tidal	Energy	9.20	$411.84	-$1,004.70
30	Methane Digesters (Large)	Energy	8.40	$201.41	$148.83
31	Insulation	Buildings and Cities	8.27	$3,655.92	$2,513.33
32	Ships	Transport	7.87	$915.93	$424.38
33	LED Lighting - Household	Buildings and Cities	7.81	$323.52	$1,729.54
34	Biomass	Energy	7.50	$402.31	$519.35
35	Bamboo	Land Use	7.22	$23.79	$264.80
36	Alternative Cement	Materials	6.69	-$273.90	N/A
37	Mass Transit	Transport	6.57	N/A	$2,379.73
38	Forest Protection	Land Use	6.20	N/A	N/A
39	Indigenous Peoples' Land Management	Land Use	6.19	N/A	N/A
40	Trucks	Transport	6.18	$543.54	$2,781.63
41	Solar Water	Energy	6.08	$2.99	$773.65
42	Heat Pumps	Buildings and Cities	5.20	$118.71	$1,546.66

	Solution	Sector	TOTAL ATMOSPHERIC CO2-EQ REDUCTION (GT)	NET COST (BILLIONS US $)	NET SAVINGS (BILLIONS US $)
43	Airplanes	Transport	5.05	$662.42	$3,187.80
44	LED Lighting - Commercial	Buildings and Cities	5.04	-$205.05	$1,089.63
45	Building Automation	Buildings and Cities	4.62	$68.12	$880.55
46	Water Saving - Home	Materials	4.61	$72.44	$1,800.12
47	Bioplastic	Materials	4.30	$19.15	N/A
48	In-Stream Hydro	Energy	4.00	$202.53	$568.36
49	Cars	Transport	4.00	-$598.69	$1,761.72
50	Cogeneration	Energy	3.97	$279.25	$566.93
51	Perennial Biomass	Land Use	3.33	$77.94	$541.89
52	Coastal Wetlands	Land Use	3.19	N/A	N/A
53	System of Rice Intensification	Food	3.13	N/A	$677.83
54	Walkable Cities	Buildings and Cities	2.92	N/A	$3,278.24
55	Household Recycling	Materials	2.77	$366.92	$71.13
56	Industrial Recyling	Materials	2.77	$366.92	$71.13
57	Smart Thermostats	Buildings and Cities	2.62	-$74.16	$640.10
58	Landfill Methane	Buildings and Cities	2.50	-$1.82	$67.57
59	Bike Infrastructure	Buildings and Cities	2.31	-$2,026.97	$400.47
60	Composting	Food	2.28	-$63.72	-$60.82
61	Smart Glass	Buildings and Cities	2.19	$74.20	$325.10
62	Women Smallholders	Women and Girls	2.06	N/A	$87.60
63	Telepresence	Transport	1.99	$127.72	$1,310.59
64	Methane Digesters (Small)	Energy	1.90	$15.50	$13.90
65	Nutrient Management	Food	1.81	N/A	$102.32
66	High-Speed Rail	Transport	1.42	$1,049.98	$310.79
67	Farmland Irrigation	Food	1.33	$216.16	$429.67
68	Waste-to-Energy	Energy	1.10	$36.00	$19.82
69	Electric Bikes	Transport	0.96	$106.75	$226.07
70	Recycled Paper	Materials	0.90	$573.48	N/A
71	Water Distribution	Buildings and Cities	0.87	$137.37	$903.11
72	Biochar	Food	0.81	N/A	N/A
73	Green Roofs	Buildings and Cities	0.77	$1,393.29	$988.46
74	Trains	Transport	0.52	$808.64	$313.86
75	Ridesharing	Transport	0.32	N/A	$185.56
76	Micro Wind	Energy	0.20	$36.12	$19.90
77	Energy Storage (Distributed)	Energy	N/A	N/A	N/A
77	Energy Storage (Utilities)	Energy	N/A	N/A	N/A
77	Grid Flexibility	Energy	N/A	N/A	N/A
78	Microgrids	Energy	N/A	N/A	N/A
79	Net Zero Buildings	Buildings and Cities	N/A	N/A	N/A
80	Retrofitting	Buildings and Cities	N/A	N/A	N/A
	Totals		**1,052.06**	**$27,378.56**	**$74,362.49**

SUMMARY OF SOLUTIONS BY SECTOR

Sector		Solution	TOTAL ATMOSPHERIC CO2-EQ REDUCTION (GT)	NET COST (BILLIONS US $)	NET SAVINGS (BILLIONS US $)
Buildings and Cities	27	District Heating	9.38	$457.07	$3,543.50
	31	Insulation	8.27	$3,655.92	$2,513.33
	33	LED Lighting - Household	7.81	$323.52	$1,729.54
	42	Heat Pumps	5.20	$118.71	$1,546.66
	44	LED Lighting - Commercial	5.04	-$205.05	$1,089.63
	45	Building Automation	4.62	$68.12	$880.55
	54	Walkable Cities	2.92	N/A	$3,278.24
	57	Smart Thermostats	2.62	-$74.16	$640.10
	58	Landfill Methane	2.50	-$1.82	$67.57
	59	Bike Infrastructure	2.31	-$2,026.97	$400.47
	61	Smart Glass	2.19	$74.20	$325.10
	71	Water Distribution	0.87	$137.37	$903.11
	73	Green Roofs	0.77	$1,393.29	$988.46
	79	Net Zero Buildings	0.00	N/A	N/A
	80	Retrofitting	0.00	N/A	N/A
		Buildings and Cities Total	**54.50**	**$4,778.30**	**$17,906.26**
Energy	2	Wind Turbines (Onshore)	84.60	$1,225.37	$7,425.00
	8	Solar Farms	36.90	-$80.60	$5,023.84
	10	Rooftop Solar	24.60	$453.14	$3,457.63
	18	Geothermal	16.60	-$155.48	$1,024.34
	20	Nuclear	16.09	$0.88	$1,713.40
	22	Wind Turbines (Offshore)	15.15	$545.28	$762.54
	25	Concentrated Solar	10.90	$1,319.70	$413.85
	29	Wave and Tidal	9.20	$411.84	-$1,004.70
	30	Methane Digesters (Large)	8.40	$201.41	$148.83
	34	Biomass	7.50	$402.31	$519.35
	41	Solar Water	6.08	$2.99	$773.65
	48	In-Stream Hydro	4.00	$202.53	$568.36
	50	Cogeneration	3.97	$279.25	$566.93
	64	Methane Digesters (Small)	1.90	$15.50	$13.90
	68	Waste-to-Energy	1.10	$36.00	$19.82
	76	Micro Wind	0.20	$36.12	$19.90
	77	Energy Storage (Distributed)	N/A	N/A	N/A
	77	Energy Storage (Utilities)	N/A	N/A	N/A
	77	Grid Flexibility	N/A	N/A	N/A
	78	Microgrids	N/A	N/A	N/A
		Energy Totals	**247.19**	**$4,896.24**	**$21,446.64**
Food	3	Reduced Food Waste	70.53	N/A	N/A
	4	Plant-Rich Diet	66.11	N/A	N/A
	9	Silvopasture	31.19	$41.59	$699.37
	11	Regenerative Agriculture	23.15	$57.22	$1,928.10
	14	Tropical Staple Trees	20.19	$120.07	$626.97
	16	Conservation Agriculture	17.35	$37.53	$2,119.07
	17	Tree Intercropping	17.20	$146.99	$22.10
	19	Managed Grazing	16.34	$50.48	$735.27

Sector		Solution	TOTAL ATMOSPHERIC CO2-EQ REDUCTION (GT)	NET COST (BILLIONS US $)	NET SAVINGS (BILLIONS US $)
	21	Clean Cookstoves	15.81	$72.16	$166.28
	23	Farmland Restoration	14.08	$72.24	$1,342.47
	24	Improved Rice Cultivation	11.34	N/A	$519.06
	28	Multistrata Agroforestry	9.28	$26.76	$709.75
	53	System of Rice Intensification	3.13	N/A	$677.83
	60	Composting	2.28	-$63.72	-$60.82
	65	Nutrient Management	1.81	N/A	$102.32
	67	Farmland Irrigation	1.33	$216.16	$429.67
	72	Biochar	0.81	$31.27	N/A
		Food Totals	**321.93**	**$777.48**	**$10,017.44**
Land Use	5	Tropical Forests	61.23	N/A	N/A
	12	Temperate Forests	22.61	N/A	N/A
	13	Peatlands	21.57	N/A	N/A
	15	Afforestation	18.06	46.51	1,014.92
	35	Bamboo	7.22	18.35	234.64
	38	Forest Protection	6.20	N/A	N/A
	39	Indigenous Peoples' Land Management	6.19	N/A	N/A
	51	Perennial Biomass	3.33	N/A	N/A
	52	Coastal Wetlands	3.19	N/A	N/A
		Land Use Totals	**149.60**	**$64.86**	**$1,249.56**
Materials	1	Refrigeration	89.74	N/A	-$902.77
	36	Alternative Cement	6.69	-$273.90	N/A
	46	Water Saving - Home	4.61	$72.44	$1,800.12
	47	Bioplastic	4.30	$19.15	N/A
	55	Household Recycling	2.77	$366.92	$71.13
	56	Industrial Recyling	2.77	$366.92	$71.13
	70	Recycled Paper	0.90	$573.48	N/A
		Materials Totals	**111.78**	**$1,125.01**	**$1,039.61**
Transport	26	Electric Vehicles	10.80	$14,148.03	$9,726.40
	32	Ships	7.87	$915.93	$424.38
	37	Mass Transit	6.57	N/A	$2,379.73
	40	Trucks	6.18	$543.54	$2,781.63
	43	Airplanes	5.05	$662.42	$3,187.80
	49	Cars	4.00	-$598.69	$1,761.72
	63	Telepresence	1.99	$127.72	$1,310.59
	66	High-Speed Rail	1.42	$1,049.98	$310.79
	69	Electric Bikes	0.96	$106.75	$226.07
	74	Trains	0.52	$808.64	$313.86
	75	Ridesharing	0.32	N/A	$185.56
		Transport Totals	**45.78**	**$15,675.92**	**$22,665.87**
Women and Girls	6	Educating Girls	59.60	N/A	N/A
	7	Family Planning	59.60	N/A	N/A
	62	Women Smallholders	2.06	N/A	$87.60
		Women and Girls Totals	**121.26**	N/A	**$87.60**

WHO WE ARE – THE COALITION

DRAWDOWN FELLOWS

Zak Accuardi, MA is a policy researcher with five years of experience addressing diverse urban sustainability challenges. He has led research and co-authored a report focused on government partnerships with emerging mobility providers like Uber.

Raihan Uddin Ahmed, MDS is an environmental specialist with more than 14 years of experience. His work focuses on impact assessments for infrastructure projects, renewable energy technologies, and climate change.

Carolyn Alkire, PhD is an environmental economist with 35 years of experience in research and analyses to advance policy improving land and resource management. She has worked with government agencies on regional transportation planning to reduce greenhouse gas emissions.

Ryan Allard, PhD is a transportation systems analyst with six years of experience examining how to improve transportation systems around the world. He has presented and published computer models on transport technology and connectivity in peer-reviewed journals and at international conferences.

Kevin Bayuk, MA works at the intersection of ecology and economy, where permaculture design meets cooperative organizations intent on meeting human needs. He is a partner with LIFT Economy, which accelerates social enterprises and facilitates investment into highly beneficial impact organizations, and is a founding partner of the Urban Permaculture Institute San Francisco.

Renilde Becqué, MBA is a sustainability and energy consultant with over 15 years of experience working internationally. She currently works with several international nonprofits on circular economy, carbon, and energy efficiency projects and programs.

Erika Boeing, MA is an entrepreneur and systems engineer with seven years of experience working with energy technologies. She has created a business to develop and commercialize a novel rooftop wind energy technology.

Jvani Cabiness, MDP is a global health and development professional specializing in family planning with five years of experience in sexual and reproductive health promotion. She had supported health systems strengthening and capacity building projects throughout Africa.

Johnnie Chamberlin, PhD is an environmental analyst with 10 years of experience working in environmental science, conservation, and research. He is the author of two guidebooks.

Delton Chen, PhD is a civil engineer with more than 15 years of experience in modelling structures, groundwater, systems, water resources, and mine plans for sustainability. Delton has investigated 'hot rock' geothermal energy and island aquifers in Australia, and is a co-founder and lead author of Global 4C, a new international policy for new climate mitigation finance.

Leonardo Covis, MPP is a program analyst and manager with eight years of experience in the economic development and environmental policy fields. His work has brought millions of dollars to low-income neighborhoods, revitalized natural habitats, and guided statewide fuel policy decisions in California.

Priyanka deSouza, MSc, MBA, MTech is a researcher in the field of urban planning with more than seven years of experience working on different energy technologies and environmental policy. She has most recently worked on setting up a low cost air quality monitoring network for schools in Nairobi.

Jai Kumar Gaurav, MSc is a research analyst with eight years of experience working in the field of climate change mitigation and adaptation. He has worked on Clean Development Mechanism and Gold Standard certified voluntary emission reduction projects. He is also working on developing a Nationally Appropriate Mitigation Action (NAMA) proposal in the waste sector.

Anna Goldstein, PhD is a science policy expert with 10 years of experience doing academic research. She has translated her scientific background into insights for management of clean energy research programs.

João Pedro Gouveia, PhD is an Environmental Engineer with more than eight years of experience working in energy systems analysis mainly in the residential sector with contributions both for research and policy. He is finishing is PhD on Climate Change and Sustainable Energy Policies at the Center for Environmental Sustainability Research at the Faculty of Sciences and Technology at Nova University of Lisbon.

Alisha Graves, MPH is a public health professional whose work focuses on improving global access to family planning. She is the Vice President of the Population Program at Venture Strategies for Health and Development (VSHD), a California-based nonprofit where she oversees the Rebirth of Population Awareness, and is the Co-founder of the OASIS Initiative, a joint project of the University of California, Berkeley and VSHD.

Karan Gupta, MPA is a high-performance building specialist with seven years of experience in the utility and building industries. He has worked with modular building systems to accelerate the market for energy efficiency in residential and commercial applications.

Zhen Han, BSc is a PhD candidate in Ecology at Cornell, with her research focusing on nutrient cycling in agro-ecosystems for which she conducted quantitative synthesis and field measurements to investigate the impact of various agricultural management practices on nitrous oxide. She has served as an Environmental Policy Fellow at UNEP, where she worked on ecosystem-based climate change adaptation and gender mainstreaming.

Zeke Hausfather, MS is a climate scientist and energy systems analyst whose work focuses on conservation and efficiency. He has worked as a research scientist at Berkeley Earth, the head of energy analytics at Essess Inc., and the chief scientist at C3, and co-founded Efficiency 2.0, a behavior-based energy efficiency company.

Yuill Herbert, MA has worked on more than 35 community climate action plans across Canada, as well as many other community planning and climate change-related projects. He is a director and founder of Sustainability Solutions Group, a workers cooperative in Canada, and previously developed the highly regarded GHGProof energy, emissions, and land-use planning model.

Amanda Hong, MPP is a public policy professional whose work includes policy suggestions for driving source reduction, recycling and composting of packaging waste in California, and a blue carbon assessment of mangrove conservation in Sri Lanka. Amanda currently works as the Organic Recycling Specialist for the US Environmental Protection Agency's Pacific Southwest Region.

Ariel Horowitz, PhD is an energy analyst with six years of experience working with energy technologies and systems. She has a doctorate in chemical engineering, with a focus on energy storage.

Ryan Hottle, PhD is a soil carbon and climate science analyst with a research focus on climate change mitigation through biological carbon sequestration. His interests include climate-smart agriculture, fast action mitigation strategies, and energy conservation and efficiency in the built environment, and he has worked as a consultant for the World Bank and Consultative Group on International Agricultural Research (CGIAR)'s Climate Change and Food Security program.

Troy Hottle, PhD is an ORISE Postdoctoral Fellow at the U.S. Environmental Protection Agency with 10 years of experience working on environmental projects and conducting research. He has worked on the application of life cycle assessment to evaluate and inform real-world systems, including biopolymer degradation, vehicle mass reduction, and the development of national energy inventories.

David Jaber, MEng is a strategic advisor with more than 15 years of experience in green building investigation, greenhouse gas analysis, and zero waste implementation. He has created dozens of greenhouse gas inventories and reduction strategies in food processing, manufacturing, and retail settings.

Dattakiran Jagu, MTech is a PhD candidate in the field of Science and Management of Climate Change with five years of experience in promoting clean energy technologies. He is a founding member of a clean energy start-up that designed the first train station in India that runs on solar energy.

Daniel Kane, MS is a PhD student at Yale University's School of Forestry and Environmental Studies with five years of experience in agricultural research. He focuses on the application of open-source tools for agricultural management and how soils can be managed to foster climate change resilience in agriculture.

Becky Xilu Li, MPP is an energy policy consultant with four years of experience in the field. She has worked with both US and Chinese governments, companies, and research institutes to promote market-driven solutions for renewable deployment.

Sumedha Malaviya, MA is a climate and energy professional with more than seven years of experience in climate mitigation, adaptation, and energy efficiency projects. She has worked with several countries on developing and implementing Low Emissions Development strategies.

Urmila Malvadkar, PhD is an applied mathematician and environmental scientist whose research and modeling focuses on water, conservation, and international development. Urmila's PhD work focused on ecological modeling, and since then her research has covered many environmental issues, including placement of dams and water intakes, managing populations under disturbance, water issues in the developing world, and the size of effective protected areas.

Alison Mason, MSc is a mechanical engineer with 16 years of experience working with solar energy. She was instrumental in launching a solar installation training and manufacturing program with the Oglala-Sioux tribe in South Dakota.

Mihir Mathur, BCom is an interdisciplinary researcher in the field of climate change with nine years of experience in finance, community engagement, and policy. He currently practices system dynamics for modeling sustainability solutions at TERI, New Delhi.

Victor Maxwell, MS is a PhD candidate in the field of environmental finance with nine years of experience working in physics and energy systems management. He has facilitated the development of decentralized sustainable energy systems for rural communities in Chile, Denmark, and South Africa.

David Mead, BA is an architect and engineer with more than 13 years of experience in the building industry. He has been involved in over 50 projects that have had high sustainability goals like LEED, living buildings, passive house, and net zero energy.

Mamta Mehra, PhD is an environmental professional with more than seven years of experience working with national and international organizations in the field of climate change adaptation and mitigation related to the agriculture sector. She is about to finish her PhD, for which she has developed a GIS framework for the delineation and characterization of resource management domain in the agriculture sector.

Ruth Metzel, MBA is an ecology and evolutionary biologist jointly pursuing a Master of Forestry from the Yale School of Forestry and Environmental Studies, and an MBA from the Yale School of Management. Her research explores the agriculture-forest interface and understanding ways in which actors from multiple sectors interact to achieve integrated landscape management objectives.

Alex Michalko, MBA is a corporate sustainability professional with more than 10 years of experience in the field across a range of industries, including technology, media/entertainment, and retail. She has worked with Disney, REI, and Amazon to advance sustainability initiatives that improve business resilience and have a positive impact on the environment and local communities.

Ida Midzic, MEng is a PhD candidate in the field of mechanical engineering with six years of experience in research and teaching. She has developed a method for mechanical engineers for eco-evaluation of conceptual design solutions in product development.

S. Karthik Mukkavilli, MS is an academic entrepreneur in energy meteorology satellite data assimilation with eight years of experience in computational science and engineering. He has developed aerosol aware solar forecasts over Australasia with hybrid atmospheric physics and artificial intelligence models.

Kapil Narula, PhD is an electrical engineer, development economist, and an energy and sustainability professional with 15 years of experience in the maritime domain. He has worked onboard ships, as a faculty at academic institutes, and as a researcher.

Demetrios Papaioannou, PhD is a civil engineer in the transportation field who specializes in mass transit, demand modeling, user satisfaction, and sustainability. His PhD studies focused on mass transit and the relationship between transit quality, user satisfaction, and mode choice, and he has presented his research at international conferences and published peer-reviewed research papers.

Michelle Pedraza, MA is a business and strategy analyst for global markets, whose work now focuses on addressing the challenges that microenterprises face in scaling their businesses. She completed an internship for the Clinton Global Initiative, reviewing and developing commitments for the Market Based Approaches and Food Systems tracks.

Chelsea Petrenko, PhD is an ecosystem ecologist with a focus on forest resources and soil carbon storage. Her PhD research measured changes in soil carbon storage after clear-cutting forests in the northeastern United States, and she has worked as a trainee in Polar Environmental Change which brought her to Greenland and Antarctica to study carbon cycling in cold environments.

Noorie Rajvanshi, PhD is a sustainability engineer with over seven years of experience in the field of environmental impact quantification using life cycle assessment methodology. She has worked with various cities in North America to evaluate technology pathways for achieving their 2050 sustainability goals.

George Randolph, MSc is an energy policy analyst with five years of experience, working most recently in electric utility regulatory affairs. He has consulted on several energy efficiency and residential rooftop solar proceedings before public utility commissions in California, Nevada, Arizona, and Colorado.

Abby Rubinson, JD is an international, environmental, human rights lawyer with more than 10 years of experience in the field. She has focused her work on the links between climate change and human rights, including in litigation and advocacy defending indigenous peoples' rights, scholarly publications, and international treaty negotiations.

Adrien Salazar, MA is a political ecologist, organizational strategist, advocate, and poet with over eight years of experience in program and campaign management for environmental and community-based organizations. He has worked with indigenous rice farmers in the northern Philippines to develop community-based evaluation indicators, in a project to support farmer empowerment and conservation of indigenous rice varieties.

Aven Satre-Meloy, BS is a Master's student in Environmental Management with five years of experience working on energy and sustainability issues. He has conducted research or worked in the field of sustainable energy on four continents.

Christine Shearer, PhD is an environmental sociologist with more than 10 years of experience doing interdisciplinary climate change and energy research. She has worked on energy policy and climate impacts and adaptation, with research published in Nature and The New York Times, among others.

David Siap, MSc is an engineer with five years of experience working on energy efficiency topics. He was the lead technical analyst on the US DOE energy conservation standards and test procedures, with a projected net present value of over $1 billion and energy savings of approximately 1 quad.

Kelly Siman, MS is a PhD candidate in the field of Biomimicry at the University of Akron with more than 10 years of experience in academia and environmental nonprofits. She is working on climate change resilience and biomimetic adaptation and mitigation applications.

Leena Tähkämö, PhD is a postdoctoral scientist with six years of experience in the field of illuminating engineering. She has studied the environmental and economic sustainability of lighting systems by life cycle assessment method to identify the areas of the greatest importance in emission reduction.

Eric Toensmeier, MA is an economic botanist with 25 years of experience investigating agroforestry systems and perennial crops. He is the author of The Carbon Farming Solution: A Global Toolkit of Perennial Crops and Regenerative Agriculture Practices for Climate Change Mitigation and Food Security.

Melanie Valencia, MPH is the Innovation and Sustainability Officer and teaches environmental sustainability at Universidad San Francisco de Quito. She co-founded Carbocycle, a startup that recycles organic waste into marketable vegetable oil substitutes.

Ernesto Valero Thomas, PhD is an architect with seven years of experience working with environmental strategies for the sustainable growth of emerging cities. He has developed methodologies to study the flow of water, food, oil, waste, telecommunications, and people in cities throughout the world.

Andrew Wade, MS is a graduate student in real estate finance and development with seven years of experience researching sustainable urban development projects in cities around the world. He has directed a panel on innovation in the real estate industry at Harvard.

Marilyn Waite, MPhil is an engineer and clean technology investment professional with more than 10 years of experience in the field. She is the author of Sustainability at Work.

Charlotte Wheeler, PhD is a tropical ecologist with six years of experience working in forest restoration and climate change mitigation. She has conducted research into the carbon sequestration potential of large scale tropical forest restoration.

Christopher Wally Wright, MPA is a researcher and analyst with over six years of experience working in the fields of public sector administration, environmental education and resource management, and social and public policy.

Liang Emlyn Yang, PhD is a geographer with nearly 10 years of experience in human-environment interactions. He has worked with long-term historical climatic and environmental impacts, natural hazards, and social and human responses in China and Southeast Europe.

Daphne Yin, MA is an environmental consultant with five years of experience in climate change, natural resource management, and development. She has co-developed methods of natural capital and social capital valuation for common lands in India, with a focus on grazing lands.

Kenneth Zame, PhD is an energy and environmental sustainability researcher and educator with over seven years of experience in research. He worked as a QESST Scholar on the sustainability of terawatt-scale PV deployment in the US, a project sponsored by the US National Science Foundation (NSF) and Department of Energy (DOE).

May Boeve is the executive director of 350.org, a climate-focused campaigns, projects, and actions organization, led from the bottom up by people in 188 countries. She became the first person in the United States to be profiled by Time magazine as part of its annual series on Next Generation Leaders.

James Boyle is the founder, CEO, and chairman of Sustainability Roundtable, a research and consulting firm dedicated to accelerating the development and adoption of best practices in more sustainable businesses, and the principal cofounder of the nonprofit Alliance for Business Leadership.

Tom Brady is a National Football League quarterback with the New England Patriots, widely considered to be among the greatest quarterbacks of all time. An advocate of environmental sustainability, Tom wanted to minimize his family's impact on the land, so he and his wife Gisele Bündchen built a new house that uses solar energy, gray water technology, and composting. Eighty percent of the construction was comprised of reused or recycled material.

Tod Brilliant is a marketing expert, writer, and photographer serving as VP of marketing and creative director for Peoples Home Equity, a mortgage lending company, and previously as creative director and social strategist at Post Carbon Institute, which aims to lead the transition to a more resilient, equitable, and sustainable world.

Clark Brockman is a longtime champion of energy-efficient, climate-responsive design and planning throughout the built environment, currently serving as a principal at SERA Architects, where he leads the firm's office in San Mateo, California. He is a founding and past board member of the International Living Future Institute, an adviser to Portland State University's Institute for Sustainable Solutions, and a member of SPUR's water and climate policy board in San Francisco.

Bill Browning is one of the green building and real estate industry's foremost thinkers and strategists, and an advocate for sustainable design solutions at all levels of business, government, and civil society. He is a founding partner of Terrapin Bright Green and has consulted for the Greening of the White House, Google, Disney, Bank of America, Starwood, Lucasfilm, Clif Bar, Grand Canyon National Park, and the Sydney 2000 Olympic Village.

Michael Brune is executive director of the Sierra Club, the United States' largest and most influential grassroots environmental organization. He previously worked for Rainforest Action Network and is the author of Coming Clean: Breaking America's Addiction to Oil and Coal.

Gisele Bündchen is a model, entrepreneur, environmentalist, and philanthropist. She has championed rainforest preservation and clean water initiatives; founded the humanitarian, educational, and environmental grant-maker The Luz Foundation; and was designated a Goodwill Ambassador for the United Nations Environment Programme in 2009. The charities she had supported include the Rainforest Alliance, Save the Children, Doctors without Borders, and many more.

Leo Burke directs the Global Commons Initiative at the University of Notre Dame's Mendoza College of Business, which offers education both within the college and in conjunction with partners such as the UN. He has also served as associate dean and director of executive education at the University of Notre Dame, and has worked at Motorola.

Peter Byck is a film director, producer, editor, and Arizona State University professor, whose first documentary, Garbage, won the 1996 South by Southwest Film Festival Jury Prize, and whose second film, Carbon Nation, highlights climate change solutions. He is currently building a series of short films celebrating ranchers who focus on soil health.

Peter Calthorpe is an urban designer, author, and leader in developing new approaches to urban revitalization, sustainable growth, and regional planning around the globe. He directs the award-winning design studio Calthorpe Associates, is the founder and the first board president of the Congress for the New Urbanism, and recently authored the book Urbanism in the Age of Climate Change.

Lynelle Cameron is president and CEO of the Autodesk Foundation and senior director of sustainability at the software corporation Autodesk. She founded both to invest in and support people using design to solve today's most difficult challenges, and under her leadership Autodesk has won numerous awards for sustainability, climate leadership, and philanthropy.

Mark Campanale is the founder and executive director of the Carbon Tracker Initiative, an independent financial think tank that provides in-depth analysis of the impacts of climate change on capital markets and investment in fossil fuels, mapping risk, opportunity, and the route to a low carbon future. With cofounder Nick Robins, Campanale conceived the "unburnable carbon" capital markets thesis, which uses the science of carbon budgets to assess investor exposure to stranded assets and the looming carbon bubble.

Dennis Carlberg, AIA, LEED AP BD+C, is an architect and the sustainability director at Boston University, where he is an adjunct assistant professor in the Department of Earth and Environment and faculty adviser for the Earth House, a living learning community at BU. He cochairs the Climate Resilience Committee at the Urban Land Institute Boston, a committee dedicated to exploring policies and solutions that address the impacts of climate change on communities.

Steve Chadima has nearly thirty years of experience in advanced energy and technology. He is senior vice president of external affairs at Advanced Energy Economy, a national association of business leaders who are making the global energy system more secure, clean, and affordable.

Adam Chambers is a scientist at the USDA's Natural Resources Conservation Service (NRCS), where he is on the Air Quality and Atmospheric Change Team, working to implement conservation measures on managed agricultural lands. Over the past two decades his work has focused on the applied sciences and reducing air pollutants and greenhouse gases in the atmosphere.

Aimée Christensen leads the Sun Valley Institute for Resilience and Christensen Global Strategies, with twenty-five years of climate experience including at the U.S. Department of Energy, World Bank, Baker McKenzie, and Google, where she served as "climate maven" at Google.org. She negotiated the first bilateral climate change agreements, including United States–Costa Rica in 1994, wrote the first university endowment investment policy on climate change (Stanford University in 1999), and was the 2011 Hillary Laureate and a 2010 Aspen Institute Catto Fellow.

Cutler J. Cleveland is an author, consultant, academic, and business executive working on research involving natural resources, energy use, and their related economies. He is the editor-in-chief of the Encyclopedia of Energy and a professor at Boston University.

Leila Conners founded Tree Media Group, setting out to build a production company that creates media to support and sustain civil society by telling inspiring stories.

John Coster serves as an independent adviser to several low-carbon or carbon-sequestering initiatives. He previously served as the green business officer at Skanska USA Building, a leading construction group that, in addition to providing construction services, develops public-private partnerships for everything from small renovations to billion-dollar projects.

Audrey Davenport is the ecology program lead for corporate real estate at Google, and previously led internal corporate sustainability efforts at Google on the Energy and Sustainability Team. She was a Fulbright scholar in Malaysia, and taught graduate courses on sustainable business strategies at Johns Hopkins University and the Presidio Graduate School.

Edward Davey is senior program manager at the Prince of Wales' International Sustainability Unit, where he leads the organization's work on forests and climate change, and is currently writing a book titled A Restored Earth: Ten Paths to a Hopeful Future. He previously served as lead adviser on environment in the Colombian presidency.

Pedro Diniz is a businessman and former Formula 1 racer. He transformed his family farm in Brazil's São Paulo state into Toca Farm, one of the country's leading producers of organic food, with a strong commitment to develop large-scale agroforestry production.

AshEL "SeaSunZ" Eldridge (a.k.a. the Uber Rapper) is the CEO of Earth Amplified Consulting, offering creative strategy for entrepreneurs, start-ups, and nonprofits, and an adjunct professor of climate justice, race, and activism at San Francisco State University. He is the founder of roots, rap, and reggae collective Earth Amplified, a vocalist with West African/West Oakland band Dogon Lights, a shamanic and plant-based health coach, Purium distributor, and teacher of meditation, creativity, and manifestation principles for activists, creatives, and entrepreneurs.

John Elkington is an entrepreneur, environmentalist, and author of seventeen books, including his most recent, *The Breakthrough Challenge: 10 Ways to Connect Today's Profits with Tomorrow's Bottom Line*. He has founded and cofounded several ventures, including Volans, a change agency aimed at looking beyond incremental change and addressing systemic challenges at scale, SustainAbility, and Environmental Data Services.

Jib Ellison is the founder and CEO of Blu Skye, a management consulting firm focused on sustainable business growth. He works with Fortune 500 companies to transform markets and create new ones, using sustainability to reveal new market opportunities.

Donald Falk is an associate professor at the University of Arizona's School of Natural Resources and the Environment, where he specializes in watershed management and ecohydrology. His research areas include fire history, fire ecology, restoration ecology, landscape ecology, and the impacts of land management and global change on ecosystems, including dynamics of abrupt change.

Felipe Faria is the CEO of the Green Building Council Brasil, an organization that has accelerated the greening of Brazil's construction industry, making the country one of the top five markets in the world for LEED, and influencing large-scale projects such as the 2014 FIFA World Cup and the 2016 Olympic Games. He previously served as volunteer of the LEED Steering Committee, a group of professionals responsible for maintaining LEED as a global leadership tool, and is currently the chair of World Green Building Council Americas Regional Network Committee.

Rick Fedrizzi is the founder and former CEO of the U.S. Green Building Council (USGBC) and CEO of Green Business Certification Inc. (GBCI). USGBC's LEED green building program has been the cornerstone of his career, and since its launch in 2000, more than 55,000 commercial projects spanning 10.1 billion square feet, and more than 154,000 residential units around the world participate in LEED.

David Fenton founded Fenton in 1982 to create communications campaigns for the environment, public health, and human rights. He has aided the rise of MoveOn.org, stimulated the increase in organic food sales, represented Nelson Mandela and the African National Congress, passed sanctions against apartheid, publicized the first gay marriages in the United States, worked with Al Gore and the UN on climate change, and led public health campaigns against tobacco and endocrine-disrupting chemicals.

Jonathan Foley spent over two decades leading interdisciplinary, university-based programs focused on solving global environmental issues before becoming the Executive Director of the California Academy of Sciences, where he has been able to incite interest and excitement in the sciences in children and adults alike. He has published over 130 scientific articles, many op-eds, and has won numerous awards and honors, including the Presidential Early Career Award for Scientists and Engineers (awarded by President Bill Clinton).

Bob Fox is one of New York City's most highly respected leaders in t he green building movement, having started the CookFox architecture firm in 2003, which is devoted to creating beautiful, environmentally responsible, high-performance buildings. The firm is best known for its design of the Bank of America Tower at One Bryant Park, the first commercial skyscraper to receive LEED Platinum certification.

Maria Carolina Fujihara is an architect who specializes in sustainable urban planning. She served as the technical coordinator for Green Building Council Brasil for five years, where she worked promoting LEED certification in the country. Maria also was the head of the technical committees that created the certification tool for the Brazilian market for homes.

Mark Fulton is a recognized economist and market strategist with a strong focus on the environment and sustainability, starting with a report he authored on climate change and markets in 1991. He has served as head of research at Deutsche Bank Climate Change Advisors, where he produced thought leadership papers for investors on climate, cleaner energy, and sustainability.

Lisa Gautier is the president and a board member of the environmental public charity Matter of Trust, which she cofounded with her husband, Patrice Gautier, in 1998. The nonprofit concentrates on eco-education, uses for man-made surplus, and naturally abundant renewable resources.

Mark Gold is the associate vice chancellor for environment and sustainability and an adjunct professor at the Institute of the Environment and Sustainability at UCLA, and has worked in the fields of water pollution, water supply, integrated water management, and coastal protection for the past twenty-five years. In addition, he has worked extensively on the development of sustainable city plans for Los Angeles and Santa Monica, and is currently leading the Sustainable LA Grand Challenge, with countywide goals of 100 percent renewable energy, 100 percent local water, and enhanced ecosystem and human health by 2050.

Rachel Gutter is the chief product officer of the International WELL Building Institute, a public benefit corporation whose mission is to improve human health and well-being through the built environment. She previously served as the senior vice president of knowledge at the U.S. Green Building Council and the director of the Center for Green Schools, where her dynamic leadership helped convene international corporations, globally recognized institutions, and government entities around the goal of putting every student in a green school within this generation.

André Heinz is a director of the Heinz Endowments, and shortly after joining its board he oversaw the creation of an environmental grant-making program in 1993. He continues to serve on the board and on the investment committee, which oversees the management of the $1.5 billion endowment, and he pursues investments in sustainable technologies through venture capital.

Gregory Heming is a municipal councillor in Annapolis County, Nova Scotia, where he chairs both the economic development and climate change committees and serves nationally on the board of directors of the Federation of Canadian Municipalities. He holds a Ph.D. in ecology with postgraduate studies in history of religions and philosophy of science, and has spoken, written, and published extensively on rural economics, ecology of place, and public engagement.

Oran Hesterman is a national leader in sustainable agriculture and food systems, and the president and CEO of the Fair Food Network. He has more than thirty-five years of experience as a scientist, farmer, philanthropist, businessman, educator, and advocate, and is a respected partner for policy makers, philanthropic leaders, and advocates.

Patrick Holden is the founding director of the Sustainable Food Trust, which works internationally to accelerate the transition to more sustainable food systems. He is patron of the UK Biodynamic Association and was awarded the CBE for services to organic farming in 2005.

Gunnar Hubbard is the principal and sustainability practice leader at Thornton Tomasetti, a global engineering design, investigation, and analysis services firm. He is a recognized leader in green building across the United States, Asia, and Europe.

Congressman Jared Huffman represents California's North Bay and North Coast in the House of Representatives as one of Congress's leading champions for clean energy, greenhouse gas reduction, and protecting our natural environment, serving on the Transportation and Infrastructure Committee and the Natural Resources Committee. Prior to Congress, he served six years in the California Assembly, where he chaired the Committee on Water, Parks, and Wildlife and authored dozens of significant laws, and was a senior attorney with the Natural Resources Defense Council.

Molly Jahn is a professor at the University of Wisconsin–Madison in the Department of Agronomy, at the Global Health Institute, and at the Center for Sustainability and the Global Environment, and joint faculty at the Oak Ridge National Laboratory. She has published more than one hundred peer-reviewed research articles and has sixty active commercial licenses from her plant-breeding programs, and her vegetable varieties are grown commercially and for subsistence on six continents.

Chris Jordan is a Seattle-based photographic artist and filmmaker whose work focuses on consumerism and mass culture. His work sends a bold message about unconscious behaviors in our individual and collective lives.

Daniel Kammen is the founding director of the Renewable and Appropriate Energy Laboratory (RAEL) at UC Berkeley, where he is a professor in the Energy and Resources Group, the Goldman School of Public Policy, and the Department of Nuclear Engineering. In 2010 he was appointed the first Energy and Climate Partnership for the Americas (ECPA) fellow by Secretary of State Hillary Clinton, and served the U.S. State Department as science envoy from 2016 to 2017.

Danny Kennedy is a clean-technology entrepreneur, environmental activist, and the author of *Rooftop Revolution: How Solar Power Can Save Our Economy—and Our Planet—from Dirty Energy* (2012). He is a cofounder of Sungevity, the managing director of the California Clean Energy Fund, and a cofounder of Powerhouse.

Kerry Kennedy is a human rights activist and lawyer, the president of the Robert F. Kennedy Human Rights organization, the author of *Speak Truth to Power: Human Rights Defenders Who Are Changing Our World*, and the author of the *New York Times* best seller *Being Catholic Now*. She served as chair of the Amnesty International USA Leadership Council for more than a decade, and was nominated by President Bush and confirmed by the Senate to serve in her current capacity as a member of the board of directors of the United States Institute of Peace.

Elizabeth Kolbert has been a staff writer at *The New Yorker* since 1999, and previously worked at the *New York Times*. She is the author of several books, including *The Sixth Extinction*, for which she won the 2015 Pulitzer Prize for General Nonfiction.

Cyril Kormos is vice president for policy at the Wild Foundation, where he researches and advocates for issues including wilderness law and policy, conservation finance, and forest policy. He also coordinates IntAct, International Action for Primary Forests, an NGO coalition promoting the protection of primary forests globally.

Jules Kortenhorst is the CEO of Rocky Mountain Institute (RMI), an independent, nonpartisan nonprofit that drives the efficient and restorative use of resources. He is a recognized leader on global energy issues and climate change, with a background spanning business, government, entrepreneurial, and nonprofit leadership.

Larry Kraft is the executive director and chief mentor at iMatter, which engages passionate youth to take climate action and hold local communities accountable for their action or inaction on climate change.

Klaus Lackner is the director of the Center for Negative Carbon Emissions at Arizona State University, which advances carbon management technologies that can capture carbon dioxide directly from ambient air in an outdoor operating environment. He has made numerous contributions to the field of carbon capture and storage since 1995, along with other scientific fields.

Osprey Orielle Lake is the founder and executive director of the Women's Earth and Climate Action Network International (WECAN). She works nationally and internationally with grassroots and indigenous leaders, policy makers, and scientists to mobilize women for climate justice, resilient communities, systemic change, and a just transition to a clean energy future.

John Lanier is the executive director of the Ray C. Anderson Foundation, a Georgia-based private family foundation honoring the legacy of the late Ray C. Anderson, Lanier's grandfather. Anderson was a globally recognized industrialist and pioneer of environmentalism, and Lanier continues his legacy today through foundation programs that seek to create a brighter, more sustainable world for the present generation and those to come.

Alex Lau is a cleantech entrepreneur, angel investor, and international renewable energy project investor based in Vancouver, Canada. He serves on the Mayor of Vancouver's Greenest City Action Team and Renewable City Action Team.

Lyn Davis Lear is an activist and philanthropist and the president of L&L Media, which aims to inspire, educate, and activate people about global environmental issues through all forms of media. She is a board member of the Los Angeles County Museum of Art (LACMA) and the Sundance Institute, where she has produced and supported several films and labs and founded the Lear Family Foundation, dedicated to supporting civil rights and liberties, the arts, and the environment.

Colin le Duc, along with David Blood and Al Gore, is a cofounder and partner of Generation Investment Management, where he coleads the firm's growth equity Climate Solutions funds. He was previously with Sustainable Asset Management in Zurich, Arthur D. Little in London, and Total in Paris, and currently serves on the board of directors of various Generation portfolio companies around the world.

Jeremy Leggett is an entrepreneur, author, and advocate. He is the founding director of Solarcentury, one of the most respected international solar companies, founder and chairman of SolarAid, a charity set up with 5 percent of Solarcentury's annual profits, and chairman of Carbon Tracker, a financial-sector think tank warning of carbon asset stranding risks to the capital markets, colloquially known as the carbon bubble.

Annie Leonard is executive director of Greenpeace USA, and has more than two decades of experience investigating and explaining the environmental and social impacts of our stuff, where it comes from, how it gets to us, and where it goes after we get rid of it. Her film and book, both titled *The Story of Stuff*, blossomed into the Story of Stuff Project, which works to empower people around the globe to fight for a more sustainable and just future.

Peggy Liu has been the chairperson of JUCCCE, one of the top environmental organizations accelerating the greening of China, since 2007. She convenes international leaders in creating systemic change through eco-city planning, clean energy, smart grid, food education, and China's sustainability marketplace.

Barry Lopez is an essayist, author, and short-story writer who has traveled extensively in remote and populated parts of the world. He is the author of *Arctic Dreams*, for which he received a National Book Award; *Of Wolves and Men*, a National Book Award finalist; and eight works of fiction.

Beatriz Luraschi is a senior program officer at the Prince of Wales' International Sustainability Unit (ISU), where she has been working on tropical forest and climate change issues since 2013, including REDD+, eliminating deforestation in commodity supply chains, and the climate policy-science interface. Before joining the ISU, she conducted research on a range of sustainability issues, and completed field work to quantify ecosystem services on coffee farms under different management systems in Central America.

Brendan Mackey is the director of the Climate Change Response Program at Griffith University, based on Australia's Gold Coast, with expertise including terrestrial carbon dynamics, the interactions among climate change, biodiversity, and land use, and the role of science in environmental policy and law. His current research is focused on coastal zone adaptation in the Pacific, information and knowledge management for adaptation and resilience planning, and the assessment and valuation of primary forest.

Joanna Macy is an activist, author, scholar of Buddhism and systems theory, and the root teacher of the Work That Reconnects. She is the author of twelve books, including *Coming Back to Life: The Updated Guide to the Work That Reconnects*.

Joel Makower is the chairman and executive editor of GreenBiz Group, as well as an award-winning journalist. He is author or coauthor of more than a dozen books, including *Strategies for the Green Economy* and *The New Grand Strategy: Restoring America's Prosperity, Security, and Sustainability in the 21st Century*.

Michael Mann is Distinguished Professor of atmospheric science at Penn State University. He is a fellow of the American Geophysical Union, the American Meteorological Society, and the American Association for the Advancement of Science, and has authored more than two hundred publications and three books, including *Dire Predictions*, *The Hockey Stick and the Climate Wars*, and *The Madhouse Effect*.

Fernando Martirena is the director of the Center for Research and Development of Structures and Materials (CIDEM) at the Central University "Marta Abreu" of Las Villas, in Cuba.

Mark S. McCaffrey is a senior research fellow with the National University of Public Service (NUPS) in Budapest, Hungary, a senior adviser for the Earth Child Institute, founder of the United Nations Framework Convention on Climate Change's Education, Communication and Outreach NGOs community, and the author of *Climate Smart & Energy Wise* (2014). He has served as climate programs and policy director at the National Center for Science Education and cofounded CLEAN, the Climate Literacy and Energy Awareness Network.

David McConville is the board chair at the Buckminster Fuller Institute (BFI) and the creative director of the Worldviews Network, a collaboration of artists, scientists, and educators integrating storytelling and scientific visualization to facilitate dialogues about social-ecological regeneration.

Craig McCaw is a telecommunications pioneer—founder of McCaw Cellular and Clearwire Corporation and now chairman and CEO of the venture capital firm Eagle River Investments LLC. He is president of the Craig and Susan McCaw Foundation which supports education access and advancement, international economic development, and environmental conservation. He has served as chairman of the board of The Nature Conservancy and founded the Grameen Technology Center.

Andrew McKenna is the executive director of the Big History Institute at Macquarie University in Sydney, an innovative center dedicated to excellence in the area of big history, or the attempt to understand in a unified and interdisciplinary way the history of the cosmos, Earth, life, and humanity.

Bill McKibben is an author, environmentalist, and activist, and is a cofounder and senior adviser at 350.org, an international grassroots climate campaign that works in 188 countries around the world. He has written fifteen books, including *The End of Nature*, published in 1989 and often regarded as the first book about climate change written for a general audience.

Jason F. McLennan is considered one of the most influential individuals in the green building movement today, serving as CEO of his own design practice, McLennan Design, and as the founder and chairman of the International Living Future Institute, an NGO focused on transforming our world into one that is socially just, culturally rich, and ecologically restorative. He is the founder and creator of the Living Building Challenge, the world's most progressive and stringent green building program, and is the winner of the prestigious Buckminster Fuller Challenge and recipient of the ENR Award of Excellence.

Erin Meezan is the vice president of sustainability at Interface, where she gives voice to the company's conscience, ensuring that strategy and goals are in sync with the aggressive sustainability vision established almost twenty years ago. She is a frequent lecturer on sustainable business to senior management teams, universities, and the growing green consumer sector.

David R. Montgomery is a professor of geomorphology at the University of Washington, in Seattle. He is a MacArthur Fellow and the author of *Dirt: The Erosion of Civilizations*, *The Hidden Half of Nature: The Microbial Roots of Life and Health* (with Anne Biklé), and *Growing a Revolution: Bringing Our Soil Back to Life*.

Pete Myers is an author and the CEO and chief scientist of Environmental Health Sciences, an organization working to close the gap between good science and great policy. He is actively involved in primary research on the impacts of endocrine disruption on human health, serves as the board chair of the Science Communication Network, and served as the board chair of the H. John Heinz III Center for Science, Economics and the Environment.

Mark "Puck" Mykleby is a founding codirector of the Strategic Innovation Lab at Case Western Reserve University, which is dedicated to developing, testing, and implementing a new grand strategy for the United States that can power a new era of prosperity, security, and sustainability. Previously, Mykleby served as a fighter pilot in the Marines and a special strategic assistant to the chairman of the Joint Chiefs of Staff.

Karen O'Brien is a professor in the Department of Sociology and Human Geography at the University of Oslo, Norway, where she works on issues related to climate change adaptation and transformations to sustainability. She has been a lead author on several Intergovernmental Panel on Climate Change reports.

Robyn McCord O'Brien has for ten years helped lead a food awakening among consumers, corporations, and political leaders. She leads a nonprofit and an advisory firm and is a best-selling author, public speaker, and strategist.

Martin O'Malley is the sixty-first Governor of Maryland and ran for President of the United States of America in 2016. He has been outspoken about the need to act on climate change and environmental issues.

David Orr is a Paul Sears Distinguished Professor emeritus and counselor to the president at Oberlin College, and author of eight books and more than two hundred articles, reviews, book chapters, and other professional publications. He holds eight honorary degrees and leadership awards from the U.S. Green Building Council and Second Nature.

Billy Parish is a cofounder and the CEO of Mosaic, a provider of consumer lending solutions for the home energy market. He previously founded and grew the Energy Action Coalition into the world's largest youth clean energy organization.

Michael Pollan is a best-selling author, journalist, activist, and professor of journalism at UC Berkeley. He focuses on issues of food, diet, and food systems, and is the author of eight books, including *The Omnivore's Dilemma*.

Jonathon Porritt is a writer, broadcaster, and commentator on sustainable development. He cofounded the Forum for the Future, a sustainability nonprofit working globally with business, government, and others to create a better future.

Joylette Portlock has worked in environmental education and advocacy for the past ten years. She is the current president of Communitopia, a nonprofit that uses new media and project-based campaigns to identify, research, and advocate for individual, community, and national climate solutions, working to give the public scientific information it can use.

Malcolm Potts is a Cambridge trained obstetrician and reproductive scientist, and has worked all over the world to give women family-planning choices. He was appointed the first Bixby Professor of Population and Family Planning at UC Berkeley in 1992, and his current focus is on population growth and climate change in the Sahel.

Chris Pyke is chief operating officer for GRESB.com, where he provides actionable environmental, social, and governance information for global property investors, and he is the vice president for research at the U.S. Green Building Council. He represented the United States for greenhouse gas mitigation issues related to residential and commercial buildings on the IPCC Working Group 3, and was chair of the EPA's Chesapeake Bay Program Scientific and Technical Advisory Committee.

Shana Rappaport has worked actively for more than a decade as a community organizer and cross-industry convener to advance sustainability solutions. As director of engagement for VERGE with GreenBiz Group, she is currently scaling the leading global event series focused on accelerating the clean economy.

Andrew Revkin has written about climate change for nearly thirty years, twenty-one of which were as a New York Times reporter and author of the newspsper's Dot Earth column. He now writes for ProPublica, focusing on long-form climate and energy reporting.

Jonathan Rose founded the multidisciplinary real estate development, planning, and investment firm Jonathan Rose Companies, which has successfully completed more than $2.5 billion of work. Along with his wife, Diana, he also cofounded the Garrison Institute.

James Salzman is the Donald Bren Distinguished Professor of Environmental Law, with joint appointments at UC Santa Barbara and the UCLA Law School. He has published eight books on environmental law, serves on both the National Drinking Water Advisory Council and the Trade and Environment Policy Advisory Committee, frequently appears as a media commentator, and remains a dedicated classroom teacher.

Samer Salty is founder and CEO of Zouk Capital and has thirty years of experience in private equity, investment banking, and technology. He designed and implemented Zouk's distinctive dual-track strategy, consisting of technology growth capital and renewable energy infrastructure.

Astrid Scholz is the chief "everything" officer of Sphaera, a cloud-based solutions-sharing platform aimed at accelerating the pace of social change by connecting the best solutions with innovative problem solvers around the world. She is the immediate past president of Ecotrust, a hybrid nonprofit with more than $100 million in assets under management.

Ben Shapiro is cofounder and nonexecutive director of PureTech Health, and its Vedanta program is developing an innovative class of therapies that modulate pathways of interaction between the human microbiome and the host immune system. Through his previous work as executive vice president of research for Merck, he led the research programs that resulted in FDA registration of approximately twenty-five drugs and vaccines.

Michael Shuman is an economist, attorney, entrepreneur, and author, and director of local economies for Telesis Corporation. He is an adjunct instructor in community economic development at Simon Fraser University in Vancouver, Canada, and in sustainable business at Bard College in New York City, and recently wrote *The Local Economy Solution* (2015).

Martin Siegert is co-director of the Grantham Institute for Climate Change at Imperial College London, and previously was director of the Bristol Glaciology Center at Bristol University as well as head of the School of GeoSciences at Edinburgh University. An expert in geophysics, he was awarded the Martha T. Muse Prize for excellence in Antarctic science and policy in 2013, and is a fellow of the Royal Society of Edinburgh.

Mary Solecki is the western states advocate for Environmental Entrepreneurs (E2), a nonprofit advocacy organization whose business members support policy with both economic and environmental benefits.

Gus Speth is a cofounder of the New Economy Law Center at the Vermont Law School and cochair of the Next System Project. He served as dean of the Yale School of Forestry and Environmental Studies, cofounded the Natural Resources Defense Council, was founder and president of the World Resources Institute, served as administrator of the UN Development Programme and chair of the UN Development Group, and authored six books.

Tom Steyer is a business leader and philanthropist who believes we have a moral responsibility to give back and help ensure that every family shares the benefits of economic opportunity.

Gunhild A. Stordalen is the founder and president of the EAT Foundation. Together with her husband, Peter, she founded the Stordalen Foundation, which she also chairs.

Terry Tamminen currently serves as the CEO of the Leonardo DiCaprio Foundation, and under California governor Arnold Schwarzenegger he was appointed as Secretary of the California Environmental Protection Agency, and later Cabinet Secretary and Chief Policy Advisor to the Governor. An accomplished author, he has written several books, including *Lives Per Gallon: The True Cost of Our Oil Addiction* and *Cracking the Carbon Code: The Key to Sustainable Profits in the New Economy*.

Kat Taylor and her husband, Tom Steyer, established the TomKat Foundation to support organizations that enable a world with climate stability, a healthy and just food system, and broad prosperity. She is the founding director of TomKat Ranch Educational Foundation, which is dedicated to inspiring a sustainable food system, and is a cofounder and co-CEO of Beneficial State Bank.

Clayton Thomas-Muller is a member of the Mathias Colomb Cree Nation and a Winnipeg-based indigenous rights activist who has campaigned across Canada and the United States in hundreds of indigenous communities to organize against encroachments of the fossil fuel industry and the banks that finance them. He serves as the indigenous extreme energy campaigner with 350.org and as an organizer for Defenders of the Land and Idle No More.

Ivan Tse is a social entrepreneur and philanthropist working to shape the new culture within the social enterprise, philanthropy, and luxury sectors. He serves as chair and president of the TSE Foundation, a Hong Kong–based philanthropic organization that promotes initiatives to unite humanity, disseminate global knowledge, and build the infrastructure of a transnational world.

Mary Evelyn Tucker teaches at Yale University, where she directs the Forum on Religion and Ecology with her husband, John Grim, with whom she wrote *Ecology and Religion*. They are coproducers of the Emmy Award–winning film *Journey of the Universe*, and have created four open online classes based on the film through Coursera.

Paul Valva is a third-generation real estate associate in the San Francisco Bay Area, specializing in both commercial and industrial properties. He is passionate about sustainability and the environment and served for four years as manager in Northern California for the Climate Reality Project, educating the public about the dangers of and solutions to climate change.

Brian Von Herzen is the executive director of the Climate Foundation, which addresses gigaton-scale carbon balance on land and in the sea while ensuring global food and energy security. Climate Foundation's marine permaculture technology has the potential to provide sustainable food, feed, fiber, fertilizer, and biofuels on a global scale, all while enabling carbon export from the atmosphere.

Greg Watson is director of policy and systems design at the Schumacher Center. He is a public voice on sustainable agriculture, renewable energy, new monetary systems, equitable land tenure arrangements, neighborhood planning through democratic processes, and government policies that support human-scale development.

Ted White is a managing partner of Fahr, the umbrella entity for Tom Steyer's business, political, and philanthropic efforts. One of the primary goals of Fahr and its affiliated entities is to accelerate the transition to a clean energy future.

John Wick is a research rancher, venture philanthropist, and cofounder of the Marin Carbon Project, which has established scientifically that durable soil carbon can be increased through the production of healthy foods and safe fibers. He and his wife, Peggy Rathmann, own the Nicasio Native Grass Ranch in Marin County, California.

Dan Wieden is an American advertising executive who cofounded Wieden+Kennedy and coined the Nike tagline Just Do It. He is also founder of Caldera, a nonprofit arts education organization and camp for at-risk youth located in Sisters, Oregon.

Morgan Williams is an ecologist and sustainable development scientist who served as New Zealand's Parliamentary Commissioner for the Environment from 1997 to 2007. He now serves as chair of the boards of WWF New Zealand and the Cawthron Foundation, which supports New Zealand's largest private research organization.

Allison Wolff is CEO of Vibrant Planet, which provides strategy, narrative, and movement building design for companies and nonprofits focused on social and environmental innovation. She has worked with Chan Zuckerberg Initiative to develop their content and movement building strategy; with Facebook and eBay on their social good and sustainability narratives, marketing, and public engagement strategies; with Google to build Google Green; with GlobalGiving on its brand identity and strategy; and at Netflix as the director of marketing.

Graham Wynne is a former chief executive and director of conservation for the Royal Society for the Protection of Birds (RSPB), and is currently senior adviser to the Prince of Wales' International Sustainability Unit, a member of the board of the Institute for European Environmental Policy, and a trustee of Green Alliance. He was a member of the Policy Commission on the Future of Farming and Food and the Sustainable Development Commission.

ACKNOWLEDGMENTS

The staff is deeply indebted to everyone who believed in and supported this project. We could write volumes about the individuals here; however, given the number, we will need to convey our specific appreciation to each person one-on-one. The book was always about the larger "we" in the world. All of you represent the goodness and benevolence that suffuses humanity. The earth and all of life now beckon us, and you have exemplified this commitment in your work and life. We say thank you from our hearts and on behalf of all living beings.

Alec Webb · Alex Lau · Amanda Ravenhill · Andrew McElwaine · Andre Heinz · Barry Lopez · Betsy Taylor Bob Fox · Byron Katie · Colin Le Duc · Cyril Kormos · Daniel Kammen · Daniel Katz · Daniel Lashof David Addison · David Bronner · David Gensler · Edward Davey · Erin Eisenberg · Erin Meezan · Gregory Heming Guayaki · Harriet Langford · Ivan Tse · Jaime Lanier · James Boyle · Janine Benyus · Jasmine Hawken · Jay Gould Jena King · Johanna Wolf · John Lanier · John Roulac · John Wells · John Wick · John Zimmer · Jon Foley Jonathan Rose · Jules Kortenhorst · Justin Rosenstein · Kat Taylor · Lisa and Patrice Gautier · Lyn Lear Lynelle Cameron · Malcolm Handley · Malcolm Potts · Marianna Leuschel · Martin O'Malley · Mary Anne Lanier Matt James · Norman Lear · Organic Valley Cooperative · Paul Valva · Pedro Diniz · Peggy Liu · Peter Boyer Peter Byck · Peter Calthorpe · Phil Langford · Ray and Carla Kaliski · Rick Kot · Ron Seeley · Russ Munsell Shana Rappaport · Stephen Mitchell · Suki Munsell · Ted White · Terry Boyer · Tom Doyle · Tom Steyer Virgin Challenge · Will Parish

Adam Klauber · Andersen Corporation · Ben Holland · Ben Rappaport · Carbon Neutral Company · Chantel Lanier Chris McClurg · Chris Nelder · Colin Murphy · Cyril Yee · Dan Wetzel · David Weiskopf · Deep Kolhatkar · Diego Nunez Ellen Franconi · Frances Sawyer · Galen Hon · Gerry Anderson · George Polk · Jai Kumar Gaurav · Jamil Farbes · Jason Meyer Joel Makower · Johanna Wolf · Jonathan Walker · Joseph Goodman · Kate Hawley · Kendal Ernst · Leia Guccione Lynn Daniels · Maggie Thomas · Mahmoud Abdelhamid · Malcolm Handley · Mark Dyson · Mike Bryan · Mike Henchen Mike Roeth · Mohammad Ahmadi Achachlouei · Organic Valley Cooperative · Nicola Peill-Molter · Nuna Teal · Robert Hutchison Sean Toroghi · Thomas Koch Blank · Udai Rohatgi · Vivian Hutchinson · William Huffman

Adam DeVito · Alicia Eerenstein · Alicia Montesa · Alisha Graves · Allyn McAuley · Anastasia Nicole · Andy Plumlee · Angela Mitcham Annika Nordlund-Swenson · Aparna Mahesh · Aseya Kakar · Aubrey McCormick · Babak Safa · Basil Twist · Ben Haggard · Betty Cheng Bill and Lynne Twist · Bruce Hamilton · Caitlin Culp · Calla Rose Ostrander · Caroline Binkley · Carol Holst · Charles Knowlton · Cheryl Dorsey Cina Loarie · Claire Fitzgerald · Clinton Cleveland · Connie Horng · Daniel Kurzrock · Daniela Warman · Danielle Salah · Darin Bernstein David Lingren · David McConnville · David Allaway · Deborah Lindsay · Diana Chavez · Donny Homer · Dwight Collins · Eka Japaridize · Ella Lu Emily Reisman · Eric Botcher · Farris Gaylon · Gabriel Krenza · Hannah Greinetz · Helaine Stanley · Henry Cundill · Jacob Bethem Jacquelyn Horton · Jamie Dwyer · Jaret Johnson · Jeff and Elena Jungsten · Jeremy Stover · Jodi Smits Anderson · Joe Cain · Jose Abad Joshua Morales · Joyce Joseph · Juliana Birnbaum Traffas · Katharine Vining · Katie Levine · Kenna Lee · Kristin Wegner · Kyle Weise · Leah Feor Lina Prada-Baez · Madeleine Koski · Matthew Emery · Matthew John · Meg Jordan · Megan Morrice · Michael Elliot · Michael Neward Michael Sexton · Michelle Farley · Molly Portillo · Nancy Hazard · Nick Hiebert · Nicole Koedyker · Olga Budu · Olivia Martin · Pablo Gabatto Ray Min · Robert Trescott · Ron Hightower · Rupert Hayward · Ryan Cabinte · Ryan Miller · Sam Irvine · Sara Glaser · Serj Oganesyan Sonja Ashmoore · Srdana Pokrajac · Sterling Hardaway · Susan McMullan · The North Face · Thomas Podge · Tim Shaw · Tyler Jackson Veena Patel · Vincent Ferro · Whitney Pollack · Yelena Danziger · Zach Carson · Zach Gold

INDEX

Page references in italics refer to captions.

235

PHOTOGRAPH CREDITS